Electronics from the Ground Up

About the Author

Ronald Quan has a BSEE degree from the University of California at Berkeley and is a member of SMPTE, IEEE, and the AES. He has worked as a broadcast engineer for FM and AM radio stations, and he also holds an Extra Class amateur radio license.

He is the author of *Build Your Own Transistor Radios* (McGraw-Hill/TAB Electronics, 2012), a book that covers theory and practice pertaining to radios, circuits, and signals and also includes a multitude of newly designed radio circuits.

For over 30 years he has worked for companies related to video and audio equipment (Ampex, Sony, Macrovision, Monster Cable, and Portal Player). At Ampex, he designed CRT (cathode ray tube) TV monitors and low-noise preamplifiers for audio and video circuits. Other designs included wide-band FM detectors for an HDTV tape recorder at Sony Corporation, and a twice color subcarrier frequency (7.16 MHz) differential phase measurement system at Macrovision, where he was a Principal Engineer. Also at Macrovision he designed several phase lock loop circuits to provide a regenerated clock signal from the raw EFM (eight-fourteen modulation) data stream from a CD player.

At Hewlett Packard, working in the field of opto-electronics, he developed a family of low-powered bar code readers, which used a fraction of the power consumed by conventional light pen readers.

Currently, he is the holder of at least 400 worldwide patents (which includes over 80 United States patents) in the areas of analog video processing, video signal noise reduction, low-noise amplifier design, low-distortion voltage-controlled amplifiers, wide-band crystal voltage-controlled oscillators, video monitors, audio and video IQ modulation, in-band carrier audio single-sideband modulation and demodulation, audio and video scrambling, bar code reader products, and audio test equipment.

In 2005 he was a guest speaker at Stanford University Electrical Engineering Department's graduate seminar, talking on lower noise and distortion voltage-controlled amplifier topologies. He has served as a mentor in three different electrical engineering lab courses at Stanford University. And in November 2010 and October 2012 he presented papers in audio amplifier distortion to the Audio Engineering Society's Conference in San Francisco's Moscone Center.

Electronics from the Ground Up

Learn by Hacking, Designing, and Inventing

Ronald Quan

New York   Chicago   San Francisco
Athens   London   Madrid
Mexico City   Milan   New Delhi
Singapore   Sydney   Toronto

Electronics from the Ground Up:
Learn by Hacking, Designing, and Inventing

2 3 4 5 6 7 8 9 0   DOC/DOC   1 2 0 9 8 7 6 5

ISBN   978-0-07-183728-6
MHID   0-07-183728-0

This book is printed on acid-free paper.

Sponsoring Editor Roger Stewart	**Copy Editor** James K. Madru
Editing Supervisor Donna Martone	**Proofreader** Claire Splan
Production Supervisor Lynn M. Messina	**Indexer** Claire Splan
Acquisitions Coordinator Amy Stonebraker	**Art Director, Cover** Jeff Weeks
Project Manager Patricia Wallenburg, TypeWriting	**Composition** TypeWriting

Contents

Preface

Learning electronics from the ground up requires a foundation of knowledge of electronic components more than resistors, capacitors, diodes, and transistors. This book also covers different types of batteries and construction techniques.

This book is roughly divided into three sections—beginner, intermediate, and advanced levels for the hobbyist. There are projects for the hobbyist in all these sections to gain some practical experience and to reinforce some basic principles in electronics.

As the title says, *Electronics From the Ground Up*, this book is intended not only to introduce the hobbyist to electronics but also to lift the hobbyist to the next level in terms of learning and building more complicated circuits. This book is geared toward the hobbyist. However, there are some advanced topics discussed as well, such as a look into harmonic distortion analysis, which is written in terms of high school mathematics.

I think this book will offer information not only for hobbyists but also for students, engineers, and perhaps even for some electronics lab courses.

Acknowledgments

I would like to thank very much Roger Stewart of McGraw-Hill, who encouraged me to write a second book. Without his support, this book would not have been written. Also, thanks to Amy Stonebraker, who coordinated my manuscript and figures.

In transforming my manuscript and figures into a wonderful book, I owe much gratitude to project manager, Patricia Wallenburg, and her team including the copy editor, James Madru, and proofreader, Claire Splan.

Writing a book generally requires a "sounding board" for feedback, and Andrew Mellows not only proofread the entire manuscript but also gave me very helpful comments on the technical matter. Thank you very much, Andrew, for your suggestions and comments.

Some material in this book pertaining to building and entry-level electronics came from working with Professor Robert W. Dutton at Stanford University. I give my greatest thanks to Bob Dutton for inviting me into the Electrical Engineering Department as a mentor in his classes and for his ongoing encouragement while I was writing this book. I was very privileged to work with him, lead teaching assistant Kevin Dade, and the very bright students in the Research Experience for Undergrads, Mechatronics, and Radio-Frequency labs that included students Kiran Rana Magar, Tyler Conklin, Luis Aguilar, Kevin Hsu, Jamie Nakamura, Rea Rostosky, Chet Gengy, Laza Upatising, Si Tan, Charles Guan, Jack Tsai, and all the others who participated. I thank you all very much.

Also, I owe immense gratitude to Professor Thomas H. Lee at Stanford University, who was responsible for my meeting with Bob Dutton in the first place and who throughout the years has encouraged me in my electronics research. And, of course, thank you to Stanford University.

Quite a few people inspired me this time around. I would like to first give great thanks to Jeri Ellsworth, whose videos and private communications provided me with some of the material I decided to include in this book, such as the Wein oscillator. Jeri not only inspired and encouraged me on this book but also on my first book, *Build Your Own Transistor Radios*, when I worked on SDRs (software-defined radios).

Another person who inspired me is Professor Robert G. Meyer at the University of California Berkeley. There are parts of this book that include looking into the large-signal behavior of transistors. His class notes, books, and lectures played an important role for me in writing the later chapters. So thank you Bob Meyer!

This book also goes into audio circuits, and I would like to thank John Curl for his insights in this area, including low-noise and low-distortion amplifier design. Larry Topping built and evaluated many of the audio circuits featured in Chapter 8, and I owe him much gratitude for that. I am also thankful to Robert K. LeBeck, Jr., who provided some much-needed information pertaining to audio circuits. Thank you to Chris Gammel for posting on his site suggestions for this book that resulted in a request for audio circuits. Also, I give my thanks to Dave Jones who provided me helpful information on the latest digital storage oscilloscopes.

With much appreciation, I also need to mention my two mentors from the industry, John O. Ryan and Barrett E. Guisinger. Both gave me a world-class education in analog and video circuits design. I learned about feedback-clamp circuits from John and resistor-capacitor image-enhancement circuits from Barrett.

During this whole process I have been indebted to the following friends for encouraging and supporting me: Alexis DiFirenzi Swale, Germano Belli, Edison Fong, George Almeida, and Jo Acierto Spehar.

I really got into my second career as an author/writer because of Paul Rako. He connected me with Roger Stewart, which started all this. Thanks to you, Paul. Also, I would like to state my appreciation to Galileo Academy of Science and Technology and its teachers, Eugene Wing and Doug Page, for their support. William K. Schwarze, who was my best teacher at Galileo, also deserves thanks for his ability to convey advanced mathematical concepts in a very clear manner.

My friend, James D. Lee, is an inspiration to me, and he also reviewed a part of my book.

Finally, I acknowledge a big debt of gratitude to members of my family, William, George, Thomas, and Frances, for their support.

And, of course, I dedicate this second book to my parents, Nee and Lai.

PART ONE

Beginning Electronics

CHAPTER 1
Introduction

When I was growing up in the Bay Area, one of the first gifts I received as a kid was a small horseshoe magnet. I later received a second one and began to experiment with them and found that like poles (e.g., north to north and south to south) repelled and opposite poles attracted. When placed near a compass, these magnets moved the needle around in "mysterious" ways—not to mention that I quickly remagnetized the compass's needle and thus ruined it.

So when my McGraw-Hill editor asked me to write another book that was different from *Build Your Own Transistor Radios*, I had to ask myself, "What type of book? What do I cover?"

After all, I grew up with *The Boys' Book of Radio and Electronics* series, for which Alfred P. Morgan wrote four masterful volumes. I remember going to the library and looking for the 621.38 section for hobbyist books in electronics. Each book in the series from the first to the fourth was packed with colorful descriptions of the history, science, and construction of projects ranging from crystal radios, phonographs, Geiger counters, low-powered broadcast stations, solar-run radios, oscillators, and so on.

By far the *The Boys' Second Book of Radio and Electronics* was the most popular in my mind. In it, Alfred P. Morgan presented some of the more interesting projects for the beginning electronics hobbyist. For example, he presented a simple crystal radio followed by a more selective tuning crystal radio and then onto adding an audio amplifier to these radios so that more stations could be heard.

Basically, Mr. Morgan's approach was to start with a simple project and later show iterations of the basic idea to improve upon the next project. Thus, this approach is what inspired me to be one of the themes of my second book, *Learning Electronics from the Ground Up*.

Goal of This Book

The main goal of this book is allow readers to advance from one hobbyist level to another. That is, if one is a beginning hobbyist, she or he will advance to an intermediate level. Likewise an intermediate-level hobbyist will learn enough to step into an advanced level. For those advanced hobbyists I will try to show some new material such as thinking out of the box or inventing.

Throughout this book, basic concepts will be reiterated to build a fundamental understanding of certain aspects of electronics. Construction techniques and their trade-offs will be presented, such as knowing the limitations of using solderless breadboards known as *superstrip protoboards* versus copper-clad or vector boards.

Essentially, the topics covered will relate to analog electronics that covers a range of circuits and signals related to optical emitters and receivers, audio, oscillators, and video or analog image processing. One of the later chapters shows how equations describing a line or a polynomial relate to electronic circuits or systems.

This book is divided into three parts:

1.   Chapters 1 to 6 for beginners will cover parts and simple circuits, including batteries, light-emitting diodes (LEDs), flashlights, and power supplies. Soldering techniques will also be shown.
2.   Chapters 7 to 9 for intermediate-level hobbyists will include audio preamp circuits and different types of oscillators.
3.   Chapters 10 to 18 for advanced-level enthusiasts will venture into AM (amplitude-modulation) and FM (frequency-modulation) radios, television circuits and systems, along with circuit analysis, and finally, closing out with inventive circuits and troubleshooting techniques.

So we will start with modifying LED flashlights and end with constructing television circuits. Thus, there will be plenty of example circuits to build from. And more importantly, we will learn electronics from the ground up.

CHAPTER 2

Components and Schematics

To begin, I will familiarize readers with various components used in electronics. For this chapter, only selected components such as wires, batteries, capacitors, resistors, inductors, and some semiconductors will be explored. Other devices not covered in this chapter will be presented in subsequent chapters. For example, in this chapter, bipolar transistors will be mentioned, but field-effect transistors (FETs) will not be covered here. However, Chapter 6 will cover FETs.

One idea in writing this chapter was to limit the scope to components and to cover the rest of the different parts elsewhere in this book. I did not plan to make this chapter a reference chapter on all types of electronic parts. Closing out this chapter will be a discussion on schematic diagrams.

Wires

We will start with *insulated* wires, which are used to connect components and circuits together. Generally, there are two types of insulated wires—multistranded wires and solid single-stranded wires (Figure 2-1).

Wire thickness in United States is specified by American Wire Gauge (AWG). The lower the AWG number, the thicker is the wire. For typical wires used in electronic circuits, AWG numbers from 20 to 24 are used. Figure 2-1 shows 22 AWG for the top two wires and 26 AWG for the single-strand wire-wrap wire.

Both stranded and solid-strand wires are fine for construction projects. Often it is much easier to use the single-strand wire-wrap wire, as shown in Figure 2-1, because of its small diameter, which allows it to fit into tight places. In "emergencies," one can use 18- to 24-gauge zip cord or speaker wire (Figure 2-2).

Note, as shown in Figure 2-2, the zip cord can be separated entirely to provide two separate pieces of wire. Speaker wire can also be used for input and output leads

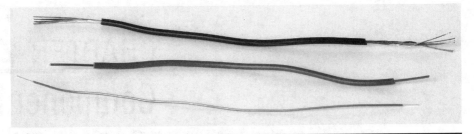

FIGURE 2-1 Stranded wire (*top*); solid single-stranded wire (*middle*); and wire-wrap wire (*bottom*).

that require hot and ground connections because one wire is normally marked with a stripe. For example, the striped side of a speaker wire can be used as the ground connection. Note that zip cord normally does not have such a marking and can result in a polarity issue, such as crossing the ground lead with the hot lead.

For audio, video, and radio-frequency (RF) signals, coaxial cables are used to shield out extraneous noise (e.g., power-line frequency hum). The outer-shield conductor is normally connected to ground, and the inner wire is the signal or hot lead (Figure 2-3).

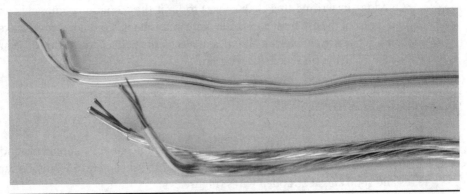

FIGURE 2-2 22 AWG solid-strand speaker wire (*top*) and 18 AWG stranded speaker wire (*bottom*).

FIGURE 2-3 Coaxial cable with an RCA phono connector for audio or video signals.

Wire Tools

When we are constructing circuits, the wires will have to be cut, and a small amount of insulation will have to be removed. Probably the most useful tool for this is the wire-stripping tool for removing insulation from insulated 30 AWG to 20 AWG wires. However, one can also use a wire cutter, diagonal cutter, or a pair of long-nose pliers with a wire-cutting feature. For "emergencies," a nail clipper or a pair of scissors can be used (Figure 2-4).

When constructing circuits with electronic components such as resistors, capacitors, transistors, and so on, wire cutters and diagonal cutters are often used to trim off the excess leads after soldering. Typical sizes of wire cutters and diagonal cutters range from 4 inches to 6 inches. Larger cutters may be used, but they may not fit in tight places where the leads have to be trimmed. For bending wires, a pair of long-nose pliers is used (Figure 2-5).

FIGURE 2-4   (*From left to right*) Wire stripper, wire cutter, diagonal cutter, scissors, and nail clipper.

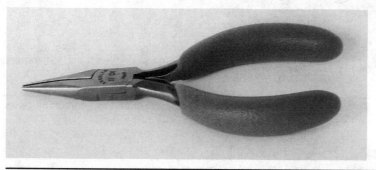

FIGURE 2-5   A pair of long-nose pliers is used for bending leads or wires.

Batteries

Probably the most common battery today is the AA cell. It is used in flashlights, radios, and electronic equipment such as voltmeters and inductance meters. However, for smaller devices, the AAA cell can be found in MP3 audio players, very small radios, and audio recorders. However, one of the most common power sources used for portable test equipment and smoke alarms is the 9-volt rectangular battery. Most batteries will have their positive (+) or negative (–) terminals marked, especially the 9-volt battery. For cylindrical batteries such as AAA, AA, C, and D cells, the positive (+) terminals are at the top and the negative (–) terminals are at the bottom.

Batteries can be categorized as nonrechargeable (e.g., carbon or heavy duty, alkaline, or lithium) and rechargeable [e.g., nickel cadmium, nickel–metal hydride (NiMh), and precharged or low-self-discharged nickel metal hydride] types. Nonrechargeable batteries are *not* to be installed in any battery charger because they can leak or burst or cause injury. Rechargeable batteries can only be installed in compatible battery chargers, or otherwise, damage can occur (Figures 2-6 and 2-7). Also, for safety reasons, do *not* install rechargeable batteries in smoke alarms.

It should be noted that as of the year 2014, many of the rechargeable cells are low-self-discharge types that can keep a charge up to one year compared with a couple of months for standard NiMh batteries. My choice is to skip the standard NiMh batteries and get the low-self-discharged types whenever possible. You can usually tell if a battery is a low-self-discharge type by looking for keywords on the battery or package, such as "low self-discharge" or "stays charged up to one year." In Figure 2-7, the energyOn and Tenergy batteries are low-self-discharge rechargeable batteries.

FIGURE 2-6 Five alkaline batteries, three lithium batteries, and two carbon batteries.

FIGURE 2-7 Regular and low-self-discharge NiMh rechargeable batteries.

The others are regular NiMh rechargeable batteries that will self-discharge within a few months.

> **NOTE**   It is recommended that you buy name-brand (e.g., energyOn, Sanyo, Ray-O-Vac, Energizer, etc.) rechargeable batteries because I found that generic no-name batteries bought on eBay had a much lower milliampere-hour capacity than stated.

See Tables 2-1 and 2-2 for battery capacities of nonrechargeable and rechargeable batteries.

> **NOTE**   There is a difference between lithium and lithium-ion batteries. The lithium batteries are not rechargeable. However, in laptop computers and other devices, rechargeable lithium-ion batteries are used.

TABLE 2-1   Data for Nonrechargeable or Otherwise Known as Primary Batteries, 2014

Battery Size and Type	mAh Rating	Voltage	Notes on Load Current
AAA carbon	540	1.5 volts	@ 25 milliamperes (mA)
AAA alkaline	1,200	1.5 volts	@ 25 mA
AAA lithium	1,200	~1.6 volts	@ 25 mA; initially, it is ~1.8 volts
AA carbon	1,100	1.5 volts	@ 25 mA
AA alkaline	2,700	1.5 volts	@ 25 mA
AA lithium	3,000	~1.6 volts	@ 25 mA, and initially, it is ~1.8 volts
C carbon	3,800	1.5 volts	@ 25 mA
C alkaline	8,000	1.5 volts	@ 25 mA
D carbon	8,000	1.5 volts	@ 25 mA
D alkaline	20,000	1.5 volts	@ 25 mA
9-volt carbon	~300	9 volts	300 mAh @ 5 mA, 180 mAh @ 15 mA
9-volt alkaline	600	9 volts	@ 25 mA
9-volt lithium	800	9 volts	@ 25 mA, and initially, it is ~10 volts

Both nonrechargeable and rechargeable batteries are rated in milliampere-hour (mAh) capacity that allows one to predict approximately their useful life under a known current drain. For example, a typical two-AAA-cell LED penlight may drain about 150 milliamperes. If a lithium AAA cell with 1,200 mAh is used, we would expect that the LED flashlight will provide about (1,200 mAh/150 mA =) 8 hours of continuous use.

The approximate service life of the battery in hours = $yyyy$ mAh/(load current in milliamps), where $yyyy$ is the rated milliampere-hours of the particular battery. For

TABLE 2-2   Data for Rechargeable Batteries, 2014

Battery Size and Type	mAh Rating	Voltage
AAA NiCad	350 to 400	1.2 volts
AAA NiMh	900	1.2 volts
AAA LSD NiMh	800	1.2 volts
AA NiCad	1,000	1.2 volts
AA NiMh	2,500	1.2 volts
AA LSD NiMh	2,100	1.2 volts
C NiMh	2,500 to 5,000	1.2 volts
C LSD NiMh	4,000 to 4,500	1.2 volts
D NiMh	2,500 to 10,000	1.2 volts
D LSD NiMh	8,000 to 9,000	1.2 volts
9 Volt NiMh	200	8.4 volts
9 Volt LSD NiMh	200	8.4 volts

LSD = low self-discharge.

example, a 1000 mAh battery with a 100 mA load current is expected to last about 1000 mAh/100 mA or 10 hours.

Some interesting factoids: New alkaline batteries' voltages start at about 1.55 volts, but they quickly drop to about 1.2 volts or 1.1 volts after use. Lithium and NiMh batteries maintain a more constant voltage over time. Lithium batteries start at 1.7 volts to 1.8 volts and drop to about 1.5 volts. NiMh batteries start at about 1.35 volts and settle to about 1.2 volts after use.

Other types of batteries are coin or button cells used for watches, car keys, LED keychain lights, laser pointers, calculators, and battery backup circuits for computers. Typically, flat coin cells have a voltage of 3 volts and are lithium types, whereas silver oxide or alkaline button cells yield about 1.5 volts. Commonly used 3-volt coin batteries are the CR2016 at 90 mAh, the CR2025 at 163 mAh, and the CR2032 at 240 mAh (Figure 2-8).

Normally, the polarity of coin or button cells is marked as shown in Figure 2-8, and the positive (+) sign is seen along with the manufacturer's name and battery number (e.g., CR2025). These batteries are generally used for continuous low-drain service or for intermittent medium-drain applications. For example, many keychain LED lights use two CR2016 batteries that drain at medium currents. However, the users of these keychain LED lights normally operate them for less than a minute at a time. Also, many watches use a single CR2016 battery for continuously draining a low current. The CR2025 battery is used in some electronic car keys. And for battery backup of data in computer motherboards, the CR2032 cell is commonly used.

Figure 2-8 Coin batteries CR2016, CR2025, and CR2032.

We have now introduced some of the more commonly used cells and batteries. Obviously, more types are available. One can find more information on various batteries at http://data.energizer.com/SearchResult.aspx.

A "First Hack" on Batteries

Technically speaking, a battery usually includes two or more basic cells. For example, a 9-volt battery is made of six 1.5-volt cells connected in series (Figures 2-9 through 2-11).

In Figure 2-9, an exhausted 9-volt carbon battery is being taken apart by hacking the bottom side. By prying open the seam with a pair of small long-nose pliers, the metal case can be removed. As a word of caution, be careful when taking apart the metal case that you do not get cut. See Figure 2-10.

Figure 2-9 A *carbon* 9-volt battery "intact" (*left*) with the battery's case being removed (*right*).

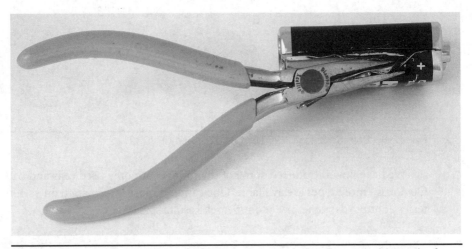

FIGURE 2-10   A carbon 9-volt battery with its metal case being removed with a pair of long-nose pliers.

Finally, Figure 2-11 shows the metal case removed, with the six cells stacked along with the connector. Note that the negative terminal with the wider mouth is connected to the top of the carbon battery and the positive terminal is connected via a flat metal ribbon to the bottom of the battery.

It should be noted that even though the cells of this battery are no longer useful, the 9-volt connector can be salvaged for future experiments.

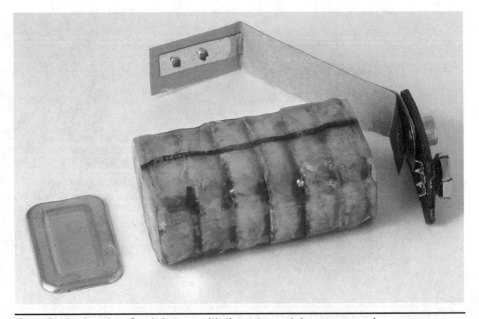

Figure 2-11   A carbon 9-volt battery with the outer metal case removed.

The Series Circuit and Battery Holders

Batteries are almost always connected in series to generate more voltage, and you see this in 9-volt batteries that are made with six cells connected in series. In a series-connected battery, the cells are connected such that the positive (+) terminal is connected to the next battery's negative (–) terminal. When batteries are connected correctly in this way, the total voltage across the batteries adds up. That is why two 1.5-volt cells connected in series correctly will yield (1.5 volts + 1.5 volts) = 3 volts, such as when used in a flashlight. Similarly, in another example, a radio may use four series-connected cells to provide 6 volts. In Figure 2-12, the radio runs off a 6-volt supply, and by connecting the four C cells in series, a 6-volt supply voltage is provided.

For some of the projects and experiments in this book, we will be using AA-cell battery holders for one, two, three, four, six, and eight cells to provide 1.5 volts, 3 volts, 4.5 volts, 6 volts, 9 volts, and 12 volts when using carbon or alkaline batteries or, alternatively, using about 20 percent less voltage output with rechargeable AA batteries. See Figure 2-13 for the various battery holders.

For the 9-volt battery, there is a connector with a matching battery holder or, alternatively, a 9-volt battery holder with built-in connector. See Figure 2-14.

The series connection of batteries is used often in flashlights, and even though most of them are connected in a straight line, other series connections are arranged

Figure 2-12   A radio that uses four cells connected in series to deliver 6 volts.

FIGURE 2-13   (*From left to right*) Battery holders for one, two, three, four, six, and eight cells.

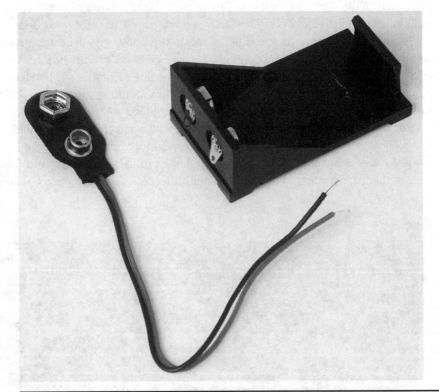

Figure 2-14   A 9-volt battery clip and 9-volt battery holder that is mountable.

in a zigzag manner (Figure 2-15). The two-cell flashlight uses a straight series connection of two D cells, while the LED flashlight connects three AAA batteries in series via a zigzag fashion.

NOTE   Of course, there have been times when the batteries are put in series the wrong way, such as connecting the positive (+) terminal of one battery to the positive (+) terminal of another battery in a two-cell flashlight

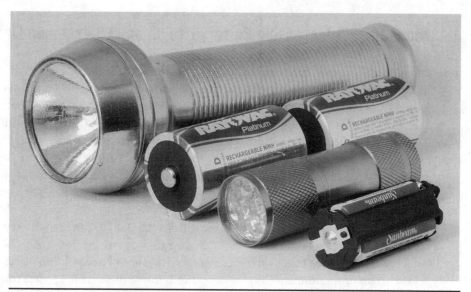

FIGURE 2-15   A two-cell regular flashlight and a three-cell LED flashlight.

that results in no light when the switch is turned on. And if the batteries are connected with the negative (–) terminal of the first battery connected to the negative (–) terminal of the second battery, the two-cell flashlight will also fail to light up when switched on. The reason is that with two batteries of the same voltage (e.g., 1.5 volts) connected back to back with the same polarity terminals, the voltage across the two batteries is (1.5 volts – 1.5 volts) = 0 volt.

You will see later that this back-to-back series connection of two equal voltage sources yields 0 volt and is a basic principle of why capacitors block direct current (DC).

CAUTION   A question does come up: Can we connect batteries in parallel? That is where the positive terminal of a first battery is connected to the second battery's positive terminal and where the negative terminal of each battery is connected together. The answer is generally and almost always no! The reason is first of all that the batteries have to have the same voltage and must be of the same type. But even if two batteries are marked of the same voltage, one battery may be more used up than the other, and what will happen is that the stronger battery will start charging the weaker battery. This charging effect may cause large currents to flow into the weaker battery, which then may cause heating or a bad chemical reaction that can lead to leaking. Therefore, for safety reasons, parallel connections of batteries are to be avoided.

A "Second Hack": Increasing Battery Life on an LED Keychain Light

Many LED keychain lights use two CR2016 batteries, which are rated at about 90 mAh. When the switch is turned on, the applied voltage to the white LED is supposed to be about 6 volts due to the series connection of the two CR2016 batteries. The LED actually does not receive 6 volts applied to it because the CR2016 batteries each have a built-in internal resistance, which causes a voltage drop. Nevertheless, the LED drains in excess of 50 mA, which then limits the battery life to about 2 hours given the 90-mAh capacity of the CR2016 batteries. Now suppose that we do not really need the LED to go to full brightness, and sufficient light is given at less than 10 mA into the LED.

If we replace the two CR2016 batteries with a single CR2032 in the LED keychain light, two things happen. One is the increased capacity of the 240-mAh CR2032, which is almost three times the capacity of a CR2016. The second thing is that the LED current drops dramatically to about 6 mA because there is only 3 volts from the CR2032 that is supplied to the LED.

In practice, only a few milliamps are needed to provide sufficient light for shining light in the dark when opening a lock or searching for items up close. The reason is that the white LED is very efficient.

Although most people may assume that you need at least 3 volts to turn on the LED, actually, a voltage of 2.4 volts is sufficient to light up the white LED. The CR2032 will give at least 80 percent more useful service life when it drops from 3.0 volts to about 2.4 volts. We can then estimate that the milliampere-hour rating is prorated to 80 percent of 240 or 192 mAh. The useful life of the keychain light is then about 192 mAh/6 mA, or about 32 hours. This is better than the 2 hours using the two CR2016 batteries, granted that the original light using two CR2016 batteries gives more light. See Figure 2-16.

From Yesteryear: A "Third Hack" by Converting a AA Cell Into a Penlight

From at least the 1950s to about 2000, the Eveready Battery Company used cardboard for the outside jacket of its carbon-zinc batteries, which were numbered 915, 1015, and 1215 for AA cells. The company also used this same material for its carbon-zinc C (#935) and D (#950) cells. Most other companies, such as Ray-O-Vac and Burgess (Gould), used a metal outer jacket instead of the cardboard tubing that Eveready used.

Today, virtually every carbon-zinc battery has a metallic outer case or is wrapped in plastic. Therefore, performing this third hack is virtually impossible today unless one can find a AA cell that has a cardboard outer casing. *Note:* eBay sometimes has

Figure 2-16 Two CR2016 batteries are removed and replaced with a single CR2032 battery to increase the service life of the keychain light.

these types of old AA batteries on auction, but their prices can be extravagant. Figure 2-17 that shows with the exception of the first battery, an alkaline power cell with a metal casing, the rest of the AA batteries have the now-obsolete cardboard outer case.

The first battery on the left in Figure 2-17 shows Eveready's first marketed alkaline AA cell, which was called an alkaline power cell and not the Energizer. And

Figure 2-17 Various Eveready AA batteries that are now obsolete.

Figure 2-18   Left is an old AA cell. Right is the result of opening the old Eveready AA cell using a knife to remove its bottom.

dating back to the early 1960s or even prior to that, Eveready had named at least some of its batteries the Eveready Energizer. The second battery in Figure 2-17, a carbon-zinc AA cell, was used for "transistor radio and electronic applications."

As a kid, I wanted to know what was inside a carbon-zinc battery but found that the metal jacket for the other batteries was harder to open as opposed to simply cutting off the bottom of an Eveready battery (see Figure 2-18).

After removing the bare battery, one will see that the cardboard tub includes a hole on the top to allow the positive (+) electrode of the bare battery to come through. This hole is then a convenient place to position a prefocus lamp such as the #112 bulb such that light shines out from the top of this cardboard tube.

Eat Your Spinach and Make a Flashlight!

All that is needed is a wire, which at that time I took from the paper-wire wrap for fresh spinach (see Figure 2-19). This paper-wire wrap is stripped of its paper, leaving only the conductive bare wire. However, the paper wire may have residual glue on it that acts as an insulator. The glue may be removed a number of ways from the wire, but a quick method is to simply use fine sandpaper or emery cloth (see Figure 2-20).

In Figure 2-20, once the wire is cleaned, a small length of the wire is then wrapped around the threads of the #112 bulb with an extra length supplied to allow "threading" the wire-bulb assembly into the cardboard tube up to the top hole.

If the original obsolete battery from Figure 2-18 is still good, I would just simply reinsert that battery back into the cardboard tubing, and I would have a working penlight. However, because the bare battery from Figure 2-18 is really exhausted due

FIGURE 2-19   Paper-wire wrap and #112 bulb.

to the fact that it is too many years old, I will need to find another bare AA cell that still has ample energy to light up the bulb.

It turns out that today's metal-case carbon-zinc AA cells can be taken apart with a small pair of long-nose pliers to pry open the seam (Figures 2-21 and 2-22). Again, the pair of long-nose pliers works as a useful tool for taking apart batteries.

With the inner plastic wrapping removed, the small diameter allows the bare battery to actually fit into the cardboard tube casing shown in Figure 2-18 on the

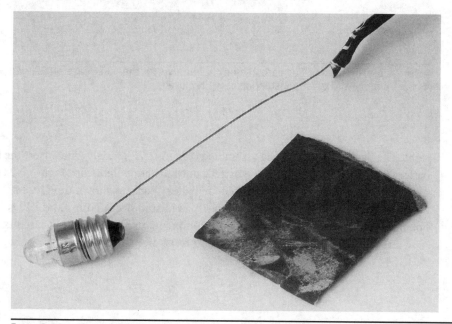

FIGURE 2-20   The wire is cleaned so as to be uninsulated via sanding off the residual glue once the paper has been removed.

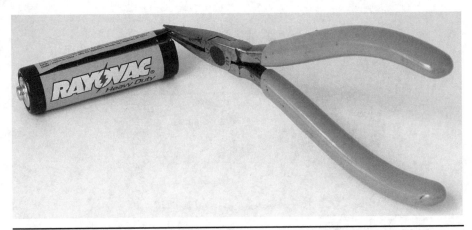

FIGURE 2-21   Prying open a present-day carbon AA cell with a pair of long-nose pliers.

FIGURE 2-22   The present-day carbon-zinc AA battery with the outer case pried open and its bare battery with plastic insulation being removed.

right. The flashlight now can be assembled (Figure 2-23). Finally, just insert the bare battery into the tube, and push the battery until the lamp lights up (Figure 2-24). To turn off the flashlight, simply pull the bare battery back, as shown in Figure 2-25.

We will return to lighting devices such as incandescent lamps and LEDs later. For now, let's explore other electronic components used for building circuits. But first, let's take a look at voltage, current, and resistance.

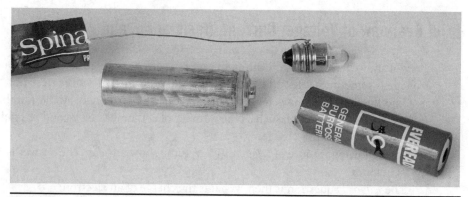

FIGURE 2-23   Wire and bulb assembly and resulting insertion into the cardboard tubing.

FIGURE 2-24   The completed flashlight.

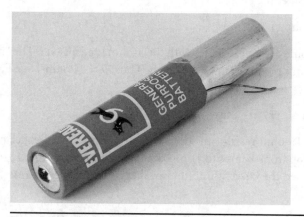

FIGURE 2-25   The flashlight is turned off by pulling out the bare battery slightly.

Brief Overview of Voltage, Current, Resistance, and Power

Previously, we have shown the different types of batteries, which are voltage sources. Voltage is sometimes referred to as the electrical *pressure* that moves electrons.

The flow of electrons generates current. *Current* is defined by the number of electrons flowing per time, such as the number of coulombs per second. A coulomb is a measure of electron charge.

The following is more of a *Jeopardy* TV game show answer: The answer is 1.6×10^{-19} coulomb.

The *Jeopardy* question is then: What is the charge of an electron?

That is, one electron charge provides 1.6×10^{-19} coulomb.

How about resistance? What is resistance? *Resistance* is the ability to restrict or reduce the current or, put another way, to reduce the flow of electrons. For example, one way to reduce the current into your 8-ohm loudspeaker and thus reduce its sound output is to place a 16-ohm resistor in series with one of the speaker cables. It should be noted that hooking resistors in series results in the addition of the resistances.

Ohm's law states that:

$$I = V/R$$

Again, the current I is measured in amperes (amps, A), and the voltage V is measured in volts.

The resistance R is measured in ohms—pretty good having the measurement of resistance named after the creator of Ohm's law, George Simon Ohm.

Going back to the loudspeaker, let's assume that V is an alternating-current (AC) voltage of 24 volts (don't worry, we will get into AC signals and circuits later). For now, just take a look at the numbers. Without placing the 16-ohm resistor in series with the 8-ohm loudspeaker the current is

$$I = V/R = (24 \text{ volts})/(8 \text{ ohms}) = 3 \text{ amps}$$

Now let's insert the 16-ohm resistor in series with the 8-ohm loudspeaker. The "new" resistance R is now 8 ohms plus 16 ohms = 24 ohms. Thus, $R = 24$ ohms. See Figure 2-26. Now we have

$$I = V/R = (24 \text{ volts})/(24 \text{ ohms}) = 1 \text{ amp}$$

Note that adding a 16-ohm resistor in series with an 8-ohm loudspeaker made the speaker's current drop from 3 amps to 1 amp. Stated in another way, the 16 ohms of added resistance caused the current into the speaker to be reduced to one-third the original current.

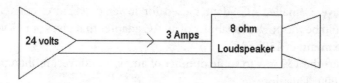

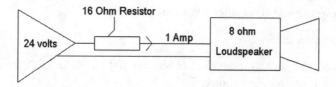

FIGURE 2-26 Amplifier with a 16-ohm resistor in series with an 8-ohm loudspeaker.

To bring the speaker's current back up to the 3-amp level requires that the voltage source V increase threefold from 24 volts AC to 72 volts AC. For $V = 72$ volts,

$I = V/R = (72 \text{ volts})/(24 \text{ ohms}) = 3 \text{ amps}$

Therefore, raising the electrical pressure threefold in V restored the 3 amps of speaker current. So one way to offset electric current constriction/reduction due to resistance is to increase the source voltage.

Now we will look at Ohm's law as applied to power. There are other types of voltage sources such as AC adapters that provide a voltage and a rated current. For example, a typical laptop AC adapter receives 110 volts AC from the wall outlet and provides about 20 volts DC with a current capability of 3 amps. In terms of power capability, this AC adapter provides a maximum 60 watts of power. So how did I get this number?

Ohm's law states that power is equal to voltage times current. Or, alternatively stated:

$P = V \times I$

This also can be written as

$P = VI$

where P is power in watts
 V is voltage in volts
 I is current in amps (A)

Thus, 60 watts is equal to (20 volts × 3 amps).

What if we have a smaller AC adapter used for lower-powered circuits? For example, what would be the power capability of an AC adapter that gives out 15 volts DC but only a maximum of 150 mA?

First, we convert the 150 mA to some number of amps. To convert milliamps to amps, divide by 1,000. Therefore,

150 mA = (150/1,000) amps = 0.150 amp

The maximum power from this smaller AC adapter is then

(15 volts × 0.150 amp) = 2.25 watts

Power produces heat such as in an electric blanket, light as in an LED, sound as in a loudspeaker, or motion as in a motor.

Resistors

Resistors are devices that limit current in a circuit, and they also can be used in voltage divider, filter, and amplifying circuits. They are available from ⅛ watt to at least 2 watts in terms of power dissipation. Resistor values have a specific tolerance from 0.1 percent to 10 percent. For this book, the most common resistors that we will use are ¼-watt 5 percent and 1 percent types. For resistors up to 2 watts, usually there are four bands painted on the resistor to denote its value. See Figure 2-27 and Tables 2-3 and 2-4. *Note that resistor values are measured in ohms.*

TABLE 2-3   Resistance Values for the First Three Bands

Black = 0	Brown = 1	Red = 2	Orange = 3	Yellow = 4	Green = 5
Blue = 6	Violet = 7	Gray = 8	White = 9	Gold = ÷10	Silver = ÷100

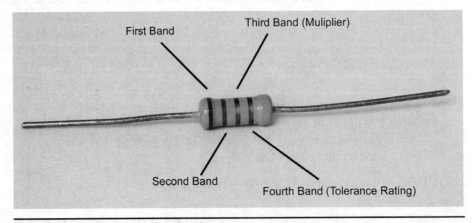

FIGURE 2-27   A resistor with four bands.

TABLE 2-4   Typical Values for Resistors

1.0	2.2	4.7	10	15	18	22	27	33	39
47	56	68	82	100	120	150	180	220	270
330	390	470	560	680	820	1K	1.2K	1.5K	1.8K
2.2K	2.7K	3.3K	3.9K	4.7K	5.6K	6.8K	10K	15K	18K
22K	27K	33K	39K	47K	56K	68K	100K	150K	180K
220K	270K	330K	390K	470K	560K	680K	1M	1.5M	1.8M
2.2M	2.7M	3.3M	3.9M	4.7M	5.6M	6.8M	10M	15M	18M

It should be noted that the gold and silver bands are usually in the third bands when denoting a value for a resistor. In the fourth band position, gold = 5 percent and silver = 10 percent. The first two bands denote a number, while the third band denotes how many zeroes are added to the first two numbers. For example, a 10,000-ohm 5 percent resistor is denoted as brown, black, orange, and gold, which is interpreted as "10 plus three zeros following = 10 000 or 10,000." With 5 percent tolerance, this means that the value of the resistor is 10,000 ohms plus or minus 5 percent of 10,000 ohms or 10,000 ohms ± 500 ohms. Therefore, this example resistor can fall within the range of 9,500 ohms to 10,500 ohms.

To shorten the number of zeroes when writing a resistor value, we assign k = 1,000 and M (or mega or Meg) = 1,000,000. For example, a 10,000 ohm resistor is the same as a 10k ohm resistor, or a 10,000,000 ohm resistor is the same as a 10 M ohm or 10 Meg ohm resistor.

Some typical values of resistor are shown in Table 2-4. More commonly used values are shown in bold fonts.

There are "in-between" values (in ohms) that are not shown for 5 percent resistors, such as 12, 13, 16, 20, 24, 30, 36, 43, 51, 75, and 91. These in-between values may be multiplied by 0.1, 10, 100, 1,000, and 10,000 to provide other resistance values. For example, multiplying by 0.1 yields values of 1.2, 1.3, and so on. If we multiply by 100, we get in-between values of 1,200, 1,300, ..., 9,100.

For 1 percent resistors, there are five bands, and the first three bands represent three digits, followed by the fourth band, which is the added number of zeroes following the three digits. The fifth band is always brown for denoting a 1 percent tolerance (see Figure 2-28).

For the resistor at the top of Figure 2-28, the first three bands denote the first three digits, while the fourth band is the multiplier. We read the resistor using the first four bands, but from which side of the resistor? The four bands that tell us the value of the resistor always have equal spacing separating these four bands. If one looks very closely, one will notice that the spacing between the fourth and fifth band in Figure 2-28 is wider than the spacing between the first and second, second and third, or third and fourth bands.

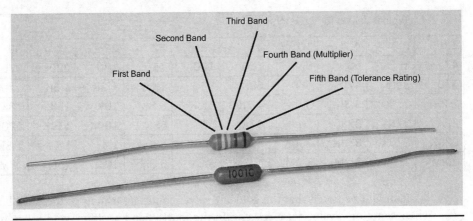

Figure 2-28   A 1 percent tolerance resistor with five bands.

The bottom resistor in Figure 2-28 has its value printed on it instead of using a color code. Here the first three numbers represent the first three digits, followed by the fourth number, which is the multiplier (or number of zeroes added to the three digits). In this example, the resistor reads "1001C," which is 100 + "1 zero" = 1,000 ohms. The C denotes the tolerance, which is coded as 0.25 percent. For precision resistors, A = 0.05%, B = 0.10%, C = 0.25%, D = 0.50%, F = 1.0%, and G = 2%. Sorry, but there is no designation for E.

It's All Greek to Me, But It's Good

Instead of writing the word *ohms*, a shorthand notation is used employing last letter of the Greek alphabet, *omega*. The upper case or capital letter omega is Ω, which denotes ohms. We can substitute Ω wherever *ohm* or *ohms* is used to describe the values of resistors. For example, 1,000 ohms is expressed as 1,000 Ω.

For resistors greater than 2 watts, we have *power resistors*. Fortunately, they have their resistance values, power rating, and tolerance printed on them. (Note that the color code is not normally used in power resistors of 3 watts or more.) See Figure 2-29.

So how do we find out the power going into a device such as a resistor? From Ohm's law,

$$P = VI$$

To determine the power absorbed or dissipated in a resistor we need to express *P* in terms of resistance. We also know that:

$$I = V/R = (V/R)$$

Let's substitute this equation into the power equation so that:

$$P = VI = P = V(V/R) = (VV/R) = (V)^2/R$$

FIGURE 2-29   These resistors are wire-wound types at 5 Ω 5% 10 watts and 10 Ω 5% 10 watts.

Thus,

$$P = (V)^2/R$$

NOTE   *P* is measured in watts when the voltage *V* is measured in volts and the resistance *R* is measured in ohms. Note that "w" denotes watt.

The equation, $P = (V)^2/R$, is probably the most common method of assessing the power absorbed into a resistor. It's also easy to just measure the voltage across the resistor. Just square the voltage and divide by the resistance to get the power dissipated by the resistor. For example, suppose that there are 25 volts across a 1K ohm (1-kΩ) resistor. The power is then:

$$P = (25 \text{ volts})^2/1 \text{ k}\Omega = (625/1{,}000) \text{ watt} = 0.625 \text{ watt} = 0.625 \text{ w} = P$$

A rule of thumb is to specify a resistor's wattage to be at least twice the actual power dissipation. So if there is 0.625 watt across the 1K ohm resistor, pick a 2-watt resistor for reliability. However, a 1-watt resistor would work too. If the power rating of the resistor is near the power dissipation, then chances are that the resistor will change value (e.g., increase in resistance) over time due to prolonged "excessive" heat.

There is another derivation for expressing the power dissipation of a resistor in terms of current flowing into the resistor. Again, let's take a look at the basic equation:

$$P = VI$$

However, we also know that:

$$I = V/R$$

which says that the current through the resistor is the voltage divided by its resistance value. Put in another way, if a current I is flowing into a resistor R, a voltage will be developed across the resistor. This means that:

$V = I \times R = IR$

Thus,

$V = IR$

The voltage V is the voltage across the resistor when current is flowing into it.
   Now let's review what we have:

$P = VI$

$V = IR$

By substitution of V, we get

$P = VI = (V)I = (IR)\ I = IRI = IIR = (I)^2R = P$

This leads to:

$P = (I)^2R$

NOTE P is measured in watts when the current I is measured in amps and the resistance R is measured in ohms.

   For example, if we know that there are 1.5 amps DC flowing into a 22-Ω resistor, then the power dissipated into the resistor is

$P = (I)^2R = (1.5)^2 22$ watts $= 2.25 \times 22$ watts $= 49.5$ watts

In this example, for reliable operation, a 100-watt power resistor should be chosen.
   A next logical component to cover is the capacitor. *Capacitors* are storage devices. When voltage is applied to a capacitor, a charge is stored inside the capacitor. The ability of the capacitor to hold a charge makes the capacitor sort of like a low-capacity rechargeable battery.

Capacitors

Whereas resistors are measured in ohms, capacitors are measured by their capacitance in farads. However, a farad is a very large amount of capacitance, so most capacitors have values of microfarads or picofarads of capacitance. A microfarad (μF) is one-millionth of a farad. And a picofarad (pF) is one-millionth of a millionth of a farad or a millionth of a microfarad. Another commonly used term to denote capacitance is the nanofarad (nF), which is one-billionth of a farad. And just in case you are wondering, the farad is named after Michael Faraday.

Capacitors are made of two plates separated by an insulator, otherwise known as a *dielectric*. A general formula for capacitance is

$C = kA\varepsilon_0/d$

The constant k is the dielectric constant. For example, the dielectric constant of a vacuum or air is about 1, whereas the dielectric constant of ceramic material is about 3 to 7. A is the area of the plates, ε_0 is the emissivity constant (don't worry about this because we will not be making any calculations), and d is the distance between the two plates.

From this formula, intuitively, it makes sense that the closer the plates are, the smaller is the number for d, which results in more capacitance. In addition, if the area of the plates increases, then A increases, which also increases the capacitance. Also, depending on the dielectric material that governs k, the capacitance can be increased or decreased.

One advantage of a capacitor is that it can be quickly charged by putting a voltage source across the terminals. The discharge rate is related to how much capacitance there is in the capacitor.

For a quick understanding of how capacitors work, for a DC voltage, think of a capacitor as a low-capacity rechargeable battery. A capacitor can hold a charge like a battery but cannot power a device for long periods like a battery. For AC voltage, a capacitor acts sort of like a resistor, except that it does not get hot. It resists or controls AC current but does not dissipate heat because during one part of the AC signal it is charging up like a battery and during the other part of the AC signal it is discharging current, and the net charge and discharge current sum to zero. This is why a perfect capacitor never gets hot!

Because of the storage capability of capacitors, they are used in power supplies. When you shut off your laptop computer and then pull the plug from the AC outlet, you will notice in some laptop AC adapters the indicator light is still on for a few seconds before it goes out. The capacitor in the AC adapter has stored enough electricity to light up the LED lamp after the AC power has been removed. Capacitors used in power supplies provide filtering at low frequencies to remove hum are called *electrolytic capacitors*.

NOTE Electrolytic capacitors in general are polarized. That is, there is usually a (–) marking or a (+) marking on the capacitor. The polarity of the capacitor must be observed and connected correctly. For example, to charge up an electrolytic capacitor with a battery, connect the (–) terminal of the battery to the (–) terminal of the capacitor and likewise with the (+) terminal.

WARNING Improperly connecting an electrolytic capacitor may lead to an unsafe situation.

Electrolytic capacitors have markings for correctly applying DC voltage to them. For electrolytic capacitors that have *axial* leads, usually the outer can or case is the (–)

polarity of the capacitor. In radial-lead electrolytic capacitors, the (–) terminal is clearly marked on one side. Figure 2-30 shows axial- and radial-lead electrolytic capacitors. Note that electrolytic capacitors have voltage ratings that always should exceed the voltage applied to them. In general, the voltage rating should exceed the supply voltage, for instance, by at least 30 percent. For example, if the circuit operates at 12 volts DC, all electrolytic capacitors should have a voltage rating of at least 16 volts. Also, electrolytic capacitors have a tolerance of at least 20 percent. Some are specified at –20 percent and +80 percent, and these are generally used in power supplies.

In some circuits involving AC (alternating current) signals where the average DC (direct current) voltage is about 0 volt, a nonpolarized electrolytic capacitor can be used. Alternatively, two equal "regular" electrolytic capacitors may be soldered back to back in series with *either* the (–) and (–) or (+) and (+) terminals. The resulting capacitance is half. For example, soldering two 100-μF electrolytic capacitors soldered back to back in series results in a 50-μF nonpolarized capacitor. See Figure 2-31 for an illustration of back-to-back connections.

Smaller-value capacitors of 2.2 μF and under are generally nonpolarized, which means that you can connect these capacitors without regard to polarity. Capacitors are generally marked with a three-digit number, where the first two numbers are significant and the third number denotes the number of zeroes added to the first two numbers in units of picofarads. The tolerance is denoted by a letter such as K = 10%, J = 5%, G = 2%, and F = 1%. For example, a 0.01-μF 10 percent tolerance capacitor is marked 103K or 10,000 pF = 0.01 μF. When in doubt about the value, you can measure the capacitance with a capacitance meter, which is usually available in a digital voltmeter costing $20 or more. The voltage rating is usually at least 25 or 50 volts, and the exact voltage rating is rarely marked on the capacitor itself, but it is known when it is ordered from a vendor (e.g., www.mouser.com or www.digikey.com). Figure 2-32 shows various types of small-valued capacitors that include ceramic disc, ceramic monolithic, film, and silver-mica capacitors.

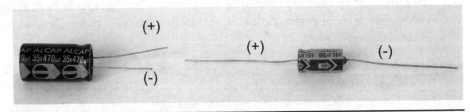

Figure 2-30 Axial- and radial-lead capacitors with polarity markings.

Figure 2-31 (*From left to right*) A nonpolarized electrolytic capacitor and two electrolytic capacitors connected with their (–) terminals and two electrolytic capacitors connected back to back with their (+) terminals.

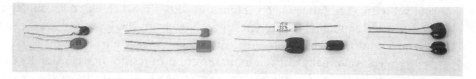

FIGURE 2-32   Ceramic disc, ceramic monolithic, film (e.g., mylar or polyester), and silver-mica capacitors.

Inductors or Coils

An *inductor* is yet another storage element that works on the principle of magnetic flux or magnetic fields. Its property is that ideally it has zero DC resistance but has an AC resistance just like a capacitor.

Whereas a capacitor's AC resistance decreases with increases in frequency, an inductor has an opposite characteristic. Instead, as the AC frequency increases, the AC resistance of an inductor increases.

In radio-frequency (RF) circuits, inductors are used in oscillators and band-pass filters. For example, inductors are used in inductor-capacitor oscillators such as a *Colpitts* oscillator. And inductors can serve as both an inductor and an antenna coil such as those used in an AM radio.

Inductors are used in power-supply circuits to filter out hum or high-frequency noise. They are used in switching power supplies to boost the voltage. For example, an inductor with a switching power-supply circuit can convert the 1.5 volts from a battery to a 4 volt output to light up white light emitting diodes (LEDs) (Figure 2-33).

Figure 2-33 shows the internal circuitry of a AAA-cell LED flashlight. Because the white LEDs need at least 2.5 volts for lighting, the 1.5 volts from the AAA cell is

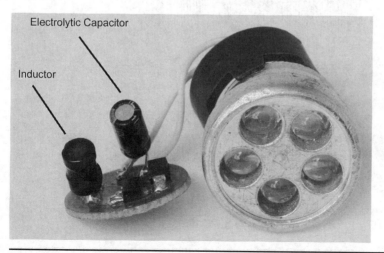

FIGURE 2-33   A miniature switching power supply inside an LED flashlight with an inductor.

not sufficient. The inductor in conjunction with the switching power-supply chip, rectifier, and aluminum electrolytic capacitor converts the 1.5 volts from the battery to about 4 volts, which is then sufficient to light up the white LEDs.

While it is normally difficult to make your own resistor or capacitor, it's easy to make an inductor. As a matter of fact, even a straight piece of wire can be used as an inductor for very high-frequency circuits. A practical way to make an inductor is to wrap wire around an empty roll of bathroom tissue, a plastic pen, or a ferrite bar. See Figure 2-34.

Resistors are measured in ohms, capacitors are measured in farads, and inductors are measured in henries (H), named after Joseph Henry. A measure of inductance is therefore the henry. Inductors that require values in the order of henries are limited to low-frequency applications such as a 60-hertz (Hz) power supply for tube amplifiers or for an audio effects circuit such as a "wah-wah" sound-effects circuit for modifying the tone of a musical instrument. Most coils have much lower inductances, especially those used in RF and audio circuits. For RF circuits, the inductances are in the nanohenry to microhenry range (one-billionth of a henry to one-millionth of a henry), and for audio circuits, typical inductances are in the millihenries range (one-thousandth of a henry).

For the coils shown in Figure 2-34, the bathroom tissue inductor has about 75 microhenries (75 µH), while the pen coil measures at 5.6 µH, and finally, the ferrite bar (antenna) coil measures the most at 180 µH. The ferrite material magnifies the inductance, so a smaller number of turns is required to provide the same amount of inductance as would be required for an air-core inductor. To reiterate, most inductors have

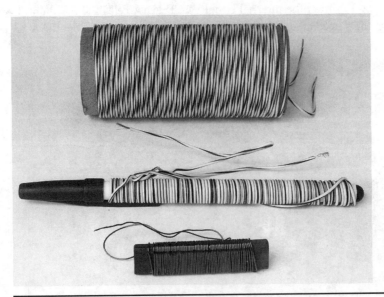

Figure 2-34 Do-it-yourself (DIY) inductors.

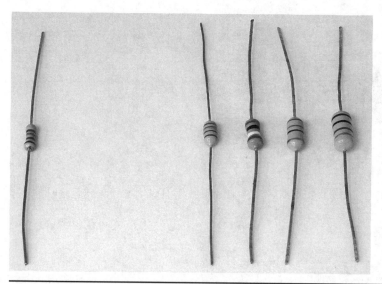

Figure 2-35   A resistor on the left and four inductors on the right.

inductances in microhenries (µH) that are used in RF (radio-frequency) circuits, while some audio circuits may use millihenry (mH) coils.

Some common inductors look like resistors and even have the same color code as resistors. However, they are distinguished in that when they are measured for resistance, the resistance value is often much lower than what the color-code marking would imply. See Figure 2-35.

The color code that is used for inductors is the same as that for resistors, and the inductance values are from the 10 percent resistor values. The first two color bands denote the first two digits, and the third band is the multiplier or number of zeroes following the two digits to determine the inductance in microhenries. The fourth band is the tolerance, usually silver for 10 percent or gold for 5 percent tolerance. For example, an inductor with yellow-violet-orange-silver bands is rated at 47,000 µH at 10 percent tolerance. To convert microhenries to millihenries, divide the microhenry number by 1,000 to arrive at the millihenry value.

From this example inductor,

47,000 µH = [47,000/1,000]mH = 47 mH = 47,000 µH

One way to identify an inductor or a resistor is to read the color code and then measure it with an ohmmeter. In the example shown in Figure 2-36, a "component" with color-coded bands of orange, white, black, and gold would normally read as a 39-ohm 5 percent resistor. But see what happens when we test it with an ohmmeter. See Figure 2-36.

The 39-µH inductor measures 0.9 ohm, which means the component is not a 39-ohm resistor. By deduction, the component must be an inductor. An inductor's

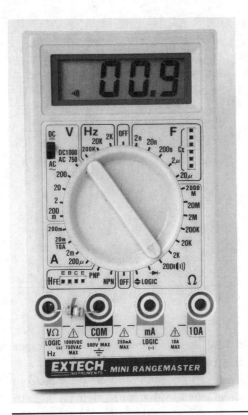

Figure 2-36 A 39-µH inductor tested for resistance.

resistance generally does not match the value as read out by its color code. By measuring the DC resistance one can then deduce whether the component is a resistor or an inductor.

Other types of inductor have markings that show a three-digit number, with the first two digits indicating microhenries and a third digit for the multiplier or number of zeroes after the two digits for the inductance value in microhenries. For example, an inductor with a marking of "470" is 47 µH (Figure 2-37).

Figure 2-37 shows an inductor with a marking of "683," which normally would mean 68,000 µH = 6.8 mH. Actually, in this case and in the last inductor with a marking of "685," that could mean 6,800,000 µH = 6.8 henries, the units of inductance are in nanohenries instead of microhenries. A nanohenry is 1 billionth of a henry and is a thousand times smaller than a microhenry. So this makes the first inductor in the figure actually a thousand-fold less than expected and thus 683 → 68 µH and 685 → 6.8 mH.

The only inductor that measures as expected is the middle coil that reads "470," which is 47 µH. Recall that the third number is the number of zeroes following the first two digits, that is,

Figure 2-37   Inductors with three-digit markings.

470 → (47 + no zero added)µH = 47 µH

The best way to verify if an inductor really is an inductor is to measure it with an inexpensive inductance meter such as the one shown in Figure 2-38, which costs about $40 on eBay. From Figure 2-38, the first inductor from Figure 2-37 with the marking "683C" measures about 64 µH, which then identifies that inductor as a 68-µH inductor. We will now venture into the world of semiconductors but will include vacuum tubes as well.

Semiconductors

There will be more detail about semiconductors in Chapters 4, 5, and 7. For now we will present a brief overview.

Diodes and Rectifiers

Diodes and rectifiers allow electricity to flow in one direction (e.g., with low resistance) while providing an open circuit (e.g., very high resistance) in the other direction. They are thus useful in converting an AC (alternating current) signal that, by definition, has part of the time a positive voltage and at another time a negative voltage to provide a voltage that is *either* positive or negative. For example, an AC adapter uses one or more rectifiers to provide either a positive or negative DC voltage. For small signals such as converting an AC AM signal to a DC signal, diodes are used. See the first two diodes (top and middle) in Figure 2-39.

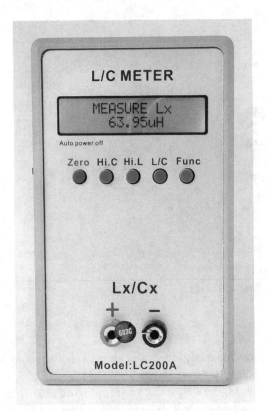

FIGURE 2-38   An inexpensive inductance-capacitance L/C meter capable of measuring both inductance and capacitance.

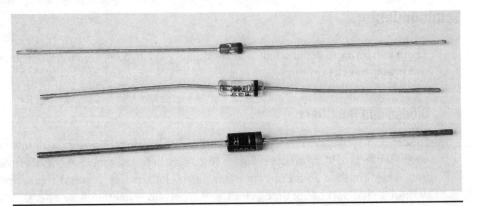

FIGURE 2-39   Diodes and a rectifier with bands to denote their cathodes.

For small-signal diodes, there are mostly standard silicon diodes such as the 1N914 or 1N4148. These diodes are generally used for switching, RF, and biasing circuits. In contrast, silicon rectifier diodes are generally not suitable for RF (radio-frequency) circuits but are suitable for higher-power applications such as power rectifiers for 50 Hz or 60 Hz AC power supplies.

The turn-on voltage of the standard silicon diode or rectifier is about 0.6 volt. For lower turn-on voltages, the Shottky diodes and Shottky rectifiers have turn-on voltages of 0.3 volt to 0.4 volt. In power supplies where the minimum losses are required, the Shottky diode is a prime source. For example, the rectifier used in the LED flashlight shown in Figure 2-33 is a power Shottky diode.

For older germanium diodes, the turn-on voltage is in the range of 0.1 volt to 0.25 volt. Common part numbers for germanium diodes are 1N34, 1N60, and 1N270. These diodes are generally used in radio circuits.

But what's a diode turn-on voltage? An idea diode when the polarity is set for forward conduction is supposed to have 0 volt. That is, for any voltage ≥ 0 volt (greater than or equal to zero volt), the ideal diode will pass the voltage through. And at any voltage < 0 volt (less than zero volt), the ideal diode will stop current flow and not pass any negative voltage through the diode. In any diode, there is a minimum threshold voltage before conduction starts. This *threshold voltage* is the turn-on voltage. There will be some experiments later on for the reader to get acquainted with diodes. For now, the diode or rectifier has two terminals—the anode and the cathode. The cathode is marked with a band. See Figure 2-39.

Figure 2-39 shows on the top, a common silicon small-signal diode such as the 1N914 or 1N4148 diode. The middle section shows a larger glass diode such as a germanium 1N34 or 1N270 diode. On the bottom, we see a power rectifier such as a 1N4003 power supply rectifier.

Generally, diodes are used for smaller signals such as AM detectors or voltage references. Rectifiers are actually diodes, but we refer to them as rectifiers for higher-power situations such as use in power supplies.

For those diodes that are made of glass, an extra lead length is required to prevent cracking of the glass body when loaded on a board. Bending the leads too close to the body causes damage at the edges of the glass diode. See Figure 2-40. Because rectifiers handle power, the leads of a rectifier act as a heat sink to allow cooling of the rectifier. Therefore, the leads are also not bent too close to the body.

Rectifiers are rated by their *peak reverse voltage* (PRV), but the older equivalent term is *peak inverse voltage* (PIV). Also they are rated for their forward-bias conduction current. Typical power diodes or rectifiers are rated at 1 amp, 3 amps, etc. For example, a 1N4001 rectifier has a PRV of 50 volts. This means that it will still conduct when the voltage applied to the anode is greater than + 0.6 volts, and will not conduct if the voltage is between −50 volts and 0 volt. If the negative voltage is more negative than −50 volts, such as −60 volts, the diode will start conducting again.

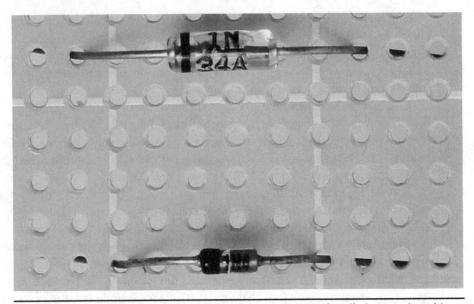

FIGURE 2-40 Examples of spreading out the leads when mounting diodes on a board to avoid cracking the glass body.

Common 1-amp rectifiers are

- 1N4001 50 PIV
- 1N4002 100 PIV
- 1N4003 200 PIV
- 1N4004 400 PIV
- 1N4005 600 PIV
- 1N4006 800 PIV
- 1N4007 1000 PIV

Amplifying Devices

An *amplifier* sometimes is defined as a device or system that makes a small or weak signal larger or stronger. This statement alone needs a qualifier in that the amplifying device or system requires power to make the signal larger. An amplifier is powered via an external DC source such as a DC power supply or a set of batteries. The external power source allows the amplifier to use a small signal at the input such as a signal from a microphone to *control* larger signals at the output of the amplifier. For example, a typical microphone does not have sufficient power output to drive a loudspeaker directly. If the microphone is connected to an amplifier that is connected to a loudspeaker, we can then drive the loudspeaker and make a whisper very loud. In FM radios, for example, at the antenna, the off-the-air signals may be as low as 10 millionths of a volt (10 μV). But by connecting one amplifier after another in the radio, the signal may be brought up to a level of 1 volt, which in this case results in a voltage gain of 100,000.

A *relay* consists of a coil that forms an electromagnet in conjunction with an actuator switch. The actuator switch controls its switch contacts to open or close, that is, to stop or allow current to flow. If there is enough current flowing through the coil, the relay switch contacts can control power by turning on very large amounts of current to a device such as a 100-watt lamp. Typically, the coil only needs a small amount of power such as 1 watt to activate the relay. In this case we have a 100:1 power gain because we have 100 watts of output for a 1-watt input signal. Thus the relay can be thought of as an amplifying device. However, on the surface, the relay only turns on or off—it's a digital signal. That is, the relay coil either pulls the contacts together to switch on or it does not.

For amplifiers, what we are commonly talking about are the ones that amplify an analog signal such as music or speech signals. Since music or speech signals are not merely on and off signals, the relay cannot normally be used as an amplifier the same way a common stereo hifi amplifier works.

Bipolar Transistors

For small-signal amplification such as radio receiver circuits, preamplifiers, video amplifiers, and so on, the plastic-case TO-92 transistor is commonly used. Small metal-case transistors (e.g., TO-18) are also available, such as the 2N2222 and 2N2907, which offer a bit more current and power capability than some TO-92 versions.

Power transistors are utilized to deliver power to a load, such as providing many watts of power to a loudspeaker. For example, a typical stereo receiver has output power transistors. For medium-power applications, some TO-5 metal-case transistors are used. For this book, we will avoid TO-5 transistors since it is easier to use the plastic-case power transistors. Also, TO-5 transistors, which were popular from the 1960s to the 1980s, have fallen out of use in favor of plastic-case power transistors.

Finally, for high-power circuits, there are TO-220 and TO-218 transistors. For even higher power, the larger plastic-case TO-247 transistor is used.

Not as common today is the TO-3 metal case power transistor. We will avoid using TO-3 transistors because the installation on heat sinks is more difficult than mounting plastic-case power transistors. Also, this metal-case power transistor is not as popular as it used to be.

A bipolar transistor has three terminals—the emitter, base, and collector. And there are two types of polarities for bipolar transistors—NPN and PNP.

Amplification occurs when a small signal is connected to the base and emitter terminals that act as the input, and the output is taken off from the collector. This type of amplifier is known as a *common-emitter amplifier*. Another form of amplification is *buffering*, where the voltage gain is 1 or less, but the output is capable of driving very high currents. In this case, the input signal is connected to the base of the transistor, and the output at the emitter of the transistor supplies amplified current. For example, certain sound cards in a computer may only have enough current to drive a set of headphones at low volume. If the output of the sound card is connected to the base of a transistor and the emitter of the transistor is connected to the same set of headphones, a louder volume will result. This type of amplifier is called an *emitter-follower amplifier*.

Of course, there are other types of amplifiers, and they will be covered in Chapter 6 and subsequent chapters. For now, we will look at different types of small-signal and power transistors (Figure 2-41).

In Figure 2-41, from left to right, are plastic-case TO-92, metal-case TO-18, metal-case TO-5, metal-case TO-3, plastic-case power TO-220, plastic-case power TO-218, and plastic-case power TO-247 transistors. For the TO-92 transistors, there are variations on the pin-out of the emitter base and collector (Figure 2-42).

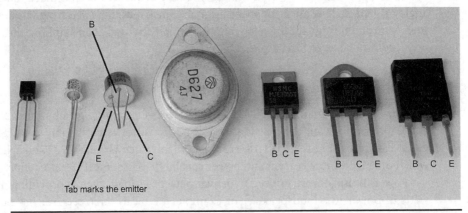

Figure 2-41 Various plastic- and metal-case transistors, including plastic power transistors.

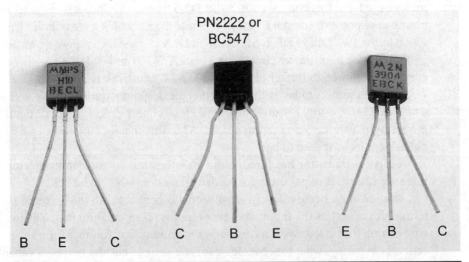

Figure 2-42 Emitter, base, and collector pin-outs for the MPSH10, PN2222 or BC547, and 2N3904 transistors.

The general pin-out sequence of a transistor is emitter-base-collector or collector-base-emitter. However, for very high-frequency transistors, generally it is desirable to minimize proximity between the base and the collector leads. This is why the MPSH10 high-frequency NPN transistor has a pin-out of base-emitter-collector. The emitter, which is often AC signal grounded, acts as a ground shield between the collector and base. Collector-to-base capacitance, as we will find out later, generally reduces the high-frequency response. The complementary PNP high-frequency transistor to the MPSH10 is the MPSH81, which also has the same pin-out as the MPSH10. For a list of some small-signal transistors with *emitter-base-collector* pin-outs, see Table 2-6. Small-signal transistors with collector-base-emitter pin-outs are listed in Table 2-7. For plastic-case power transistors, see Table 2-8. The pin-out is base-collector-emitter, with the metal tab connected to the collector terminal. See Figure 2-41 that shows B C E.

TABLE 2-6   Various NPN and PNP Small-Signal Transistors

NPN	PNP
2N2222	2N2907
2N3904	2N3906
2N4124	2N4126
2N4401	2N4403
2N5089	2N5087
2N5551	2N5401

TABLE 2-7   Small-Signal Transistors with Collector-Base-Emitter Pin-Outs

NPN	PNP
BC547	BC557
BC548	BC558
BC338	BC328
PN2222	PN2907

TABLE 2-8   Plastic-Case Power Transistors

NPN	PNP
TIP29	TIP30
TIP31	TIP32
MJE3055	MJE2955
TIP3055	TIP2955

The four transistors listed as TIP29 to TIP32 are medium-power transistors capable of about 3 amps of collector current. However, the "2955" and "3055" transistors can deliver up to 15 amps.

Vacuum Tubes

One of the first vacuum tubes was merely a light bulb with an additional metal plate inside the glass bulb. This two-element device (filament and plate) was invented by Thomas Edison and is also known as a *vacuum-tube diode*.

To operate this device, first apply a filament power supply to light up the bulb. The *Edison effect*, as it became known, occurred when the inventor noticed that if a DC voltage source (negative polarity) was connected to one of the filament's terminals (either would do) and the other terminal of the DC voltage source (positive polarity) was connected to the plate, current would flow via the plate. However, if the DC voltage source had its terminals reversed, no current would flow via the plate. You may say this was the first semiconductor, even though it was done with a hollow-state device.

Around 1907, Lee De Forest decided to build on Edison's idea and put a see-through metal fence or grid between the filament and the plate, making this a three-element device (filament, grid, and plate) known as the *triode* (e.g., "modern" tubes of the 1940s and 1950s, the 6J5 and 6C4). By adding the grid, De Forest was able to control the flow of electrons to the plate of the tube by using the principle of repulsion.

We know from static electricity that like charges repel and unlike charges attract. Since electrons are negatively charged, varying a negative voltage at the grid and one terminal of the filament then controlled the current flow to the plate. The more negative the voltage is at the grid, the fewer electrons reach the plate because the grid becomes a repeller of electrons and bounces the electrons back to the filament. If the grid voltage is less negative, then fewer electrons are repelled, and more electrons get to reach the plate, thereby increasing *plate current*.

In the intervening years, approximately 10 years apart, second and third grids were eventually placed inside the bulb. The second added grid, a screen grid, lead to the creation of the *tetrode* in 1919, a four-element tube with a filament, control grid, screen grid, and plate. The *pentode* was invented in 1926 and was a five-element tube with a filament, control grid, screen grid, suppressor grid as the third grid, and plate. The major advantage of the tetrode and pentode over the triode is higher gain or higher amplification. Examples of tetrodes are the 12DS7 and 6CL8. As for pentodes, they are still made today for guitar and high-fidelity amplifiers such as the power-output tubes 6L6, 6550, and EL34. Smaller-signal example pentodes are the 12BA6, 12SK7, and 6CB6.

Eventually, engineers got creative and placed more grids, leading to the seven-element *heptode*, a five-grid tube commonly called the *pentagrid converter* (e.g., 12BE6, 12SA7, and 1R5), and the eight-element *octode*, a six-grid tube that can be called a *hexagrid converter* (e.g., 6A8). These tubes were used almost exclusively in radios and in audio-processing amplifiers such as an audio amplitude compressor or automatic gain control amplifier.

Because there are so many types of diodes, triodes, pentodes, heptodes, and so on that are housed in seven-, eight-, and nine-pin tubes, there is no set rule as to which pin is always assigned as the plate or grid. Also, many vacuum tubes have multiple sections, such as the twin triode (e.g., 12AT7, 12AU7, 12AX7, and 12BH7) or triode and pentode combination tubes (e.g., 6AW8, 6GH8, 6U8, and 7199).

There are many different types of tubes in terms of pin numbers. For this book, we will concentrate on the seven- and nine-pin miniature and the eight-pin octal tubes such as those shown in Figure 2-43. And the pin-number assignments as viewed from the top are shown in Figure 2-44.

FIGURE 2-43 Vacuum-tube examples with seven, nine, and eight pins.

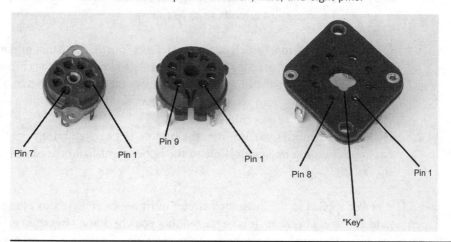

FIGURE 2-44 Pin assignments for the seven- and nine-pin miniature tubes and the eight-pin octal tube are shown via tube sockets.

With vacuum tubes, the pin-out convention is as follows:

1. For seven- and nine-pin tubes, *from the top view* as in Figure 2-44, start from the lower-right side as pin 1 and go counterclockwise for each succeeding pin number. By the time you reach the last pin on the lower-left side, that pin will be pin 7 for a seven-pin tube and pin 9 for nine-pin tube.

2. For an octal tube, again *from the top view*, start at the lower-right pin for pin 1, as shown in Figure 2-44 at the pin just right of the "key" or "index" marking. Again, count each number as you go counterclockwise. You should end with pin number 8 for the pin that is to the left of the "key" or "index" marking.

Integrated Circuits as Amplifiers and Logic Gates

Integrated circuits contain a number of transistors and often also contain resistors and capacitors. They provide a system function such as an amplifier, voltage regulator, radio receiver circuit, digital logic circuit, and all the way to a complicated mixed digital and analog system on a chip (SoC) or to a central processing unit (CPU) that is the heart of a laptop or desktop computer. What used to be built on hand-wired chassis or printed circuit boards are now condensed into *chips* or *integrated circuits*.

For example, look at the USB ATSC television tuners that are available today. Just plug the chip tuner into the USB port of your computer and view your favorite shows. A television tuner built in the early 1980s would have required printed circuit boards, transistors, RF coils, high-frequency transformers, capacitors, mechanical switches, etc. It would have been relatively large—about the size of a 23-watt compact fluorescent bulb. For this chapter, we will just concentrate on 8-, 14-, and 16-pin integrated circuits for through-hole mounting. See Figure 2-45.

The pin-outs for integrated circuits work as follows:

1. Oriented from the top, a small dot or notch faces "north." Note that the closest pin to the dot is pin 1. Starting from the upper left-hand side of the chip is pin 1; now count one pin at a time in a *counterclockwise rotation* such that by the bottom of the left column you are at pin 4 for an 8-pin chip, pin 7 for a 14-pin chip, or pin 8 for a 16-pin chip.

2. Continue to count as you go from the bottom left to the bottom right, and continue counting one pin at time to the upper right-hand corner, which is the last pin.

The power applied to the integrated circuit must not be reversed so as to avoid permanent damage. Therefore, it is suggested that you check the pin-out of an integrated circuit twice just to make sure that the power pins are connected properly.

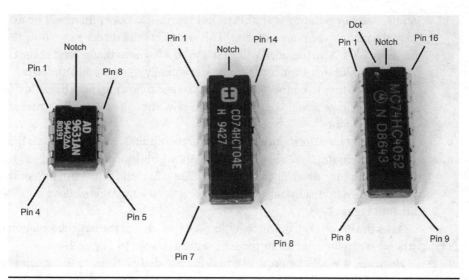

FIGURE 2-45   Integrated-circuit packages for 8, 14, and 16 pins.

Depending on which integrated circuit you are using, the power pins may be in different locations. Therefore, always download or get a copy of the data sheet for each integrated circuit that you use. For example, in a single op amp such as the LM741, the plus supply is at pin 7, whereas in a dual op amp integrated circuit such as the LM1458, the plus supply is at pin 8.

Also, the plus and minus power pins of all integrated circuits should have small decoupling capacitors about 0.1 µF to each power pin and ground. For example, the LM741 has a plus power pin at pin 7. Connect one lead of the 0.1-µF capacitor to pin 7 and the other lead of the 0.1-µF capacitor to ground. Then locate the minus supply pin of the LM741, which is pin 4. Take a second 0.1-µF capacitor and connect one lead to pin 4 and the other lead of this capacitor to ground.

Connecting decoupling capacitors to the power pins of integrated circuits is necessary to avoid noisy or oscillating signals from the output of the integrated circuit.

Other types of integrated circuits come in TO-92 packages, such as voltage regulators and radio circuits. Also, there are power regulators and amplifiers in the TO-220 case form. These and other integrated circuits will be described in more detail as needed in the subsequent chapters.

Schematic Diagrams

If one wants to communicate the essentials of a circuit, one way is to literally illustrate by hand each component. Alternatively, one can build the circuit in an organized manner in which all the wires and connections are well in view. For simple circuits, one can get away with this technique. In this book, there will be examples where the circuits are photographed. This will allow the reader to see how the parts were assembled. Also included will be *schematic diagrams* that represent the circuits.

As one builds more and more circuits, eventually she or he will understand a rule of thumb as to how to connect the parts when just following the schematic diagram. Figure 2-46 shows actual components and how they fit within the corresponding schematic diagrams.

This chapter will introduce the reader to two circuits, a video low-pass filter, and an amplifier. Figure 2-47 shows a video bandwidth low-pass filter, which has two inductors and four silver mica capacitors. The schematic representation of this circuit will evolve from a literally drawn schematic to a more refined version. We will start with Figure 2-48.

After the rough sketch, which looks pretty much like the basic orientation of the parts, we try to further redraw the schematic as shown in Figure 2-49.

However, it would be great to add reference designations to the schematic diagram since there will be at time parts that have the same value. In this case, both inductors are 18 μH. For inductors, we choose L as the designator for an inductor or coil. And for a capacitor, we can take the obvious letter C. See Figure 2-50.

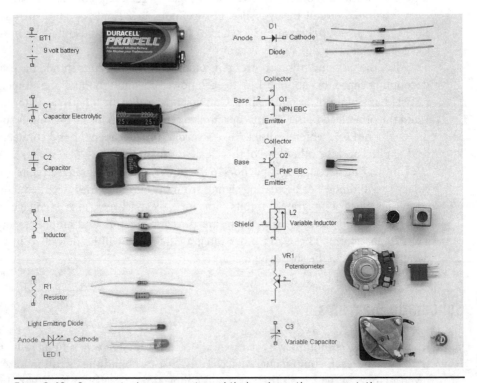

FIGURE 2-46   Some actual components and their schematic representations.

FIGURE 2-47   Video bandwidth low-pass filter.

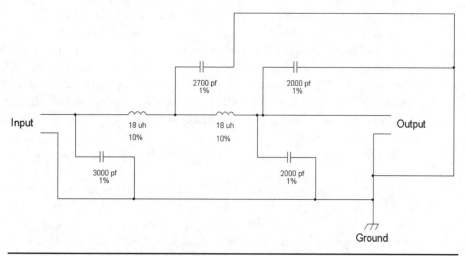

FIGURE 2-48   First rough-sketch schematic of the circuit in Figure 2-47.

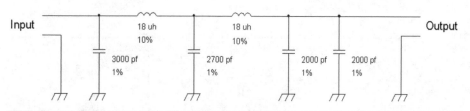

FIGURE 2-49   Redrawn schematic of the video low-pass filter.

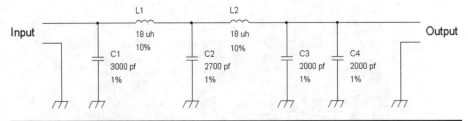

FIGURE 2-50   Schematic of the video low-pass filter with reference designations.

Finally, we can reduce the amount of information on the schematic and move that information elsewhere. For example, since all capacitors are 1 percent and all inductors are 10 percent for the schematic, we can just draw the component, its reference designation and value (e.g., C1 3,000 pF), and other pertinent information can be footnoted outside the schematic diagram (see Figure 2-51). This figure shows the result after a few iterations from the original rough-sketch schematic.

We can draw this schematic with a computer-aided design (CAD) program such as the free ORCAD Capture Lite (version 16.x), which is downloadable at www.cadence.com/products/orcad/pages/downloads.aspx. Alternatively, you can also download LT SPICE from Linear Technology at www.linear.com/designtools/software/. However, for the reader, it may be easier to keep a notebook and draw the schematics by hand. Using a CAD schematic program is a learning process that can take some time to master.

For the second example, Figure 2-52 shows the one-transistor amplifier from Chapter 1.

Figures 2-53 through 2-55 show how schematic drawings are improved from one iteration to another. Figure 2-53 shows the circuit with no ground yet. However, it points out the common lead or common wire that may be construed as a (chassis) ground. The "final" schematic is not really the last one. A different person may have another way to redraw it. But this schematic is close enough for accurate documentation.

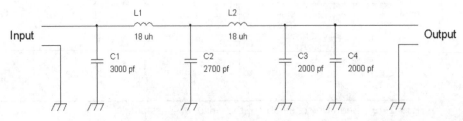

Note: All inductors, L1 and L2, have 10% tolerance; and all capacitors, C1- C4, have 1% tolerance.

FIGURE 2-51   Schematic of the video low-pass filter with footnotes.

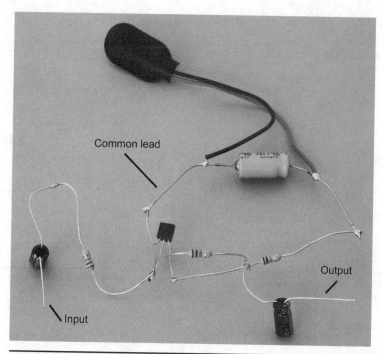

Figure 2-52   A one-transistor amplifier circuit.

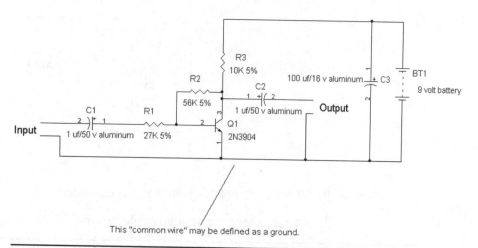

Figure 2-53   First schematic diagram of the one-transistor amplifier shown in Figure 2-52.

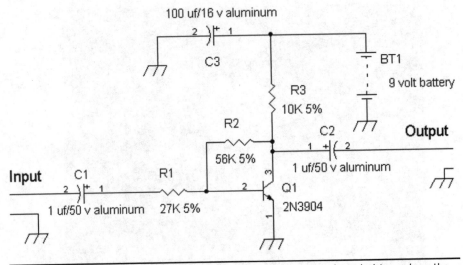

FIGURE 2-54 Schematic diagram further refined to show a ground symbol to replace the common-wire lead. The schematic looks a bit less cluttered.

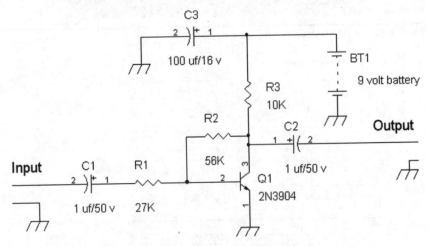

Notes: 1) All resistors are 1/4 watt and 5% tolerance.

2) All capacitors are aluminum electrolytic.

FIGURE 2-55 "Final" iteration of the one-transistor amplifier schematic.

NOTE Other electronic components such as variable inductors, variable capacitors, and variable resistors or potentimeters will be covered in the later chapters. For now, this chapter has covered the basic components for simple projects and experiments.

References

1. Robert L. Shrader, *Electronic Communication*. New York: McGraw-Hill, 1967.
2. Mouser Catalog, www.mouser.com.
3. "Eveready Battery Applications and Engineering Data," Union Carbide, Bound Brook, NJ.
4. Jameco Catalog, www.jameco.com.
5. RadioShack, *Enercell Battery Guidebook*, 2nd ed. Niles, IL: Master Publishing, 1990.
6. Digikey Catalog, www.digikey.com.
7. "Triode," wikipedia.org.
8. "Tetrode," wikipedia.org.
9. "Pentode," wikipedia.org.
10. "Vacuum tubes," wikipedia.org.

CHAPTER **3**

Construction
Techniques and
Simple Test Equipment

In this chapter, we will look at a few different types of circuit construction techniques. With the overview of certain electronics components introduced in Chapter 2, this chapter will show how they can be assembled to make a useful circuit.

For starters, solder-less circuits will be examined such as *superstrip* prototype boards or circuits assembled via twisting the wire leads and taping down the components onto an index card. Alternatively, a giant loopstick antenna coil radio is shown to work even though the wires leading up to the tuning capacitor are at least a foot long. Later we will cover assembling circuits via soldering. There will be examples of soldered circuits in copper-clad and vector boards. Also, did you know that it is possible to solder electronic components that hang in thin air and still work? A first project is included at the end of the chapter that shows how to make a simple battery tester.

Solderless Circuit Construction Techniques

Let's Do the Twist—Twisting Wires to Build Circuits

The easiest way to confirm whether a circuit works or to build a temporary circuit is to use just wires twisted together. Figure 3-1 shows the schematic diagram and Figure 3-2 shows the prototype circuit. There are three resistors, which are 20 kΩ 5 percent resistors (red-black-orange-gold), a precision 10 kΩ 1 percent resistor (brown-black-black-red-brown), and a 12 kΩ 5 percent bridging resistor (brown-red-orange-gold).

This resistor network has a 9-volt battery for its power source. The object is to find the voltage across R5, the 12 kΩ resistor. One way to find out is to measure the voltage, which was found to be 0.633 volt across the 12 kΩ resistor with the 9-volt battery measuring 9.24 volts. The other way is to calculate the current flowing

53

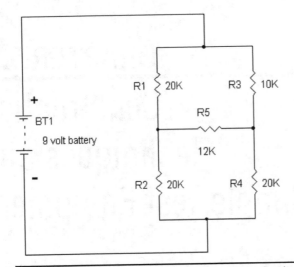

FIGURE 3-1   Schematic diagram of a five-resistor circuit.

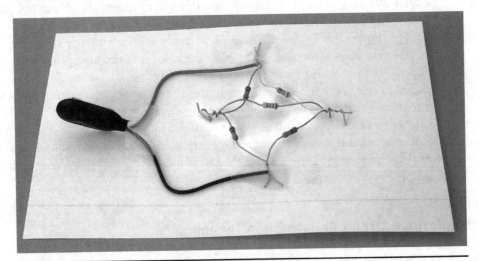

FIGURE 3-2   Solderless prototype build on an index card.

through the 12 kΩ resistor. For now, it's easier to do a direct measurement than to go through the pain of circuit analysis. Normally, the circuit would be analyzed with five equations. However for the advanced readers, by using *Thevenin* equivalent circuits that are mentioned in Chapter 15, this circuit is simplified to one equation. The general answer is the current, I_{R5}, flowing through R5 from left to right is:

$$I_{R5} = BT1\{[R2/(R1 + R2)] - [R4/(R3 + R4)]\}/[R1\|R2 + R5 + R3\|R4]$$

My previous book, *Build Your Own Transistor Radios* (McGraw-Hill), shows a detailed analysis of converting a bridge circuit to a simpler one.

Note that the construction of this circuit is relatively easy, just requiring adhesive tape (e.g., transparent, packing, masking, or duct tape) and an index card or even a piece of paper or cardboard. One advantage of this type of construction is fast verification of simple circuits, say, fewer than 20 components, depending on the size of the index card or cardboard. However, including integrated circuits into this type of prototyping technique is difficult because of their short leads. Most through-hole lead semiconductors will work fine, including transistors, light-emitting diodes (LEDs), photodiode, diodes, and rectifiers. Even vacuum tubes inserted into sockets will work with this approach.

Let's now look at an extreme radio circuit put on a 24 inch by 36 inch piece of drafting paper. All of the connections are twisted together. Figures 3-3 and 3-4 show the schematic and prototype. Note in Figure 3-4 the size of the antenna coil (top of the figure) in comparison with the AA battery on the bottom left and the tuning capacitor on the bottom right.

Believe it or not, this radio actually works. It pulled in about seven to nine AM radio stations. However, I would not recommend building radios with lead lengths this long. This extreme radio was built to show that one can get away at times with ridiculously long wire lead lengths and still have a working circuit.

When building prototypes, one may ask, "Will the circuit still work if the wire leads are long?" The answer is that generally anything working at under 2 MHz (AM band or below) can tolerate long leads and will still work. However, keeping lead lengths less than a few inches long is a general rule of thumb. For circuits working at higher frequencies or for avoiding picking up extraneous noise (e.g., 50 or

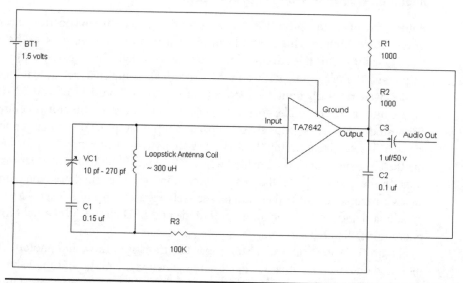

Figure 3-3 Schematic of a tuned radio-frequency (RF) radio using a 30-inch (2½-foot) loopstick antenna coil.

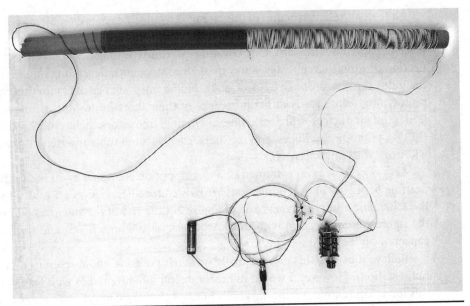

FIGURE 3-4   Extreme radio built with a 30-inch antenna coil with wires exceeding one foot.

60Hz hum), one should build circuits with leads that are less than an inch or half an inch long.

Knock on Wood—A Wooden Breadboard

Putting circuits onto a piece of wood is one of the oldest ways to construct electrical or electronic circuits. This method can be seen in books dating back to the nineteenth century with the advent of the incandescent light bulb, batteries, the telegraph, and the telephone. Radios in the early 1900s using *coherers* and later *galena crystals* were routinely built on wooden platforms or just on slabs of wood. Now see Figure 3-5. In this example, #6 and #4 sheet metal screws are fine for fastening the Fahnestock clips. These clips are often used as input, output, and power-supply terminals. To assemble a circuit, the #6 screws are used as anchor points or for mounting the clips onto the wood. I drilled with a $7/64$-inch drill bit to allow easier installation of the screws. Note that Fahnestock clips can be purchased on Amazon (www.amazon.com) and other sites on the Web. Figures 3-6 and 3-7 show a wooden breadboard and a schematic for a photodiode and LED circuit. When light hits the photodiode, the LED turns on.

Figure 3-7 shows how a resistor, capacitor, transistor, LED, and photodiode are mounted on a wooden board. The #6 screws anchor the parts, and the Fahnestock clips provide the power terminals for the 9-volt battery. As seen in this figure, the parts are put in a north-south-east-west structure. That is, the parts are preferably

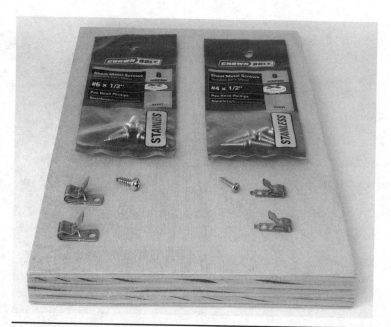

Figure 3-5 Wooden breadboard using Fahnestock clips and screws.

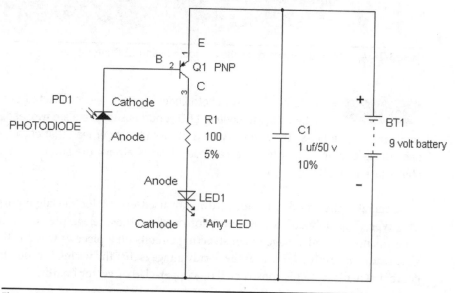

Figure 3-6 Schematic of a photodiode and LED circuit.

oriented at 90-degree angles, although at times diagonal orientations may be fine. Transistor Q1 may be a silicon transistor such as the 2N2905, 2N5087, or 2N3906 or even an old PNP germanium transistor, as shown in Figure 3-7. Note that the transistors today have shorter leads for the emitter (E), base (B), and collector (C).

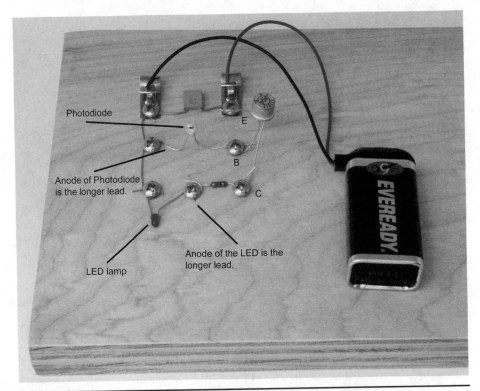

FIGURE 3-7   A wooden breadboard for the photodiode and LED circuit.

NOTE   Be sure that the photodiode and LED are connected correctly. When a photodiode or LED is purchased, there are two leads—the anode and the cathode. The anode's wire is always the longer lead that comes with a photodiode and an LED, while the cathode's lead is always the shorter wire in the photodiode and LED.

Generally, the wooden breadboard is not used except for nostalgic purposes. This type of breadboard is used often for crystal radios, telegraphs, or electricity experiments. One advantage of constructing circuits on a piece of wood is that the *stray capacitance* is low. However, the disadvantage is that the wiring is generally longer than when a circuit is constructed on copper-clad or vector boards.

Superstrip Prototype Boards

This is probably the most popular method of building circuits today. No soldering is needed, and the circuits can be quickly assembled and evaluated. The superstrip board has rows of connections conveniently spaced at 100 mils (0.100 inch), which

Figure 3-8   Precut wire with pins on each end and 22 AWG solid wire.

accommodates through-hole integrated circuits. Many of these types of boards can be purchased with binding-post connectors for hooking up one or more power supplies, and precut wires may be bought as well. Generally, insulated wires are used [e.g., 22 or 24 American Wire Gauge (AWG)], and preferably the insulation is removed by about ¼ inch. However, precut wires with pins on each end for the superstrip board can be purchased as well. See Figure 3-8.

There are many sizes of superstrip boards to choose from, including the board shown in Figure 3-9. This board is actually split into four independent sectors, which

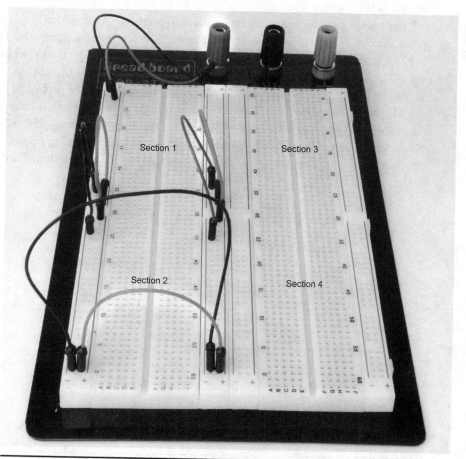

Figure 3-9   Interconnecting power-rail buses for sections 1 and 2.

allows prototyping of four independent circuits with separate positive (+) and negative (–) power-supply buses. Each section has its own (–) and (+) power rails to the left and right of the main sections A to E and F to J as seen at the bottom of the figure. To connect the power rails for sections 1 and 2, jumper wires are required (see Figure 3-9).

By interconnecting the (–) and (+) buses around "30," sections 1 and 2 now have power rails for the plus and minus supplies going down the whole left section of the prototype board. By further connecting (–) and (+) wires at "0" and "60" in the lower-left portion of the board, the plus and minus supplies are connected at both the left and right portions of the left-hand superstrip prototype board sections 1 and 2. Figure 3-10 shows a schematic of the one-transistor amplifier from Chapter 2, and Figure 3-11 shows an example of the one-transistor amplifier assembled on a superstrip board.

In this circuit, there is only a single supply, so the (–) supply is ground, and the (+) supply is +9 volts. Thus the two binding posts, Va for (+) and Vb for (–), are connected to the plus and minus buses of the board.

For another example of using the superstrip prototype for constructing an integrated-circuit (IC) audio amplifier, see Figure 3-12 for the schematic and Figure 3-13 for the assembled circuit. In this circuit, a smaller superstrip prototype board is used. The LM386 integrated circuit is inserted as shown. Pin 1 of the LM386 is connected to "10e to 10a," and pin 8 is connected to "10f to 10j" of the prototype board.

This audio amplifier should have worked on this prototype board but instead had problems with parasitic oscillations. The capacitances between the adjacent rows, such as rows 11 and 12 or rows 12 and 13, caused the LM386 to operate in an unreliable manner. *Therefore, **it is not recommended to the reader that this circuit***

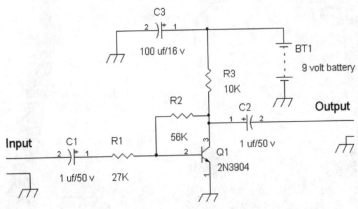

Notes: 1) All resistors are 1/4 watt and 5% tolerance.

2) All capacitors are aluminum electrolytic.

Figure 3-10   Schematic diagram of the one-transistor amplifier featured in Figure 2-52.

FIGURE 3-11   One-transistor amplifier assembled on a superstrip prototype board.

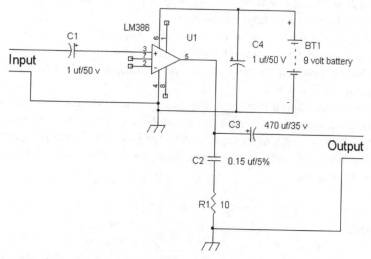

Note: For lower noise, pin 2 may be grounded.

FIGURE 3-12   Schematic of LM386 integrated-circuit audio amplifier.

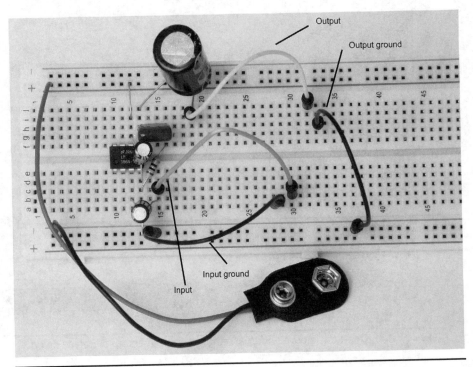

FIGURE 3-13 Audio amplifier prototype.

be built with this type of prototype board. It was found later with other types of methods, such as vector board or three-dimensional (3D) wire constructions, the LM386 operated properly without oscillations.

The superstrip board in general is fine for low-frequency circuits generally less than 2 MHz. I would not recommend using it for radio circuits. The parasitic or stray capacitances between the rows are sufficient to throw off the performance of radio-frequency (RF) circuits.

For certain audio applications and low-power circuits of less than half a watt (< 0.5 watt), this board is okay. Just be aware that if one builds a circuit on this board and it does not quite perform to expectations, the hobbyist should try a different construction technique, such as using vector or copper-clad boards, which generally require some soldering.

Soldering Tools

In general, soldering is a way of making a breadboard more permanent compared to super-strip boards. By soldering, the circuit can be transported without fear that the connections will come loose or fall off.

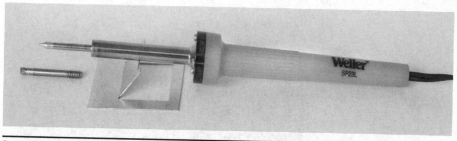

Figure 3-14   A Weller SP-23L soldering pencil with extra tip and holder.

The most commonly used soldering tool is the *soldering pencil*. It is light and can be more precisely manipulated into tight locations of the circuit board than using a soldering gun. Because it is lighter than a soldering gun, it causes much less fatigue than holding a soldering gun over a long period of time.

Generally, for small circuit boards, a 25 watt to 35 watt soldering pencil will do. For soldering more than just capacitors, resistors, inductors, small transistors, etc., a 45 watt soldering pencil allows reliable soldering to electronic parts that soak up the heat, such as open-frame variable capacitors, large power resistors, and mounted electrolytic capacitors for power supplies. Soldering pencils have replacement tips, which can be the conical or flat-blade chisel types. See Figure 3-14.

In the figure, the soldering pencil is resting on a soldering pencil holder to keep the tip from inadvertently moving around or from moving away. The holder is used so that the tip does not accidentally heat another object that can cause a fire or injury.

NOTE   For safety reasons, always rest the soldering pencil on a holder.

For the general method of soldering, normally, the lead or connector is heated with the soldering tip, and solder is added such that the heated lead or part melts the solder. This is the standard way that soldering is taught. However, I have found that placing the solder tip over the solder to melt the solder onto the wire and then moving the solder tip to heat the wire will produce a satisfactory solder joint.

Another type of soldering pencil includes a solder station, such as the one shown in Figure 3-15. The sponge normally holds some water to keep it moist so that the hot tip can be cleaned from time to time. Every now and then while soldering, the tip will accumulate dirt or other types of debris, which can be cleaned off by wiping the hot tip on the wet sponge.

Note that the AC power plug is three pronged with a ground plug. This means that if your circuit is still on, the tip of this soldering pencil that is soldering a part in the circuit will cause a short to ground. Therefore, be sure to turn off your circuit before you do some soldering on it. However, the tip of the soldering pencil can be isolated from ground with a 120-volt three-pin-to-two-pin ground-isolation adapter.

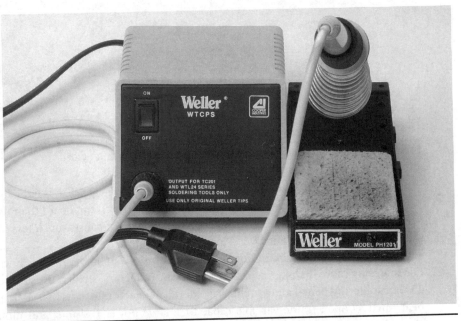

FIGURE 3-15 A solder station with a holder and sponge.

At this point, what type of solder is suitable? For electronics work, use *rosin-core* flux solder. There are other types of solder, such as bar solder, solder without flux; and then there's *acid core* solder, normally used for plumbing or pipes, which should ***never*** be used in electronic circuits.

Solder with a composition of 60 to 63 percent tin and 40 to 37 percent lead usually works best. For large jobs, the diameter of the solder can be on the order of 0.062 inch to 0.080 inch. Normally, solder of this diameter is used for large wires, cables, or electronic components. For most electronic circuits, tin-lead (Sn-Pb) rosin solder with about a 0.032-inch diameter should work fine. See Figure 3-16.

Tin-lead rosin-core solder is the preferred type for electronics work. There are other types, such as lead-free and 2 percent silver solder, but both are not recommended. Lead-free solder tends not to flow as well as the tin-lead solder, and 2 percent silver solder is expensive and does not have any significant advantage for general electronics work. Therefore, I would just recommend using 60/40 or 63/37 tin-lead rosin-core solder. Also note that solder can be bought in the ½ or 1 pound spool.

FIGURE 3-16 A tube of 0.032-inch-diameter tin-lead rosin solder.

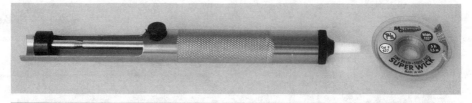

Figure 3-17   Desoldering pump for vacuuming excess solder and flat braided wire (Super Wick) for removing solder.

NOTE   Before shutting off a soldering pencil or gun, melt some solder onto the tip to prevent the tip from oxidizing or corroding. If the tip is oxidized, let the tip cool, and use fine sandpaper or emery cloth to remove the oxidation. Then turn on the soldering pencil or gun and melt some solder on the tip.

What if too much solder is applied, and there is a short circuit between connections? Is there a way to clear or remove the short circuit? Yes, there is a way to clear the excess solder by using a desolder pump. Alternatively, one can use bare copper braided meshed wire to soak up the excess solder and clear the short circuit. See Figure 3-17.

This desoldering pump has a trigger button shown in the middle of it. The desoldering pump must be loaded by compressing the plunger. Once loaded, place the tip of the pump in the desired location to remove excess solder. Then use a soldering pencil to heat the excess solder, and depress the trigger button of the solder pump to suck out the excess hot solder.

To remove solder using the "Super Wick," place the braided wire over the location of the excess solder. Use a soldering pencil to heat up the braid so that the excess solder flows into the braided wire. Remove the braided wire while the solder is still hot, and the excess solder should be cleared. If not, repeat the process with a clean patch of the braided wire. If there is a great amount of solder that is shorting the connections, a desoldering pump is preferred.

The Art of Pre-Tinning Wires or Connections

Sometimes it's hard to find a "third hand" while soldering two pieces of wire or connections together. Ideally, one makes a mechanical fit between the two connections by wrapping or bending the leads together. However if this is not possible, then pre-tinning the wires is a good practice. *Actually, in general **pre-tinning wires or connections should be done** whenever possible.*

FIGURE 3-18 Pretinning wires using two hands.

Figure 3-18 shows an example of pre-tinning leads. By using one hand to hold the electrolytic capacitor and the other to hold the soldering pencil, the negative lead of the capacitor is pretinned.

When two different leads are pretinned, one can then solder the two leads without having to feed solder. Once the two leads are secured mechanically, one can touch up the connection with a little extra solder.

Examples of Various Solder Joints

Figure 3-19 shows a series of nine solder joints on a copper-clad board. Solder joints numbered 1, 2, 3, and 5 are unreliable. Joints numbered 1 and 2 are cold solder joints that have the solder dabbed onto the copper instead of the solder melting and adhering to the copper for reliable electrical conduction. Solder joints 3 and 5 lacked sufficient heat and did not adhere to the copper very well. A telltale characteristic of soldering with insufficient heating is sharp or jagged edges on the solder joint.

Soldering examples from numbers 4 and 7 show good solder joints, with the solder adhering well to the copper. Solder joints number 6 shows a bubble or beading on the surface of the copper, which does not bond to the copper with a good

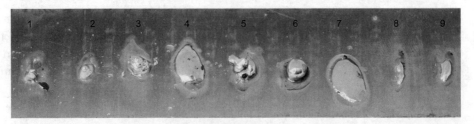

FIGURE 3-19 Examples of cold and good solder joints.

electrical connection. Joints numbered 8 and 9 may work, but more solder should have been used.

Prototype Circuits via Soldering

An example is a 3D breadboard of the circuit shown in Figure 3-13. Recall that using the superstrip prototype board that inherently has parasitic capacitances led to unreliable operation of this audio amplifier. Figures 3-20 and 3-21 show the schematic and the 3D wired prototype circuit. In Figure 3-21, we start with the LM386 integrated circuit placed upside down. With the (upside down) LM386's notch or dot pointing north, pin 1 starts at the upper righthand side (RHS), and the pins counts downward in a clockwise manner. Note that when the LM386 (or any 8 pin IC) is placed right side up and viewed from the top, pin 1 starts from the upper-left side and increments via a counterclockwise rotation.

First, solder the decoupling capacitor, with the negative terminal of the electrolytic capacitor soldered to pin 4 and the positive terminal of the capacitor soldered to pin 6 (see the left side of Figure 3-21). Then, as shown on the right side of the figure, add the 9-volt battery-connector lead, with the red positive lead soldered on the decoupling capacitor's positive lead and the black lead of the 9-volt battery connector soldered to the negative lead of the decoupling capacitor. Then start soldering the other parts as shown, such as the input and output capacitors, and observe the

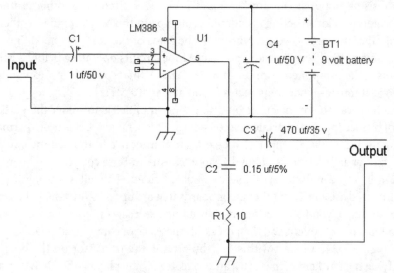

Note: For lower noise, pin 2 may be grounded.

FIGURE 3-20   Schematic of the LM386 audio amplifier.

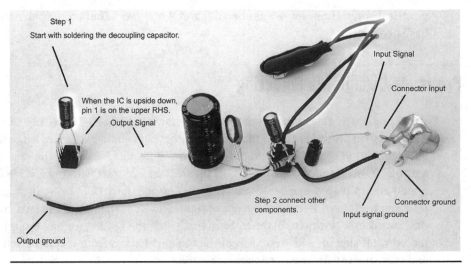

FIGURE 3-21   Prototype circuit wired in a 3D manner.

polarities of these capacitors as well. Finally, Figure 3-21 also shows examples of pre-tinning the input signal/ground leads and the input connector.

This amplifier as shown worked reliably and well. Who would have figured that a messy 3D prototype circuit can sometimes work better than a neatly assembled circuit on a superstrip board? But this happens often. The 3D circuit has minimal parasitic capacitances as one of its advantages over the superstrip board.

One of the first soldered prototype circuits was introduced in Chapter 2 as the one-transistor amplifier (Figures 3-22 and 3-23). This circuit is another example of a 3D prototype. Constructing a soldered prototype in this way provides a quick way to verify that the circuit works. Normally, the leads of the components do not need to be trimmed. Full-length leads can be used for circuits operating below 2 MHz. If the leads of the components are cut short (<½ inch), this type of 3D prototype circuit will work at much higher frequencies (up to at least 10 MHz).

Another way to construct the one-transistor amplifier is to solder the parts on a piece of blank copper-clad printed circuit board. This arrangement is more permanent (see Figure 3-24). Blank copper-clad boards are common, and they can be bought on the Web or at an electronics hobbyist's store. Normally, I use copper-clad boards that are 22 mils to 32 mils thick, single or doubled sided. This allows using a pair of medium-sized scissors to cut them. For boards that are up to 50 mils thick, I use metal cutting shears. Beyond 50 mils, one usually has to use a saw or score the board with an Exacto knife or equivalent and then break off the piece of copper clad board.

In Figure 3-24, we start with step 1 by placing the transistor on the board and identifying the emitter (E), base (B), and collector (C) by placing the 2N3904 transistor flat side up. In step 2, you know that the emitter is tied to "common," so we will use the copper as the common, ground, or ground plane. Therefore, the emitter of

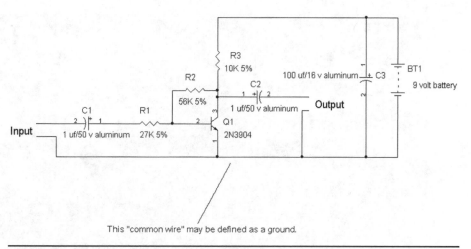

FIGURE 3-22   Schematic of a one-transistor amplifier.

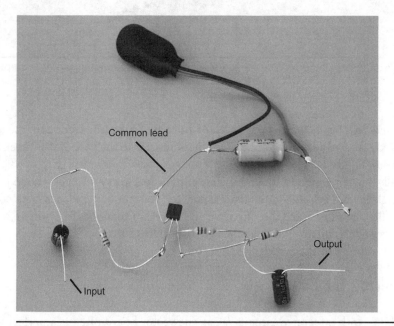

FIGURE 3-23   A 3D circuit of a one-transistor amplifier.

the 2N3904 is soldered to the copper ground plane. We also will need a decoupling capacitor, and thus its negative terminal is also soldered to the ground plane. Again, observe the polarities of electrolytic capacitors.

Using the positive terminal of the decoupling capacitor as a convenient support, solder one lead of the 10 kΩ resistor (brown, black, orange, and gold bands) to it, and solder the second lead of the 10 kΩ resistor to the collector of the 2N3904 transistor. Also, you can now solder in the two other resistors, a 56 kΩ resistor (green, blue,

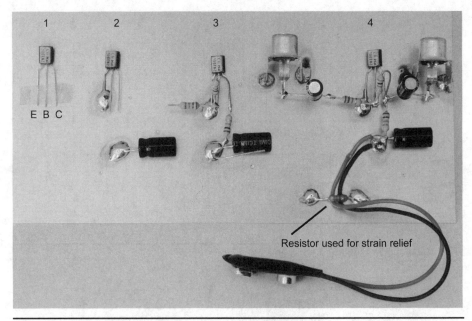

FIGURE 3-24   Copper-clad prototype version of a one-transistor amplifier shown in steps from left to right.

orange, and gold bands) across the collector and base of the transistor and one lead of the 27 kΩ (red, violet, orange, and gold) input resistor to the base of the transistor. See step 3 in Figure 3-24.

Finally, in step 4 we add the input and output capacitors and RCA phono jacks (left to right) and the 9-volt battery connector. To avoid the battery leads from coming off if the circuit is flexed too often, we add a resistor that goes to nowhere, with both leads soldered to ground to anchor the battery clip's two power leads.

Vector or Perforated Boards

For building with integrated circuits (ICs), the vector board often is a perfect way to mount them. Vector or perforated boards come with 100-mil spacings and thus allows easy installation of integrated circuits, certain connectors, etc. The leads of ¼ watt resistors can be bent to a 400- mil or 500-mil spacing. Radial-lead capacitors are often made with 100-mil spacings or multiples thereof, such as 200 mils, 300 mils, etc. See Figure 3-25.

Certain vector or perforated boards have solder pads on one side or solder pads on both top and bottom sides, and some of the more expensive ones have a ground-place mesh on one side. However, for most hobbyists, a perforated or vector board without the solder pads is more commonly used.

Figure 3-25   A perforated board without solder pads and one with solder pads on one side.

Building the LM386 Audio Amplifier on a Vector Board

For this circuit, a vector board without solder pads is chosen. To allow for flexibility, an 8-pin socket is used for the LM386. Instead of soldering the integrated circuit (IC) into the vector board directly, a socket allows changing the IC just in case there is a problem, such as a blown chip. Because there are no pads on the vector board to solder on, the socket is taped onto the board, and the power-supply decoupling capacitor is placed close to the socket. The socket has a notch, which is normally pointed north. Figures 3-26 and 3-27 show the schematic and the vector board.

On the left, the IC socket is taped to the vector board for an initial step in the construction. An electrolytic capacitor is placed near pins 4 and 5 of the IC socket.

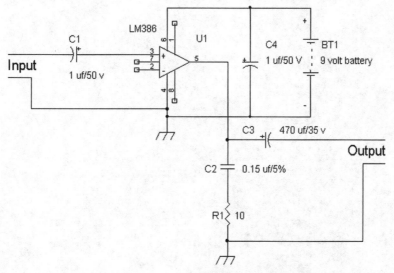

Note: For lower noise, pin 2 may be grounded.

Figure 3-26   Schematic diagram of the LM386 amplifier.

FIGURE 3-27   Vector board first try. The IC socket is taped to the board.

The socket is oriented to point its notch up or north. On the right side is a completed version of the audio amplifier, which works fine. Figure 3-28 shows the underside of the vector board. Once the decoupling capacitor's leads are soldered, the socket is secured to the vector board, which means that the tape on the top side

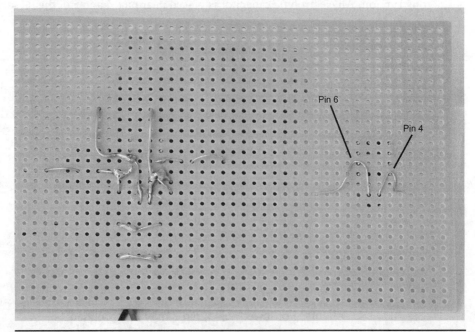

Figure 3-28   Bottom side of the vector board showing the decoupling capacitor's leads to be soldered at pins 4 and 6 of the IC socket.

can be removed, and the other parts may be installed and soldered to the other pins of the socket.

There are times when the battery can be reversed by accident, which can cause permanent damage to the LM386 chip. To prevent any momentary reversal of supply voltage into the IC, a protection diode (e.g., 1N4001-1N4005) is placed such that the cathode of the diode is connected to pin 6, the (+) supply terminal, and the anode of the protection diode is connected to pin 4, the (–) or ground pin of the LM386. Figure 3-29 shows the improved circuit with reverse battery protection, and Figure 3-30 shows the schematic diagram. Recall that the diode has a band on its body to denote its cathode.

As shown in the previous examples, different types of prototype constructions have advantages and disadvantages. For quick verification, the 3D and superstrip prototype methods work fairly well. For high frequencies, building on copper-clad boards is preferable. And for building something a little more permanent, a vector board is a good choice.

When the hobbyist starts building circuits or systems, eventually, the hobbyist, she or he, will begin to see a pattern such as starting with placing the decoupling capacitors near a chip or a main power-supply line. Eventually, a part of the circuit is memorized by she or he, and the hobbyist need not look at every part and connection one at a time to build the circuit. That is, a portion of the schematic is visualized in the hobbyist's head. Once the hobbyist has built maybe a dozen circuits, other skills will fall into place, such as remembering part of the resistor color code for often-used resistors. As I have mentioned to some people, building circuits is sort of like the way we learned to first copy a word letter by letter, then in groups of letters, and then finally just looking at the word and writing it down because we have learned

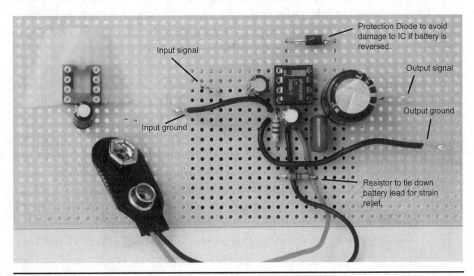

Figure 3-29 Completed LM386 audio amplifier with a protection diode to avoid accidentally damaging the IC.

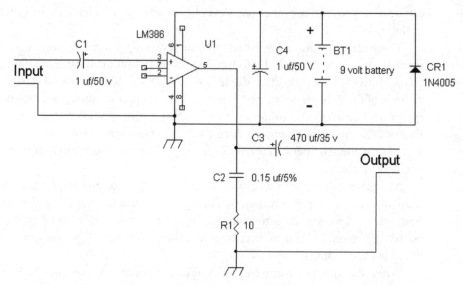

Note: For lower noise, pin 2 may be grounded.

Figure 3-30 Schematic diagram of the LM386 amplifier with CR1 a 1N4005 rectifier/diode for reverse battery protection.

that word. Eventually, an experienced hobbyist will be able to build circuits by soldering a few components at a time before looking at the schematic to connect the next few parts.

For building complicated circuits with many parts, one way to keep track of what has been wired or soldered is to make a copy of the schematic. Then, as the circuit is being built, use a yellow, green, or red highlighter (marker) to check off or to highlight over the area of the schematic that has been built.

Simple Test Equipment: Battery Testers and VOMs

In terms of a simple piece of test equipment, a battery tester is a prime example. An ideal voltmeter measures the voltages of circuits or batteries and draws very little current from the circuit or battery it is measuring. For testing batteries, we are interested really in one characteristic—what is the battery voltage tested under *load*? By *load*, we mean that batteries normally provide power to lamps, radios, motors, and other devices. These devices draw current.

A battery tester is a voltmeter that also includes a load, which is a resistor to draw current that is typical of a device's power consumption. For example, in flashlights, the typical current draw is between 250 mA and 1000 mA. And the current drain for radios ranges from about 10 mA to 100 mA.

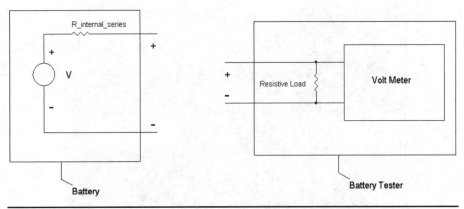

Figure 3-31   Model of a battery (*left*). Block diagram of a battery tester (*right*).

Any battery can be modeled or characterized as a perfect voltage source V with an internal series resistance R_internal_series. As the battery is being used, the battery voltage V drops, and the internal series resistance R_internal_series increases. See Figure 3-31).

An older battery tester (Eveready) is shown in Figure 3-32, and it has a selector switch for testing different types of batteries made for high and low current drains. This includes for testing at high currents that simulate the load from flashlights and testing at lower currents typical of the current draw from radios. This battery tester also measures the performance of photoflash batteries under "1.5 V PHOTOTEST FOR AN INSTANT ONLY." Dating back to at least the late 1950s, special batteries with higher current output (e.g., lower internal series resistance R_internal_series) were manufactured. I remember that these photoflash carbon batteries were available in D and AA sizes. The higher output current was needed to ensure that *flash-bulbs* were ignited reliably. However, these types of batteries were discontinued with the advent of inherently high-current alkaline batteries.

For this older type Eveready battery tester in Figure 3-32, we shall only concern ourselves with the tester's selector switch, set at the D-C-AA for toys, clocks, flashlights, and movie camers. Yes, in the old days, before video recorders, people shot 8-mm or super 8-mm home movies on film. The other selection we will concern ourselves with is the 1.5-volt and 9-volt setting for "transistor," meaning for use in transistor or solid-state radios (as opposed to the really old vacuum-tube portable radios).

Before we look at how this battery tester works, I tested three 9-volt carbon batteries (all of them had been used to some degree). See Table 3-1 for a summary.

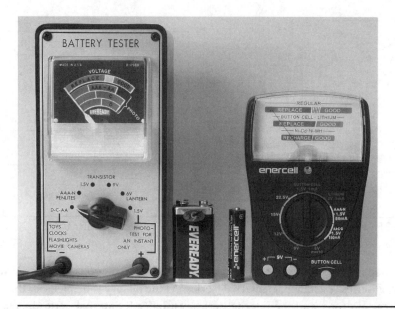

Figure 3-32   An older type ("Eveready") battery tester from the 1960s or 1970s (*left*), and a newer type (RadioShack's Enercell) battery tester from 2013 (*right*).

TABLE 3-1   Test Results for Three 9-Volt Carbon Batteries Using the Older Eveready Battery Tester In Figure 3-32

Battery Sample Number	Reading on Battery Tester	Voltmeter Reading (No Load)	Internal Series Resistance
1	GOOD—up to the first *O* in GOOD	9.24 volts	42 ohms
2	REPLACE—up to the *P* in REPLACE	8.90volts	445 ohms
3	REPLACE—one-quarter of the way up to the *R* in REPLACE	7.08 volts	2,950ohms

From Table 3-1, we see that as the battery is being more and more used up, the voltage drops, and the internal series resistance increases. In general, the voltage will drop a little, as evidenced by looking at samples 1 and 2 with voltages of 9.24 volts and 8.90 volts (showing a 3.7 percent drop), while the internal resistance increases dramatically from 42 ohms to 445 ohms, an increase of over 10 fold or over a 900 percent increase.

 Chapter 15 will explain Thevenin series resistances, which may be used for describing the internal resistance of batteries.

Now let's go back and analyze the battery tester. With a power supply, I found the voltage ranges for "REPLACE," "IN BETWEEN," and "GOOD." Also, with an ohmmeter, I measured the resistive load of the battery tester for 1.5 and 9 volts. The results are shown in Table 3-2, which also includes the current draw from the battery tester based on the nominal voltages of 1.5 and 9 volts. The current draw is calculated by Ohm's law. That is, $I = V/R$, where R = resistive load, and V = nominal voltage of a fresh battery, either 1.5 or 9.0 volts. For example, the current draw for a 1.5-volt battery with a 3 Ω load is $I = V/R = 1.5$ volts/3 Ω = 0.50 amp = 500 mA.

TABLE 3-2   Voltage Ranges for "GOOD" to "REPLACE" Resistive Loads and Current Draws for the Various Settings for the Older Eveready Battery Tester Shown in Figure 3-32

Selector Setting	"REPLACE" Voltage Range	"IN BETWEEN" Voltage Range	"GOOD" Voltage Range Resistive	Load and Current Draw
1.5-volt D-C-AA (toys, etc.)	0 to 0.75 volt	0.76 volt to 0.87 volt	0.90 volt to 1.30 volts	3 Ω, 500 mA
1.5-volt transistor	0 to 0.75 volt	0.76 volt to 0.87 volt	0.90 volt to 1.30 volts	10.2 Ω, 147 mA
9-volt transistor	0 to 5.7 volts	5.8 volts to 6.65 volts	6.70 volts to 9.88 volts	263 Ω, 34 mA

For 1.5-volt batteries, about 0.9 volt has been a historical cutoff voltage for a battery that is considered no longer good, which includes the "IN BETWEEN" range. At 0.90 volt, most flashlight bulbs or toy motors will operate very dimly or run very slowly. Batteries that measure above 0.90 volt and typically at 1.2 volts or more will be deemed good or sufficient to power a device.

Prior to the introduction of LED flashlights, when incandescent lamps were used, the nominal current draw from a two-D-cell flashlight with a PR-2 bulb was about 500 mA. Ergo, the Eveready battery tester via data from Table 3-2 has a 3 Ω resistive load for a nominal drain of 500 mA for a "fresh" 1.5-volt flashlight battery.

The 9 volt battery's usable performance has a cut-off voltage at about 6 volts. This was based on that by the time the 9-volt battery dropped to 6 volts, a radio was probably not playing very loudly anymore. If one looks carefully, the cutoff voltage is about 66 percent of the nominal voltage for both 1.5-volt and 9-volt batteries. That is, it's about 1.0 volt for the 1.5-volt battery and about 6 volts for the 9-volt battery.

For assessing what constitutes a GOOD or REPLACE rating, the actual voltage range and resistive loads are based on a judgment call. Other battery testers will have different ranges and load resistances. For example, available in the year 2013 is a RadioShack Enercell battery tester, as shown in Figure 3-32, that has a different take on the performance of batteries by testing batteries at lower currents. For instance, the highest current tested for 1.5-volt batteries is 150 mA versus 500 mA with the older battery tester. Also see Table 3-3 and compare it with Table 3-2 in terms of load currents and what voltages constitute "REPLACE" and "GOOD" readings on the meters.

TABLE 3-3 Voltage Ranges for GOOD to REPLACE, Resistive Loads, and Current Draws for the Various Settings for Testing Batteries of the Newer RadioShack Enercell Battery Tester Shown in Figure 3-32

Selector Setting	"REPLACE" Voltage Range	"IN BETWEEN" Voltage Range	"GOOD" Voltage Range	Resistive Load in ohms and Current Draw
1.5-volt AA-C-D high current	0 volt to 1.0 volt	1.0 volt to 1.1 volts	1.1 volts to 1.6 volts	10.2 Ω, 147 mA (150 mA)
1.5-volt AAA-N lower current	0 volt to 1.0 volt	1.0 volt to 1.1 volts	1.1 volts to 1.6 volts	30.0 Ω, 50 mA
9-volt transistor	0 volt to 6.0 volts	6.0 volts to 6.77 volts	6.8 volts to 10.0 volts	910 Ω, 9.9 mA

Using a Standard Volt-Ohm-Milliamp Meter (VOM) as a Battery Tester

We shall look at the VOM's voltage measuring capability and using it as a battery tester. Typically, digital VOMs are *auto-ranging*, which only requires an operator to select for DC voltages. VOMs with manual selections of the DC voltages are set for a maximum of 200 mV, 2 volts, 20 volts, 200 volts, etc.

For a 1.5-volt battery tester, we could choose a resistive load as shown in Tables 3-2 or 3-3. I will choose two examples from Table 3-2, a battery test for 1.5 volts with a 3 Ω load and for 9 volts with an approximately 263 Ω load.

The most common resistors are ¼ watt resistors, which will be used for the load resistors. However, for the two chosen examples above, we should first calculate the power dissipations to the resistive loads. The power into a 3 Ω resistor with 1.5 volts is

$$P = [V^2]/R$$

which leads to:

$$P = [(1.5 \text{ volts})^2]/3 \ \Omega = [2.25/3]\text{watt} = 0.75 \text{ watt, or } ¾ \text{ watt}$$

Thus, a 3 Ω, ¼ watt resistor will be destroyed if the 1.5-volt battery is applied to this resistor. One way to increase the wattage is to just string three ¼ watt 1 Ω resistors in series to provide a 3 Ω resistor at ¾ watt (3 × ¼ watt). However, this is running the power dissipation to the maximum. A safer approach may be to increase the wattage by connecting five 15 Ω, ¼ watt resistors in parallel to provide a 3 Ω resistor rated at 5 × ¼ watt, or 1.25 watts. Of course, other resistance values and combinations of series or parallel connections can be used. For example, ten 30 Ω, ¼ watt resistors connected in parallel will yield a 3 Ω resistor at 10 × ¼ watt, or 2.5 watts. Or six 0.51 Ω, ¼ watt resistors connected in series provide a 3.06 Ω resistor at 6 × ¼ watt, or 1.5 watts.

Now let's take a quick look at the 9-volt example, where the resistive load is 263 Ω. The closest standard value 5 percent resistor is 270 Ω. Therefore, the power dissipation is

$$P = [(9\text{Volts})^2]/270\ \Omega = [81/270]\text{watt} = 0.30\ \text{watt}$$

which exceeds ¼ watt. We can use a number of resistor combinations to provide a 270 Ω resistor that exceeds 0.30 watt. For example, stringing three 91 Ω resistors in series would work, providing a 273 Ω, ¾ watt resistor. Alternatively, connecting three 820 Ω, ¼ watt resistors in parallel will also give a 273 Ω, ¾ watt resistor.

We are now ready to use a VOM as a battery tester. Simply connect the resistive load across the VOM's test leads, and measure the voltage of the battery. Then refer to Table 3-2 to determine whether the battery is good or bad. For example, with a 273 Ω resistive load as described connected to the test leads of the VOM, measure the voltage of a 9-volt battery. If the resulting voltage reading is between 0 and 5.7 volts, the 9-volt battery is bad. If it reads between 5.8 volts and 6.65 volts, the battery is weak and should be replaced. And if the VOM reads 6.7 volts to 9.88 volts (or more), then the 9-volt battery is good.

Figures 3-33 through 3-35 illustrate a method of using a digital VOM as a battery tester. The resistive loads, as shown in Figure 3-34, are five 15 Ω, ¼ watt resistors connected in parallel to provide a 3 Ω, 1¼ watt resistor and three 820 Ω, ¼ watt resistors wired in parallel to provide a 273 Ω, ¾ watt resistor. Figure 3-31 shows an example of how a 1.5-volt battery is tested with a digital voltmeter (VOM) set at 2 volts (2,000 m for 2,000 millivolts) full scale.

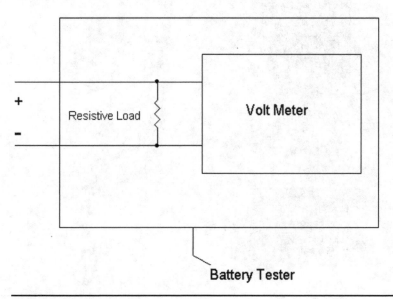

FIGURE 3-33   Block diagram of a (digital) voltmeter used to test batteries via a resistive load.

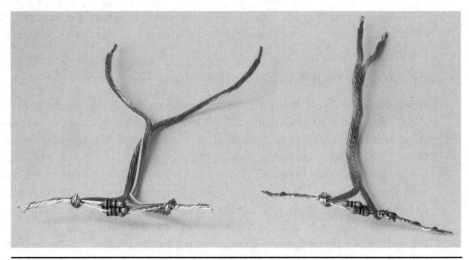

Figure 3-34 Five 15 Ω resistors connected in parallel to a zip-cord wire (*left*). Three 820 Ω resistors connected in parallel to another zip-cord wire (*right*).

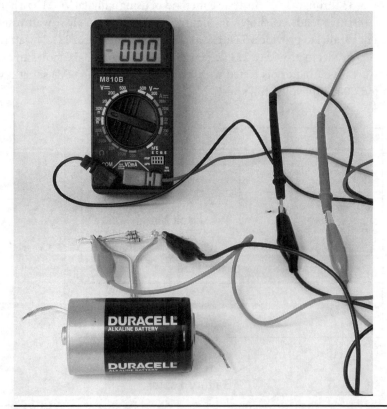

Figure 3-35 Example of hooking up the battery to the resistive load and then to the digital VOM.

Parts List

- Any digital VOM
- Two alligator jumper clip leads, preferably red for (+) and black for (−) leads
- Four wires or two short zip cords cut to about 4 inches to 6 inches
- Five 15 Ω, ¼ watt resistors connected in parallel, preferably soldered to one set of wires for testing 1.5 volt batteries
- Three 820 Ω, ¼ watt resistors connected in parallel, again preferably soldered to another set of wires for testing 9 volt batteries

Connect the battery terminals to the wire leads as shown in Figure 3-35. Test the battery for about 3 to 5 seconds to get a steady reading on the digital VOM, and use Table 3-2 to determine the voltage range for a good and bad battery.

 NOTE   Make sure that the respective resistive load is tested with the correct battery voltage.

References

1. Alfred P. Morgan, *Wireless Telegraphy and Telephony*. New York: Norman W. Henley Publishing Company, 1922.
2. *Eveready Battery Applications and Engineering Data*. New York, NY: Union Carbide, 1968 or 1971.
3. RadioShack, *Enercell Battery Guidebook*. Niles, IL: Master Publishing, 1990.
4. Ronald Quan, *Build Your Own Transistor Radios*. New York, NY: McGraw-Hill, 2013.

CHAPTER 4

Light Emitters and Receivers

We will explore light-emitting devices that turn electric current into light, such as incandescent lamps and light-emitting diodes (LEDs). The relative efficiency of these two types of light-emitting devices will be compared, and some experiments and simple projects will be introduced. Later on, photonic receiving devices that convert light into electricity will be examined. These receiving devices include solar cells and photodiodes. We shall see later that light emitting diodes and small signal germanium or silicon diodes convert light into electric current.

Incandescent Lamps

An incandescent lamp is made with a wire filament enclosed in a bulb without oxygen and glows as the filament is heated. Less than 10 percent of the electrical power into an incandescent light bulb is converted into light, and the rest is converted into heat. Lamps of this type are still used, but they are being replaced with fluorescent lights or light emitting diodes. The incandescent lamp therefore is a resistor that just happens to give out light. But what type of light?

White light is measured by its color temperature in degrees Kelvin (K). Typically, when we look outside on a sunny clear day, the Sun along with the blue sky provides a color temperature of about 4,500 to 5,500 degrees Kelvin. As the sun starts to go down in the afternoon, the color temperature drops to about 3,000 to 4,000 degrees Kelvin. Finally as the sun sets, we can clearly perceive the sunlight with a yellow to red tint, which means the sun's color temperature has dropped below 3,000 degrees Kelvin. Human eyes adapt to the color temperature for the most part from about 3,000 to 5,000 degrees Kelvin and perceive light in this range as "white," albeit at 3,000 degrees Kelvin, it has a warm tone. A standard incandescent bulb for room lighting such as a 100 watt bulb provides light at about 2,700 degrees Kelvin, which provides warm white light.

For studio or movie lighting, generally the color temperature is a bit whiter (between 3,200 and 3,500 degrees Kelvin, and sometimes up to 4,000 degrees Kelvin). Halogen lamps or white photoflood lamps provide light in this color temperature range. Incandescent lamps exceeding 4,000 degrees usually are specially made and they are often coated in blue. For standard low-power lamps such as flashlight bulbs or indicator lights, the color temperature is somewhere between 2,000 and 3,000 degrees Kelvin.

One can find out how any of these lamps look color temperature–wise by setting a digital camera's "White balance" setting to "Sunlight" and taking pictures while a flashlight or other incandescent lamp is on. Then view the picture on your monitor and you will notice a yellow or orange tint. The human eyes and brain adjust to these lower color temperatures and balance what is really a yellow or orange light as a warm white tone.

Incandescent bulbs can be raised in color temperature by using krypton or halogen lamps. Another way of raising the color temperature of an incandescent bulb is to manufacture a lamp to normally operate at a lower voltage, such as 110 volts instead of 120 volts, and then to run the bulb at 120 volts to provide a higher color temperature. The downside is that the life of the bulb is shortened considerably. For example, a 5,000 degree photoflood lamp for photography runs at 120 volts and has a guaranteed life of only 6 hours. The filament in this bulb is run much hotter than a standard 120 volt lamp one would buy for bathroom or living room lighting.

If raising the filament current beyond its standard operating level causes a dramatic shortening of its life, is the opposite true? That is, does lowering the current below its normal level increase its life? The answer is yes—and by a lot. A standard 120 volt AC 100 watt bulb normally lasts about 750 hours, but a 125 volt or 130 volt 100 watt bulb will last thousands of hours at 120 volts.

So if you want to extend the life of an incandescent bulb, just lower the voltage by about 5 percent to 10 percent. For example, if you have a typical flashlight bulb such as the PR2, rated at 2.4 volts, run it at about 2.2 volts, which will give a dimmer and more yellowish light, but it will last a lot longer. By the way, most flashlight bulbs are specified for about 10 hours of life, which may seem short, but since we generally use a flashlight intermittently, about a few minutes at a time, its bulb will last for years. Table 4-1 lists incandescent lamps and their standard operating voltages and currents.

An Experiment with a #222 Penlight Bulb

In this experiment, we will look at the relationship between current draw and voltage across the #222 lamp, a bulb that has been used in many penlights using two AA or AAA batteries. We will measure the bulb's current draw with 1.5 volts and 3.0 volts for the supply. Also, the bulb's filament resistance is measured with an ohmmeter.

TABLE 4-1   Incandescent Lamps for Flashlights and Indicators

Part Number	Voltage (volts)	Current (amp)	Description
14	2.47 volts	0.300 amp	Screw-base round-bulb flashlight
47	6.3 volts	0.150 amp	Bayonet-base pilot lamp
49	2.0 volts	0.060 amp	Bayonet-base pilot lamp
112	1.2 volts	0.250 amp	Screw-base prefocused penlight
222	2.2 volts	0.250 amp	Screw-base prefocused penlight
327	28.0 volts	0.040 amp	Single-contact flanged pilot lamp
1828	37.5 volts	0.050 amp	Bayonet-base pilot lamp
PR2	2.4 volts	0.50 amp	Flashlight bulb, two D cells
PR4	2.3 volts	0.27 amp	Flashlight bulb, two C or AA cells

After each reading for voltage and current, we shall calculate the resistance of the bulb by using Ohm's law, that is, $R = V/I$ (see Figures 4-1 and 4-2).

Parts List

- Double AA cell battery holder with two AA cells
- Two digital volt-ohmmeters (VOMs), but one VOM is workable
- Jumper clip leads
- #222 lamp
- Lamp holder for miniature bulbs

Alternatively one can wrap a wire around the #222 bulb's outer screw contact and solder a wire to the middle contact to bring out the two connections of the bulb.

To apply two different voltages to the lamp, we use a battery holder for two cells. A digital VOM is set to measure current at 10 amps full scale, and it is placed in

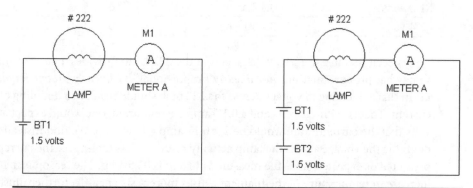

FIGURE 4-1   Schematic diagram of the test circuit with one and two cells providing current to a #222 bulb.

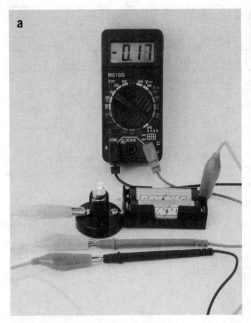
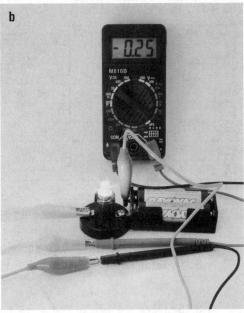

FIGURE 4-2 (*a*) A three-dimensional (3D) wired prototype test circuit for measuring current draw and voltage with one cell. (*b*) Current draw measured with two cells.

series with the battery. Due to the resistances in the digital VOM and the jumper wires, another VOM (not shown in the figure) measures the DC voltage across the lamp. A jumper lead selects power from one or two cells. The results are shown in Table 4-2.

TABLE 4-2 Current Draw of a #222 Light Bulb with Various Voltages Across It and Their Equivalent Resistances

Voltage (V)	Current (amp)	Equivalent Resistance V/I
1.2 volts	0.17 amp	7.0 Ω
2.4 volts	0.250 amp	9.6 Ω

As seen in Table 4-2, the voltage across the 222 lamp provides a current that is somewhat proportional but not *linearly* proportional. By *linearly proportional*, we mean that a doubling in voltage across the 222 bulb should result in a doubling of the current. That is, with 1.2 volts and a 0.17-amp current drain, one would expect at 2.4 volts that the current drain would be 2 × 0.17 amp = 0.34 amp. We find instead that doubling the voltage across the lamp actually results in less than twice the current as predicted or expected since the measured current is 0.25 amp. The reason is that the increase in temperature of the filament causes more resistance. To further illustrate this point, with the bulb unlit, the measured resistance (with a digital VOM) is about

1.0 Ω, which is much lower than the equivalent resistances of the partially and fully lit bulb at 7.0 Ω and 9.6 Ω.

What do we take away from this experiment on incandescent bulbs? The resistance is lower when the filament is colder than when filament is hotter. Another important characteristic of incandescent lamps is that they are not polarity sensitive. You can reverse the battery voltage into them, and they will light up the same. Moreover, as we also know, they work fine with AC voltages, which are voltages that alternate in polarity.

Light-Emitting Diodes

Light-emitting diodes (LEDs) are semiconductors that give out light when current is applied in one polarity (forward bias), and no light is given off when in the opposite polarity (reverse bias). Because LEDs are diodes also, electricity only conducts in one direction when the anode of the LED is coupled via a resistor or directly to the positive terminal of a supply and the cathode is connected to the negative terminal.

For example, in the LED keychain light from Chapter 2, the positive terminal of the battery is connected to the anode lead of the LED, while the cathode of the LED is connected to the negative terminal of the battery. However, reversing the LED leads will not allow the LED to be lighted. Figure 4-3 shows how to identify the anodes and cathodes of LEDs, and Figure 4-4 provides a close-up view of a LED. When you buy a LED, the short lead is the cathode, and long lead is the anode.

Unlike incandescent lamps that generally give out a full light spectrum from infrared to blue light, most LEDs give out light at only portions of the spectrum. These LEDs come with infrared, red, green, blue, yellow, violet, and ultraviolet emitted light. But what is the sequence of colors starting from the longest to the shortest wavelengths?

To answer this question, we go to our "friend," the resistor color code, which is based on the order of colors seen in a rainbow. Highlighted in bold are the common

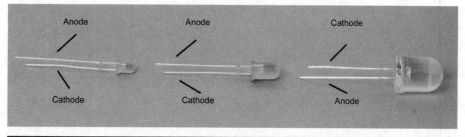

Figure 4-3 3 mm, 5 mm, and 10 mm LEDs with lead identifications.

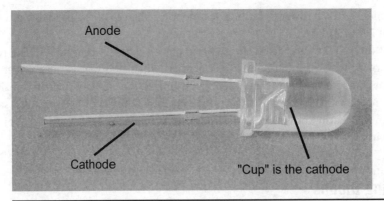

Anode

Cathode

"Cup" is the cathode

FIGURE 4-4   Close-up view of a LED, where identification of the cathode is via the cup.

visible colors available in LEDs. Recall that for resistors, part of the color code includes

- 0 = black
- 1 = brown
- 2 = **red**
- 3 = orange
- 4 = **yellow**
- 5 = **green**
- 6 = **blue**
- 7 = **violet**

Starting from the invisible longest wavelength, infrared, the color code tells us correctly that the sequence of visible colors will be red, yellow, green, blue, and violet from the order of the longest to the shortest of wavelengths for visible light.

In solid-state physics, which we do not have to know, there is a *bandgap* voltage that is related to the turn-on voltage of a LED. The higher the frequency or shorter the wavelength, the higher is the bandgap or turn-on voltage. Normally, a straight voltage source such as a battery or power supply is not connected to the LED unless we know that the applied voltage source is safe enough to do so without damaging the LED. How do we know this? First, we need to know the turn-on and turn-off characteristics of LEDs. The safest way to light up an LED is to include a series resistor. See Figure 4-5.

Now let's take a look at the minimum voltage needed to light up various LEDs. Table 4-3 lists typical turn-on voltages for LEDs, denoted by *VF*.

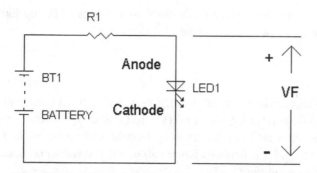

FIGURE 4-5 LED driven via a series resistor R1.

TABLE 4-3 Turn-On Voltage VF for Various LEDs

LED Type	VF (The Forward Bias or Turn-On Voltage)
Infrared (IR)	~1.5 volts
Red	~1.7 volts
Yellow	~2.0 volts
Green	~2.1 volts
Blue	~2.8 volts to 3.0 volts
White	~2.8 volts to 3.0 volts

But what about the white LED? It turns out that the white LED is really a blue LED that "cheats" by having the blue light excite a yellow phosphor to provide a white light that mimics a full-color spectrum like incandescent or fluorescent lamps. To see if this is true, see Figure 4-6. Here we will take a lit blue LED on the right and shine it into an unpowered white LED that has the yellow phosphor shown on the left to emit white light.

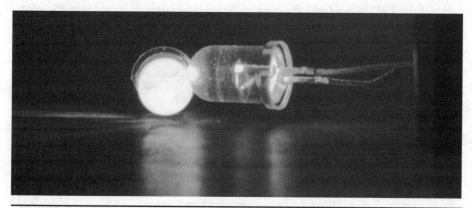

FIGURE 4-6 Shining blue light from the right into an unlighted white LED on the left produces white light.

 Since a white LED is really made from a blue LED, the turn-on voltage for a blue and white LED is the same.

Driving LEDs

Once an LED has a voltage applied beyond the turn-on voltage, the LED current will increase exponentially. Theoretically, about every 0.100 volt added to the turn-on voltage will result in an increase in LED current by tenfold. *Depending on the LED, the actual number varies, and it is between 0.060 volt and 0.120 volt of increase for a change in LED current by tenfold.*

For example, consider connecting a voltage source such as BT1 in Figure 4-5 directly to a white LED (i.e., there is no series resistor, or R1 = 0, in Figure 4-5) and turning it on initially with 2.60 volts that results in 0.01-mA of LED current. If we increase the LED voltage by 100 millivolts to 2.70 volts, the LED current will increase 10× to 0.10 mA. And if we increase the voltage by another 100 millivolts, resulting in 2.80 volts, the white LED current will increase another tenfold to 1.00 mA.

Put another way, every increase of 200 millivolts results in an increase in a particular LED current by a hundredfold. This is actually true until we take into account of the internal (series) resistance inside the LED. However, the internal resistance values in LEDs vary, and we cannot depend on these internal resistance values to drive a LED reliably. Therefore, to be on the safe side, always add an external series resistor to the LED or have the LED current driven. A current-driving circuit will be shown in Chapter 6. Generally, the external resistor value is larger than the internal resistance of the LED. For example, an external resistor may be 470 Ω, which is greater than the most internal resistances, which are typically less than 50 Ω. To determine the driving current of a LED, consider the following: given a voltage source BT1 that is greater than the LED turn-on voltage VF and series LED resistor R1, the voltage across the resistor is BT1 – VF (see Figure 4-7).

Also, we know that the current flowing into the LED resistor goes into the anode of the LED. Therefore, the current flowing through the resistor is the same as the

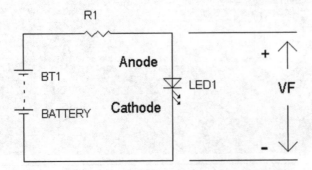

Figure 4-7 A voltage source BT1 driving an LED via R1.

LED current since the LED acts as a termination load to the resistor. Thus, the LED current is

$$I_{LED} = (BT1 - VF)/R1$$

For example, given a 5 volt supply for BT1 and a 1,000 Ω resistor for R1, for a typical white LED, the turn-on voltage VF is 2.8 volts, and the resulting LED current is $(5.0 - 2.8)$volts/1,000 Ω = 2.2 mA. For a yellow LED, where VF is 2.0 volts, the resulting current will be $(5.0 - 2.0)$volts/1,000 Ω = 3.0 mA.

For indicator lamps, typical LED currents are 1 to 10 mA. In LED flashlights, the current range is 5 to 50 mA for each white LED. Specialized LED bulbs can handle 1 watt or greater. For example, a 2 watt white LED has a turn-on voltage of about 3 volts and requires about 670 mA to deliver 2 watts (i.e., $P = VI = 3 \times 0.67 \approx 2$ watts).

Light Output Specifications for LEDs

The standard unit for measuring light output from a LED is the millicandela (mcd) for a specified LED current such as 20 mA. Until about 1983, the typical LED light output was about 10 mcd. However, by the mid-1980s, ultrabright LEDs began to emerge. These were brighter red-orange (660-nanometer wavelength) LEDs that were headed for the automobile market. In fact, an LED lamp assembly was to replace the incandescent third tail light in a car. During those days in the mid-1980s, the rated output of ultrabright red LEDs was 500 mcd, then 1,000 mcd, and then up to 2,000 mcd. The light output just kept on increasing as time passed. Also, in the middle to late 1980s, blue LEDs were just beginning to be manufactured, and the price was very high (in the hundreds of dollars for just one).

At the year 2013, one can routinely buy inexpensive white LEDs that provide about 24,000 mcd for 20 mA of current. In a dark room, an LED like this one was found to give off light with 0.8 μA, which is less than 1 μA (microamp). At 10 μA, this white LED was clearly visible in a normally lighted room. With efficiency like that, it is easier to make an LED flashlight with longer battery life, or make an emergency lamp that runs off a hand-squeezed generator. For example, with just 1.5 mA into a 24,000 mcd white LED would provide sufficient light for searching in the dark.

An improved longer-life LED keychain flashlight was featured in Chapter 2 (Figure 2-16), where the two original 2016 batteries were replaced with one 2032 battery. To further improve this flashlight, we can increase the light output by replacing the original LED. Note that the lead of the cathode of the LED has an insulating sleeve to prevent shorting out the (+) and (−) terminals of the battery. Be sure to remove the insulating sleeve from the original white LED and insert it into the cathode of a newer, brighter white LED of 24,000 mcd or more. Figure 4-8 shows where the light output is compared between the newer higher-light-output LED with the original LED. Each LED is driven with a 2032 battery.

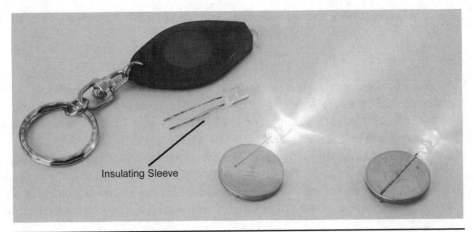

Insulating Sleeve

FIGURE 4-8   Comparison of a newer, brighter LED with the dimmer original LED.

LEDs Compared with Incandescent Lamps

For 120-V AC lamps, the compact fluorescent lamp is about 3 to 4 times more efficient than an incandescent lamp for the same amount of wattage. The LED lamp is about twice (2×) as efficient as the compact fluorescent lamp, which means that the LED lamp is about 2 × (3 to 4) or about 6 to 8 times more efficient than the incandescent light bulb.

Note that an LED projects light out from a two-dimensional surface, which is different from a regular bulb that emits light in all directions three-dimensionally. Another difference between an LED and an incandescent lamp (light bulb) is that as the voltage or current is reduced in LED, the tint of the light stays the same, as opposed to the light bulb that has a deeper color (e.g., from yellow to orange or orange to red) when the filament has reduced current. Therefore, the color temperature is essentially constant with an LED independent of current drive. Put another way, dimming an LED maintains the original color with just less light output.

Experiments with LEDs

LED Experiment 1

For providing small amounts of light, such as an indicator light to show that a device is on, typical currents for an incandescent lamp are 60 mA to 150 mA. To show just how efficient LEDs are in terms of converting electricity into light, we will show that an LED can be powered with a 220 µF capacitor driving the white LED at about 0.138 mA or less.

Given that R1 = 47 kΩ, BT1 = 9 volts, and VF ≈ 2.5 volts, the LED current is

$I_{\text{LED1}} = (\text{BT1} - \text{VF})/\text{R1} = [6.5 \text{ volts}/47 \text{ k}\Omega] = 0.138 \text{ mA}$

In Figures 4-9 and 4-10, the 9-volt battery charges capacitor C1 for a second or two, and the white LED will stay on for a while. See Figures 4-9 and 4-10.

The circuit consists of a capacitor as a temporary storage "battery" that is charged with a 9-volt battery. The 9-volt battery's (+) and (−) terminals are connected to the (+) and (−) terminals of the 220 μF, 16 volt electrolytic capacitor for about one or two seconds, and then it is disconnected from the capacitor. The capacitor delivers 9 volts to the 47 kΩ resistor, which provides about 138 uA to the white LED. The LED will stay on for seconds with the 220 μF capacitor, but the reader is encouraged to try a different capacitor, such as a 47 μF, 100 μF, 470 μF, or 1,000 μF capacitor, all being at least 16 volts in rating. Just be mindful that these electrolytic capacitors are polarized and that the battery charging the capacitor *must* be applied appropriately to the (+) and (−) terminals of the capacitor.

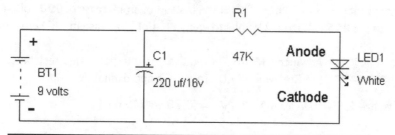

FIGURE 4-9   Schematic diagram of the circuit.

FIGURE 4-10   3D wired prototype of the circuit showing the LED being lighted by only the capacitor.

Parts List

- 9 volt battery
- 47 kΩ resistor (¼ watt, 5%)
- White LED (preferably at least 24,000 mcd, but a lower-light-output LED will also work)

LED Experiment 2

Earlier, it was noted for an ideal LED without an internal series resistor, every 100 mV of increase added to the turn-on voltage results in a tenfold increase in LED current. However, all LEDs include an internal series resistor. For many white LEDs after a few hundred microamps of current are applied, more than 100 mV is needed to generate a tenfold increase in LED current. For example, for the C503D-WAN white LED from Cree, Inc., at 2.9 volts, the LED current is 5 mA, and at 3.65 volts, the current increases to 50 mA. Therefore, instead of an increase of 0.10 volt, it took an increase of 0.75 volt from 2.9 volts to yield a tenfold increase in LED current.

 NOTE   **Advanced Math:** The equivalent internal series resistance from the Cree white LED example is approximately:

$$(3.65 \text{ volts} - 2.9 \text{ volts})/(50 \text{ mA} - 5 \text{ mA}) = [0.75 \text{ volt}/45 \text{ mA}]$$
$$= 16.67 \ \Omega$$

This is internal series resistance of this Cree white LED. The specification sheet for this Cree white LED states that usable light is provided with 3 volts or less for *VF*.

The next experiment is designed to determine whether a 24,000-mcd white LED will provide enough light with just two AA cells. The measured current with two NiMh batteries was 1.5 mA at 2.8 volts, and 9.8 mA at 3.2 volts with two alkaline batteries. Although the alkaline batteries provide much brighter light, the NiMh batteries supply 1.5 mA of current for useful lighting in the dark , which will last at least 1,000 hours, or 41 days, continuously. This makes a good emergency light! See Figure 4-11.

Parts List

- Two NiMh AA cells
- Two alkaline AA cells
- Two 10 mm (or 5 mm) white 24,000 mcd LEDs
- Two double AA battery holders with 9 volt battery connectors or equivalent

For other LEDs, the internal series resistance will vary. For example, see the 9 LED flashlight in Figure 4-12. After measuring the current for the 9 LEDs, it was deter-

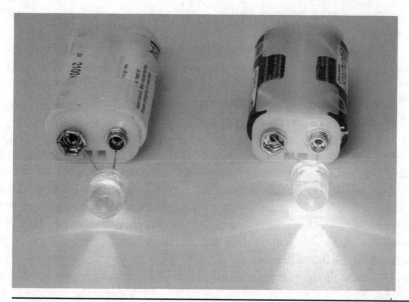

FIGURE 4-11 Two white LED experiments with NiMh cells on the left and alkaline AA cells on the right.

mined that the 9 LEDs wired in parallel give an equivalent internal resistance of about 3.3 Ω, which then results in each LED having an internal series resistance of 3.3 Ω × 9 = 30 Ω. That is, nine 30 Ω resistors wired in parallel is equal to 30 Ω/9 = 3.3 Ω.

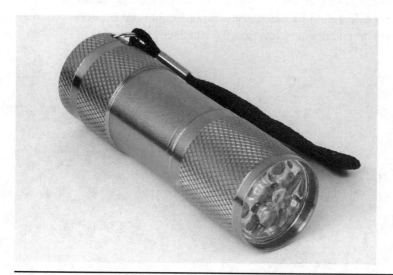

FIGURE 4-12 A nine-LED flashlight.

Photoreceivers

Devices that convert light into electricity can be classified into two categories:

1. *Solar cells*, which act as a voltage source or battery
2. *Photodiodes*, which convert light into an electric current

Figure 4-13 shows a rechargeable LED light on the left and on the right, after hacking into the LED light, its 9-cell panel solar cell array. These cells are "wired" in series. Each cell generates about 0.6 volt to develop a total of 9 × 0.6 volt = 5.4 volts to charge the 3-volt rechargeable battery, which then can be used to power the three LEDs (see Figure 4-14). A steering diode D1 provides charge only to the battery BT1 so that the rechargeable battery does not discharge through the solar cells when there is little or no light falling on them.

Note that the LEDs are connected in parallel and that they are connected straight to battery BT1, which this design can "get away" with because the voltage is known to be about 3 volts, and the battery has some internal resistance that prevents over-driving the LEDs. Normally, driving LEDs include series resistors, but in this case, the device works just fine.

A solar cell is essentially a voltage source very much like a battery that is modeled as a voltage source and an internal series resistance VR1. The solar cell's internal series resistance varies with the amount of light cast onto it. With more light, the lower the value of VR1, the internal series resistance, which allows more current output capability from the solar array. With less light falling on the array, the internal series resistance VR1 increases, which then lowers the output current capability of the solar cell. It should be noted also with less light onto the solar cell, the output voltage will drop and will be less than 0.6 volt. (see Figure 4-15).

Figure 4-13   Solar cell LED light on the left and its solar panel on the right.

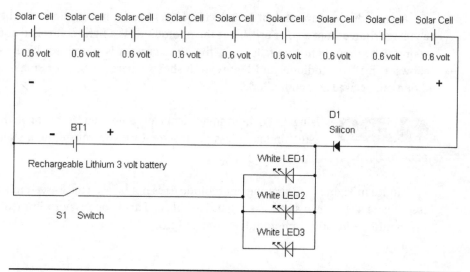

Figure 4-14 Schematic of the solar cell LED light.

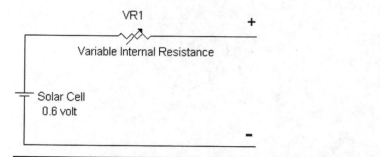

Figure 4-15 Schematic representation of a solar cell with a variable internal resistance, VR1, which is dependent on the amount of light received.

The most common photoreceivers today are the photodiode and phototransistor. A photodiode connected to a transistor results in a phototransistor that provides for amplified photocurrent. Figure 4-16 shows a schematic of a photodiode and phototransistor (photo NPN transistor).

Figure 4-16 Photodiode and photo transistor (NPN) schematic diagrams.

A photodiode converts light into electric current and is not considered to be like a voltage source from a solar cell. Thus the photodiode is a light-controlled current source. The more light that falls on it, the more current is generated. Figure 4-17 shows a "real" photodiode, and Figure 4-18 shows where an LED or small-signal diode can be used as a photodiode.

NOTE   Using LEDs and small-signal or zener diodes as photodiodes results in much less photocurrent when compared to a standard photodiode.

Notice in Figure 4-17 that the photodiode does not have a cup, as seen in a LED (see Figure 4-4), but instead has a flat dark plate. This is one way to immediately distinguish or identify a photodiode from an LED.

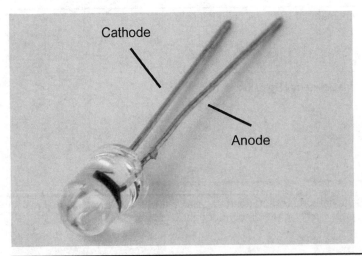

FIGURE 4-17   Photodiode with its anode (longer) and cathode (shorter) leads identified, which are the same for an LED.

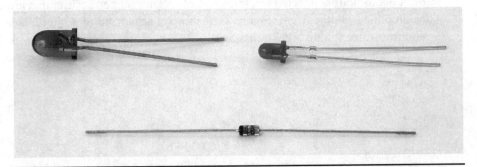

FIGURE 4-18   Two LEDs and a small-signal glass silicon diode used as photodiodes, which produce much less current than a standard photodiode.

So what exactly *is* a current source? First, let's look at the concept of an *ideal voltage source* or something close to it. For example, let's look at a 1.25 volt NiMh AA cell. If we put the two leads of a 12 Ω, ¼ watt resistor across the (+) and (−) terminals of the AA cell, we will read 1.25 volts across this AA cell, with about 100 mA of current draw via the 12 Ω resistor.

Recall that $I = V/R = (1.25 \text{ volts}/12 \text{ } \Omega) \approx 100$ mA. Now try the same experiment with a 120 Ω and 1,200 Ω resistors for current draws of 10 mA and 1 mA, and you will find that the voltage across the battery is still 1.25 volts. That is, regardless of the current draw from an "ideal" voltage source, the voltage is still the same.

An ideal current source supplies a specified current to a load or resistor regardless of the value of the resistor or load. Also, the output of the current source does not change with voltage across the current source. A photodiode then "dumps" current into a resistor based on the amount of light it receives. For example, if the photodiode generates 1 μA and is connected to a 100 kΩ resistor, 0.1 volt will be generated across the 100 kΩ resistor. That is, by Ohm's law, 1 μA × 100 kΩ = 0.1 volt. And, if the resistor is changed to 10 kΩ, the photodiode current is still dumping 1 μA into the 10 kΩ resistor to provide 0.010 volt across the 10 kΩ resistor. And again by Ohm's law, 1 μA × 10 kΩ = 0.01 volt.

Photodiode Experiment

We will look at measuring currents generated from LEDs, photodiodes, and silicon small-signal diodes. A zener diode is a voltage reference diode that can be used as a photodiode and will be covered in Chapter 5. Briefly, as a voltage reference diode, the zener diode works in reverse-bias mode (positive supply via a resistor to the cathode and negative supply to the anode) and provides a reference voltage from about 1.2 volts to over 100 volts depending on the zener diode's part number. In the forward-bias mode, the zener diode provides about 0.6 volt (positive supply via a resistor to the anode and negative supply to the cathode). The experiment will use whatever LEDs, diodes, or photodiodes the reader has at hand (see Figure 4-19).

In Figure 4-19*a*, the measured voltage is 17.9 mV across the LED. The digital VOM "automatically" has a 1 meg ohm (1 MΩ) input resistance. Thus the photocurrent is [17.9 mV/1 MΩ =] 17.9 nA. If a photodiode were used instead, the voltage measured would be much higher because of its greater efficiency to convert light into current when compared to an LED or diode used as a photodiode. If the reader measures a photodiode's voltage as shown in Figure 4-19*a*, the maximum voltage would limit out at about 0.4 volt DC because the photocurrent will develop a forward-bias voltage across its own photodiode. To more accurately measure the photocurrent, the photodiode is connected in reverse-bias mode by hooking up the positive terminal of a battery or supply to the cathode of the photodiode. This is shown in Figure 4-19*b*, and the schematic is shown in Figure 4-19*c*.

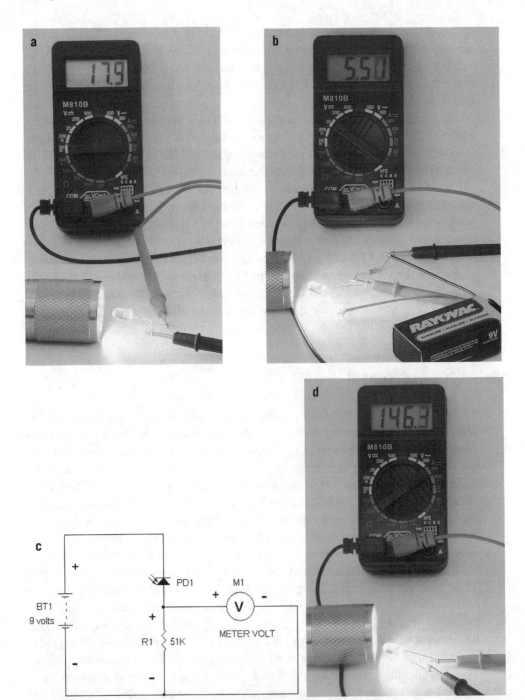

Figure 4-19 (*a*) Flashlight shining into the LED to measure voltage across the two LED leads on the 200-mV scale of the digital VOM. (*b*) Reverse biasing a photodiode via a 9-volt battery to allow a higher voltage to be read via measuring 5.5 volts across a 51 kΩ resistor, which results in 107.8 uA of photocurrent. (*c*) Schematic for measuring voltage across the 51 kΩ resistor. (*d*) Flashlight shining into a photodiode to measure current.

In Figure 4-19c, the photodiode's current from its anode dumps positive current into R1 to develop a voltage across R1. Via Ohm's law, this voltage is equal to the resistance value of R1 multiplied by the photodiode current. The reader is then encouraged to try different values for R1. *Without* moving the LED flashlight and the photodiode, change the resistor value of R1 by using jumper leads to parallel another resistor to R1. For example, take another 51 kΩ and parallel it to R1, and the voltage should drop by half because two 51 kΩ resistors in parallel provide ½ × 51 kΩ. Because the photocurrent does not change, the voltage drops by half because the new resistance value drops by half. Recall that $V = IR$. A more direct way of measuring photocurrent is simply to connect the photodiode to the digital VOM and measure for current on the 200 µA scale (see Figure 4-19d). In Figure 4-19d, the digital VOM is set to the 200 µA scale, which reads 146 µA of photocurrent generated from the photodiode.

Parts List

- Digital VOM (any)
- A 9-LED flashlight with fresh batteries or any very bright flashlight
- Alligator-clip leads
- LEDs, glass diodes, or photodiodes
- 9 volt battery and its connector
- 51 kΩ resistor(s)

Experiment for Listening to a Remote Control's Signals

For a second experiment on photodiodes, did you know that you can "listen in" on your remote control? A remote control sends out a series of light pulses whose frequencies are in the audio range. All we need to do is to connect a photodetector such as a 1N914 or 1N4148 signal diode to a simple amplifier that drives a loudspeaker (see Figures 4-20, 4-21a, and 4-21b).

The schematic in Figure 4-20 shows that an LED or photodiode may replace D1. Note that if a LED is used as a photodetector, it is generally only sensitive to the wavelength it emits. For example, a red LED can sense red light but has poor sensitivity to other than red light, such as green, blue, yellow, or infrared light. The remote control provides infrared (IR) light, so you should use an IR LED as the photodetector in this experiment. Other types of LEDs most likely will not produce sufficient photocurrent from the remote control. Note that the remote control's own LED is placed very near the photodetector (D1, LED1, or PD1) for this experiment.

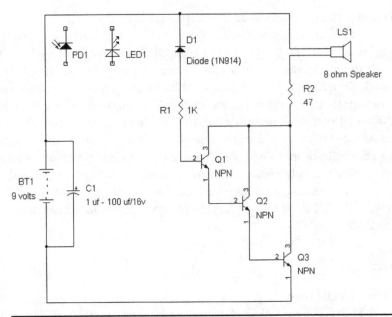

Figure 4-20   Schematic including a 1N914 (D1) or IR photodiode (PD1) or IR LED (LED1) connected to a three-transistor loudspeaker amplifier.

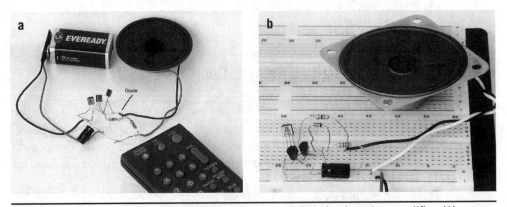

Figure 4-21   (a) Prototype board in 3D style of the photodiode loudspeaker amplifier. (b) Photodiode receiver using a superstrip board.

Parts List

- 9 volt battery
- Three NPN transistors (e.g., 2N3904, 2N4124, 2N2222, or 2N4401) of virtually any silicon type
- 1 kΩ resistor
- 47 Ω resistor

- 1µF to 100 µF, 16 volt electrolytic capacitor
- Loudspeaker (4 Ω to 100 Ω)
- IR LED or photodiode (e.g., RadioShack Part 276-0142)
- Silicon glass diode (e.g., 1N914, 1N4148, or glass zener diode > 10 volts)
- Any remote control with a LED emitter

The photodiode D1 is a common small-signal silicon diode. D1 is connected in reversed-bias mode with the cathode being tied to +9 volts. The anode is connected to a 1 kΩ series resistor R1 that acts to protect diode D1 and transistor Q1 should D1 be connected incorrectly. The other lead of the 1 kΩ resistor is connected to the base of Q1, which amplifies the photocurrent from D1. The emitter output of Q1 is connected to the base of Q2 to further amplify the photodiode's current. Finally, the emitter output of Q2 is connected to the base of Q3 to yet amplify the photodiode's current again. The collector output of Q3 then supplies sufficient pulsating current to the loudspeaker via the 47 Ω series protection resistor (see Figures 4-20 and 4-21*a*).

To allow a change to various other types of photoreceivers, such as LEDs, germanium diodes, photodiodes, zener diodes, etc., one can build the amplifier on a superstrip and simply replace each photoreceiver/diode on the board (see Figure 4-21*b*). Just be aware of the polarity of the diodes. That is, the cathode is tied to the +9 volt supply. If the LED used as a photodiode lights up, the leads need to be reversed. Also note that a 9 volt supply may provide too much reverse bias to an LED used as photodiode, and therefore, the supply voltage may be reduced from 9 volts to 4.5 volts.

We have seen in this chapter that LEDs give out light and can be used as photosensors as well. LEDs also have turn-on voltages for specific colors or wavelengths that can be used as voltage reference diodes. In Chapter 5, we will look into other types of diodes that are used for rectification and voltage references. Chapter 6 will cover transistors and show you that LEDs are also useful as voltage references for transistor circuits.

References

1. "Cree 5 mm C503D-WAN LED Data Sheet," Cree, Inc., Durham, NC.
2. "Osram SFH 229 Photodiode Data Sheet," Osram GmbH, Munich, Germany.
3. Hewlett-Packard, *Optoelectronics Fiber-Optics Application Manual*, 2nd ed. New York: McGraw-Hill, 1981.
4. Infrared Emitter and Detector, Part 276-0142 (includes a data sheet for the IR detector and emitter), RadioShack, Fort Worth, TX.
5. Assorted LEDs, Part 276–1622 (includes data sheets for the red, green, and yellow LEDs), RadioShack, Fort Worth, TX.

CHAPTER 5

Diodes, Rectifiers, and Associated Circuits

Diodes are used almost everywhere for converting alternating current (AC) to direct current (DC), signal detection, switching or diverting signals, or even controlling the gain of signals. For example, in every AC adapter that generates a DC power supply, diodes are used whether the AC adapter is a conventional iron-core linear adapter or a highly efficient switching-power-supply circuit. In this chapter, we will explore the turn-on voltage characteristics of various types of diodes and some applications of diodes in electronic circuits.

Diodes and rectifiers are devices that conduct electricity in one way, or have polarity. Generally, rectifiers are relegated to applications that provide power or large signals, while diodes are often referred to as *small-signal devices*. Sometimes a device may be called a *diode-rectifier*, meaning that the diode can handle large currents or signals. Either way for this book, a *diode* is a device that handles both small and large signals, and a *rectifier* normally will just be used for large signals, such as a power-supply rectifier. In addition, zener diodes that are generally used for defining reference voltages will be briefly covered.

Characteristics of Diodes and Rectifiers

An ideal diode is a device that conducts with zero resistance in one direction and provides an open circuit in the other direction. For instance, if CR1 is an ideal diode instead of a 1N4002 or 1N4007 device and its anode is connected to a positive voltage, as shown in Figure 5-1, we would expect the voltage across resistor R1 to be 9 volts for the forward-bias mode and 0 volts for the reverse-bias mode when the battery's negative terminal is connected to the anode.

However, with any manufactured diode, there is always a voltage loss across the anode to cathode when the diode is conducting (e.g., in the forward-bias mode). This forward voltage loss is sometimes called the *forward voltage drop VF*. The for-

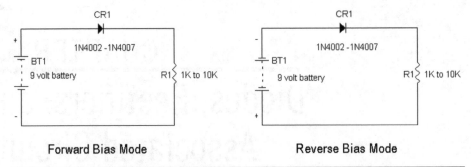

Figure 5-1   Forward- and reverse-bias connections of a diode.

ward voltage drop depends on the type of material in the diode, such as silicon, germanium, or silicon with aluminum (a Schottky diode). To investigate the forward voltage drop of a diode, we can forward bias a power rectifier such as the 1N4002 and compare the voltage drop for various currents to a small-signal diode such as a 1N914 or 1N4148 (Figure 5-2).

Parts List

- 9 volt battery and connector
- 1 kΩ, 10 kΩ, and 100 kΩ ohm resistors, ¼ watt, 5%
- 1N914 or 1N4148 diode or any silicon small-signal diode
- 1N4002 to 1N4007 rectifier or any silicon power rectifier
- Digital volt-ohmmeter (VOM)

For the experiment, the following results are shown in Table 5-1. Note that the reader may have slightly different results.

The diode current is calculated by finding the voltage across R1 (V_{BAT1} – VF), and using Ohm's law, the diode current is

$$I = (V_{BAT1} - VF)/R1 \qquad \text{Note: } BT1 = V_{BAT1} = 9 \text{ volts}$$

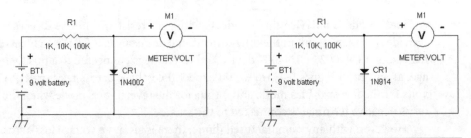

Figure 5-2   Experiments in measuring the forward voltage drop across the anodes and cathodes of a rectifier and a diode.

TABLE 5-1   Measured Forward Voltages of Diodes

Diode CR1	Resistor R1	Forward Voltage (VF)	Diode Current
1N4002	100 kΩ	0.420 volt	0.0858 mA
1N4002	10 kΩ	0.520 volt	0.848 mA
1N4002	1 kΩ	0.635 volt	8.365 mA
1N914 or 1N4148	100 kΩ	0.469 volt	0.0853 mA
1N914 or 1N4148	10 kΩ	0.585 volt	0.8415 mA
1N914 or 1N4148	1 kΩ	0.699 volt	8.301 mA

Small-signal diodes have appreciable internal resistances, while power diodes, by their nature to handle high currents inherently have low internal resistances. As shown in Table 5-1, at about 10 mA, VF, the forward-bias voltage, is higher with a 1N914 when compared to a 1N4002 power rectifier. Also note that for about every tenfold (10×) increase in current, the 1N4002 shows about a 100 mV to120 mV increase in voltage drop. Thus, for a hundredfold (10×) increase in diode current, the diode voltage increases somewhere in the range of 200 mV to 240 mV. From this observation, we see that a diode has a similar voltage-to-current relationship as an LED. For a tenfold (10×) increase in LED current the result is an increase in forward LED voltage (across the anode and cathode) in the range of 60 mV to 120 mV. Typical turn-on voltages are 0.2 volt for germanium diodes, 0.4 volt for Schottky (silicon-aluminum) diodes, and 0.6 volt for silicon diodes.

Converting AC to DC for Power Supplies

Probably since the beginning of electronics, one of the most common uses of diodes or rectifiers was to convert an AC voltage into a DC voltage. Vacuum-tube diodes served this purpose before the invention of solid-state diodes. The galena crystal diode, iron oxide or "rusty razor blade" diode, copper oxide, and selenium rectifier preceded the "modern" germanium and silicon diode.

There were early uses for converting AC signals to DC signals with diodes or rectifiers in the early part of the twentieth century. One was in the rectification of an amplitude-modulated (AM) signal for crystal and tube radios. A second use was just converting the AC power from the household power lines into a usable DC voltage supply to power tube amplifiers or radios.

Let's first take a look at an AC signal. An AC signal holds positive and negative voltage (e.g., +1 volt and –1 volt) values. But just as a person cannot be in different places at the same time, a signal cannot be at two different values at the same time. A person can be at two different places at different times, and similarly, a signal or voltage can assume a different value at different times. Figure 5-3 shows a familiar

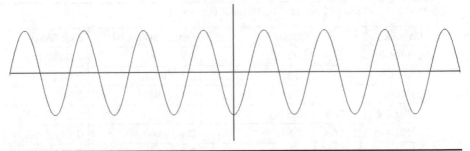

Figure 5-3 Sine-wave signal with vertical axis showing amplitude and horizontal axis showing time.

sine-wave signal that oscillates between positive and negative peak values. Values above the horizontal axis are positive, and those below this axis are negative. An AC signal can be provided by an AC adapter that converts 110 volts or 220 volts from the wall socket to a lower AC voltage such as 10 volts AC.

Before delving into rectifier circuits, we should know the following about sine waves:

1. An AC voltage is usually rated in RMS (root-mean-squared) voltage. For this chapter, we will need to convert this voltage to its positive or negative peak value.

2. The positive and negative peak values of an AC signal are just $1.41 \times$ RMS voltage. For example, if an AC adapter provides 10 volts AC (VAC), the positive peak value is $+10 \times 1.41$ volts peak, which is $+14.1$ volts peak. Similarly, the negative peak value for this example is -10×1.41 volts peak, or -14.1 volts peak.

3. To convert back to RMS voltage when the peak voltage is known, the RMS volts = (peak voltage)/1.41. Using the preceding example, a 14.1 volt peak sine wave is = (14.1 volts/1.41) RMS = 10 volts RMS, which is what we would expect.

NOTE When you use a digital or analog VOM to measure AC voltage, the VOM is measuring only the RMS value.

Basically there are two types of common rectifier circuits, half wave and full wave. We shall examine the half-wave rectifier circuit first. See Figure 5-4.

The left side of Figure 5-4 shows a rectifier circuit to provide a positive pulsating DC voltage, while on the right side a circuit is shown to provide a negative pulsating DC voltage. Both circuits as shown will generate *average* positive and negative DC voltages. Figure 5-5 shows rectification of the positive and negative cycles of a sine-wave voltage source with "ideal" diodes that have VF = 0.

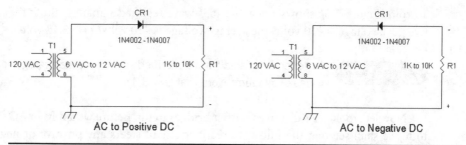

AC to Positive DC   AC to Negative DC

Figure 5-4   Half-wave rectifier circuits to provide positive and negative DC voltages.

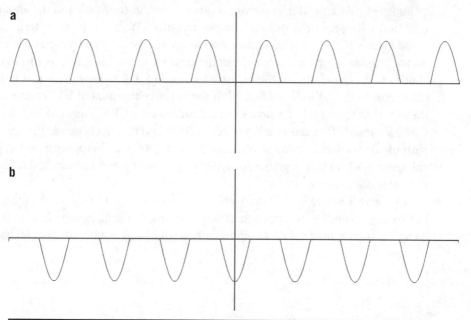

a

b

Figure 5-5   (*a*) Pulsating positive DC voltage, where the vertical axis is the amplitude and the horizontal axis denotes time. (*b*) Pulsating negative DC voltage, where the vertical axis is the amplitude and the horizontal axis denotes time.

If the peak positive voltage is +10 volts, the average voltage is +10 volts/π = +3.18 volts DC. One may ask why is it such a low number for the average voltage? After all, it's about a third of the peak voltage. The reason is that unlike a +10-volt battery that gives out +10 volts DC all the time, the voltage shown in Figure 5-5*a* only peaks out at 10 volts for a short time, and at other times, the voltages are less than 10 volts. Also, the positive sine-wave cycles are "on" only half the time, while the other half of the time the waveform is flat-lined to 0 volt. Therefore, we know that the average voltage has to be less than 5 volts because half the time there is no voltage and the other half of the time the positive-cycle voltage of the sine wave is generally less than

10 volts. Figure 5-5*b* shows a negative half-wave rectified signal. Again, if the negative peak voltage is –10 volts, the average voltage is –10 volts/π = –3.18 volts.

NOTE In general, for a half-wave-rectified sine wave, the average voltage = peak voltage/π. Note that π ≈ 3.14.

Now let's look at a full-wave rectifier bridge circuit. Before we apply an AC signal to it, we shall see that the full-wave rectifier circuit steers any positive or negative voltage to a positive voltage (Figure 5-6).

In Figure 5-6*a*, the output voltage into the load resistor R1 is supplied by the cathodes of CR1 and CR3 to provide a positive output terminal, and the anodes of CR2 and CR4 provide a negative output terminal. With the positive terminal of 9-volt battery BT1 connected to the cathode of CR2, we get a reverse-bias condition. So the only path for conduction from the positive terminal of BT1 is through the anode of CR1. Similarly, for the negative terminal of BT1 to flow current, it must be connected to the cathode of CR4. With the negative terminal of BT1 connected to the anode of CR3, we get a reverse-bias condition for CR3. Thus diodes CR2 and CR3 are turned off or essentially not connected. And therefore, we see that the only path of conduction is through diodes CR1 and CR4, which appropriately apply a voltage of positive and negative polarities across the top and bottom of R1. See the right side of Figure 5-6*a*.

In Figure 5-6*b*, we reverse the 9-volt battery's connection to see the resulting output voltage. With the battery connections reversed, the negative terminal of BT1 is connected to the anode of CR1 and the cathode of CR2. The only path of conduction

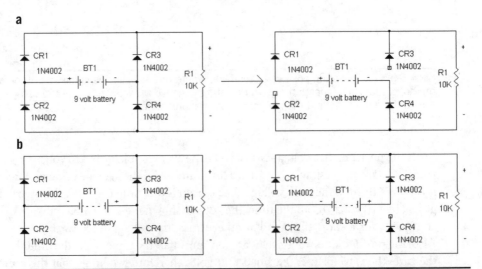

FIGURE 5-6 (*a*) Full-wave rectifier circuit with a 9 volt DC source attached to two of the inputs. (*b*) Full-wave rectifier circuit with the 9 volt battery reversed, with the same output voltage.

is via the cathode of CR2. Diode CR1 is running at reverse-bias mode and is therefore not conducting current. Now, looking at the positive terminal of BT1, we see that it is connected to the anode of CR3 and the cathode of CR4. Again, the only path of conduction is via the anode of CR3 to the positive terminal of the battery. Having the positive terminal tied to the cathode of CR4 results in a reverse-bias condition and no conduction of current. This essentially means that CR4 is an open circuit.

Therefore, CR1 and CR4 are open circuits with the only path of conduction being via CR2 and CR3. From the right side of Figure 5-6b, we see that again the voltage across R1 has the positive polarity on the top and the negative polarity on the bottom.

A full-wave bridge rectifier circuit then steers the input signal, whether positive or negative, to provide an output voltage of the same polarity. That's pretty cool! This circuit is very useful not only for providing rectification of AC signals but also as a failsafe circuit so that if the battery or DC source is hooked up backwards at the input, the DC output voltage from the full-wave rectifier circuit always provides the (same) correct polarity.

Now let's look at the full-wave rectifier circuit connected to an AC source. See Figure 5-7. The voltage across R1 is another pulsating DC waveform. Figure 5-8 illustrates the output of a full-wave rectifier circuit with "ideal" diodes where the forward bias voltage VF is zero.

For a full-wave signal as shown in Figure 5-8, how would we guess what the average voltage is if we know the peak voltage? We know from the half-wave rectifier signal that the answer is that:

Average voltage = (peak voltage)/π

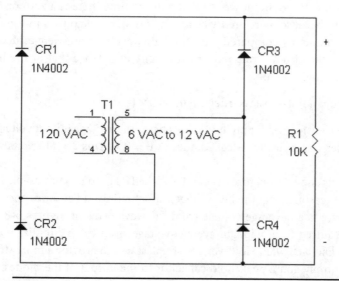

Figure 5-7 An example of a full-wave rectifier with an AC input source.

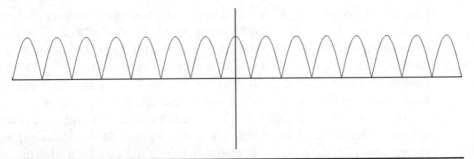

FIGURE 5-8   Full-wave rectifier output signal.

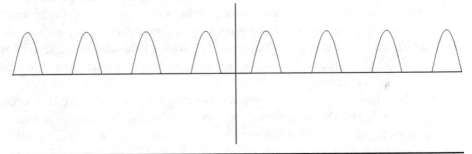

FIGURE 5-9   Half-wave rectifier output signal.

But we also know that we have twice the amount of half-cycles in the full-wave signal than in the half-wave version. Figure 5-9 (and Figure 5-5a) shows the half-wave rectifier version.

For a half-wave rectifier circuit we see that half the time there is a zero or flat-lined voltage. Thus, because there are twice as many (half) cycle pulses in the full-wave rectifier output signal compared with the half-wave version, we can deduce that the average voltage from a full-wave rectifier circuit is twice that of the half-wave circuit. Accordingly, then:

Average voltage from a full-wave rectifier circuit is 2 × (peak voltage/π)

For example, for a transformer supplying 10 volts peak, the average output voltage of a full-wave rectifier circuit using "ideal" diodes (VF = 0 volt) is 2 × (10 volts/π) = 6.36 volts DC.

For both circuits shown in Figures 5-4 and 5-7, the DC output is not usable for a power supply. The reason is that the DC voltage is fluctuating and at times goes to 0 volt. Although there is an average voltage from the waveforms in Figures 5-8 and 5-9, a circuit is required to provide an average voltage from these circuits. Such a circuit would be a *low-pass filter*. However, the simplest way to provide a more steady DC voltage is to simply add a peak-hold capacitor to the output of the diodes. See Figure 5-10.

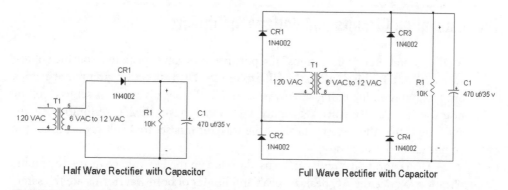

FIGURE 5-10   Half- and full-wave rectifier circuits with capacitors to convert the pulsating DC voltage into a more steady DC voltage.

In both circuits, the voltage output is about the peak voltage. The primary winding of the transformer takes in 120 volts AC, and the secondary winding delivers 6 volts AC to 12 volts AC into the rectifier circuits. At 6 volts AC, the peak voltage is 6 volts × 1.41 = 8.46 volts, which is approximately the voltage at C1 in both circuits.

Again, for a sine-wave voltage source, the peak voltage $V_{peak} = 1.41(V_{RMS})$. Note that V_{RMS} is the rated AC voltage from the transformer. The actual DC output voltage is less. In the half-wave rectifier circuit, there is about a 0.6 volt loss due to the 1N4002 diode. Thus the output is (8.46 – 0.6) volts = +7.86 volts into C1.

In both rectifier circuits, the working voltage for C1 is 10 volts or greater. Usually it is best to pick a capacitor with at least a 50 percent margin over the peak voltage or, as a rule of thumb, just have the voltage rating of the capacitor at twice the peak voltage. For example, with 8.46 volts peak, choose a capacitor with a 16 volt or more rating.

For the full-wave circuit, the conduction path always goes through two diodes in series. See Figure 5-6 for the conduction paths. Thus in the right side circuit of Figure 5-10, there is a 2 × 0.6 volt loss, or 1.2 volts, which means from the full wave rectifier circuit, the output voltage with 6 VAC RMS, is 8.46 volts – 1.2 volts = +7.26 volts DC into C1.

Because the full-wave bridge rectifier circuit causes more loss at the output, why use it? The answer is that because the output pulses occur twice as often when compared with a half-wave circuit, the *ripple voltage* is half that from a full-wave circuit versus a half-wave circuit given the same load current and capacitor C1. Put in another way, for the same amount of ripple voltage, a full-wave rectifier circuit only requires half the capacitance value of the half-wave circuit.

There is a way to calculate what value C1 should be that depends on the current load. But, in general, for load currents up to about 0.5 amp, C1 should be on the order of 470 μF to 2,200 μF. For larger load currents of about 1 amp or more, I would recommend C1 at 4,700 μF to 10,000 μF.

DC Restoration Circuits and Voltage Multipliers

A DC restoration circuit "clamps" the positive or negative peak to a specific voltage such as 0 volt. In this section, we will explore the DC restoration circuit that clamps or repositions an incoming sine wave such that the negative peak is set to 0 volt or near 0 volt (e.g., –0.6 volt). See Figure 5-11. What makes the DC restoration circuit amazing is that the negative peak of the waveform is set to 0 volt regardless of the amplitude of the sine wave.

A DC restoration circuit can be used to step up the DC voltage further by multiplication. Notice the AC peak-to-peak amplitudes of both waveforms are the same, but the absolute peak values are different between the top and bottom waveforms. In fact, the top waveform's positive peak is twice the value of the bottom waveform's positive peak. If each waveform were half-wave rectified and connected to a capacitor (see left side of Figure 5-10), the top waveform would provide twice the DC voltage as the bottom waveform and thus provide voltage multiplication.

Note that the input signal has a voltage swing of $\pm V_{peak}$, and the DC restoration circuit outputs a signal from 0 to $+2V_{peak}$. For example, if the input signal is 10 volts AC RMS, the input signal has a swing from –14.1 volts to +14.1 volts. The DC restoration circuit then provides an output signal that swings from 0 volt to +28.2 volts. *In essence, the DC restoration circuit has added a DC voltage of value $+V_{peak}$ to the AC signal source.*

Now let's look at a typical DC restoration circuit (Figure 5-12). The left side of Figure 5-12 shows a 10 volt AC signal from pins 5 and 8 of T1 connected to C1 and

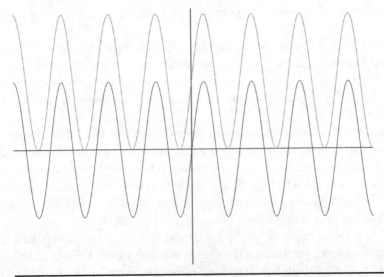

Figure 5-11   The bottom waveform is the input signal, and the top waveform is from the output of a DC restoration circuit.

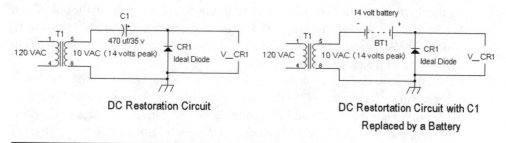

FIGURE 5-12   A DC restoration circuit (*left*). Replacing C1 with a battery for analysis purposes (*right*).

the anode of CR1. Therefore, the voltage from pins 8 and 5 of T1 swings from −14 volts to +14 volts and vice versa. CR1 and C1 form a half-wave rectifier and capacitor circuit, which allows C1 to be charged to 14 volts. That is, during the part of the AC signal when the voltage from pin 8 to pin 5 is +14 volts, CR1 conducts in the forward-bias mode and supplies +14 volts in the (+) and (−) terminals of C1. Once C1 is charged to 14 volts, we can consider it as a 14-volt battery, which is shown on the right portion of Figure 5-12.

If we now look at the output voltage of the DC restoration circuit at V_CR1, note that the 14-volt battery BT1 on the right side of Figure 5-12 adds to the 10 volt AC signal from T1. Thus BT1 adds a level-shifting voltage equal to V_{peak} of the 10 volt AC signal (14 volts). At this point, one can say the output at V_CR1 is a level-shifted AC signal.

To further illustrate how the DC restoration circuit works, see Figure 5-13. In Figure 5-13, BT2 represents the peak voltage from pins 5 and 8 of T1 at two different times. On the left side, the voltage referenced to ground is +14 volts, which is the peak voltage of T1 during its positive cycle. The voltage at V_CR1 for the diagram on the left is just the addition of two 14-volt batteries in series (BT1 and BT2) that results in a +28 volt output.

On the right side of Figure 5-13, the polarity of BT2 is reversed or alternated, which leads to two 14-volt batteries connected back to back with their (−) polarities. Thus the net voltage from BT1 and BT2 is 0 volt. Recall that in a two-cell flashlight,

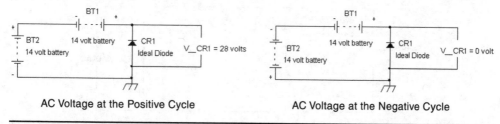

FIGURE 5-13   Using battery BT2 to denote the ±14 volt nature of the 10 volt AC signal for positive and negative cycles.

putting in the two batteries back to back results in no voltage going to the bulb. From Figure 5-13, we see that the range of voltages at the output V_CR1 is 0 to 28 volts, which is the AC waveform that has its negative peak clamped to 0 volt that results in a sine wave with a range of 0 to 28 volts.

To finalize making the voltage-doubler circuit, we simply add another diode and capacitor to peak hold the +28 volt portion of the level-shifted sine wave . See Figure 5-14. From Figure 5-14 the added second diode and capacitor CR2 and C2 allow capturing and storing of the +28 volts into C2, which results in Vout1 having +28 volts DC. Note that even though C2 is rated at 35 volts, it is preferable to have the working voltage increased to 50 volts or more. That is, it is better for C2 to be a 470 µF, 50 volt capacitor for an increased safety margin should there be a surge in the 110 volt AC line.

The next question is, can we repeat the process of DC restoration and voltage doubling to make a 3× voltage multiplier? The answer is yes. See Figure 5-15.

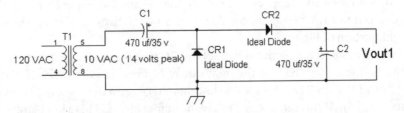

FIGURE 5-14   Voltage-doubler circuit completed by adding CR2 and C2.

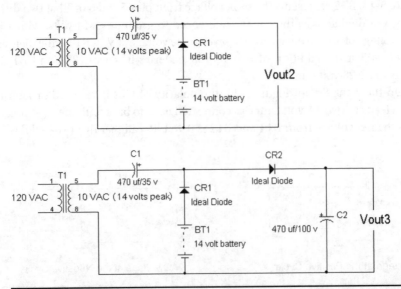

FIGURE 5-15   A DC restoration to clamp the negative peak to +14 volts (*above*). A voltage-tripler circuit (*below*).

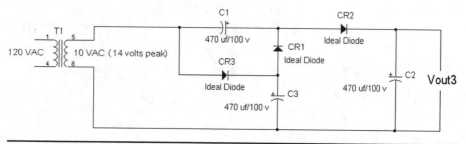

FIGURE 5-16   A voltage-tripler circuit with CR3 and C3 providing the 14 volt source to BT1.

Suppose we want to generate even more voltage via a DC restoration circuit? We do not have to necessarily clamp the negative peak of the incoming sine wave to 0 volt. Instead, we can clamp the negative peak to a positive voltage. In Figure 5-15, the top circuit shows that the anode of CR1 is now connected to a 14-volt battery. This means that when the AC voltage across pins 8 to 5 swings to its peak voltage of +14 volts, the total voltage sent to C1 via CR1 is 14 volts plus 14 volts, or 28 volts. Note that although C1 is rated at 35 volts, using a 100 volt version is better. With 28 volts stored in C1, the incoming 10 volt AC voltage is level shifted by 28 volts instead of the 14 volts that is from the voltage-doubler circuit. Therefore, the signal at the cathode of CR1 of Figure 5-15 ranges from (+28 volts + 14 volts) to (+28 volts − 14 volts), which is 42 volts to 14 volts. By inserting a 14 volt source into the anode of CR1, the negative peak of the 10 volt AC signal is clamped to +14 volts. And because the 10 volt AC signal has a total of 28 volts peak to peak, the voltage range of the sine wave at CR1 is then +14 volts to +42 volts. If we harvest the +42 volt peak at CR1's cathode via CR2 and store the voltage into C2, we get a +42 volt DC source at Vout3, which is triple the voltage from a simple half-wave or full-wave bridge circuit that gives out only +14 volts DC. The resulting circuit is shown in Figure 5-16.

As shown in Figure 5-16, by using simple half-wave rectification via CR3 and using C3 as a peak-hold capacitor for the 14 volt peak voltage of the 10 volt AC source, a +14 volt source via C3 is then connected to the anode of CR1. Vout3 then provides about +42 volts DC. The reader is encouraged to build any of the circuits with 1N4004–1N4007 rectifiers and 100 volt electrolytic 470 µF capacitors in Figures 5-14 and 5-16. Just remember to observe the polarities of the diodes and capacitors and use safety glasses, just in case.

Diodes as Switches and Gain-Control Devices

A diode bridge circuit, as shown in Figure 5-17, may be used to gate or switch analog signals. Before the advent of CMOS (complementary metal oxide silicon) switches, the diode bridge circuit was commonly used (since the 1920s).

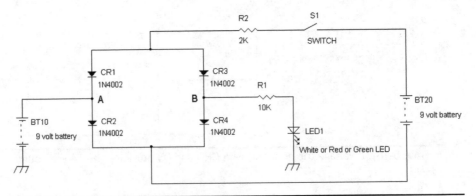

Figure 5-17   Diode bridge circuit for gating a signal through from points A to B.

Figure 5-17 shows a four-diode (CR1–CR4) switch circuit that is an open circuit when switch S1 is open because the diodes have no way to conduct via BT10. At first glance, CR2 may provide a conduction path from point A since the anode is connected to the (+) terminal of BT10. However, CR4 stops the current from flowing to point B since its cathode is tied to the cathode of CR2. When S1 is closed, current flows via BT20 and R2 to turn on all of the diodes, and conduction occurs from points A to B.

With the diodes in the forward-bias mode, each of them can be thought of as a low-power battery at 0.6 volt, as seen in Figure 5-18. Following the voltages from A to B via BT2 to BT4 or BT1 to BT3, we have voltages that are connected back to back. This back-to-back connection such as the voltages in series from BT2 and BT4 results in +0.6 volt (via BT2) and –0.6 volt (via BT4), which results in 0 volt between points A and B. Also see Figure 5-18. Note that a perfect wire with no resistance has no voltage across it from end to end when current flows through it [recall that V = IR, and if R = 0, then V = I (0 Ω) = 0 volt], so therefore the path from point A to point

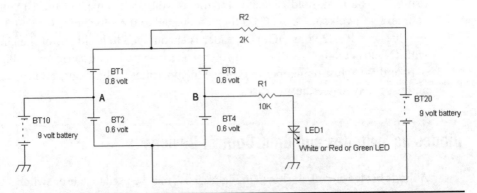

Figure 5-18   Bridge circuit with current flowing through the diodes, where the diodes' turn-on voltages are represented as 0.6 volt sources.

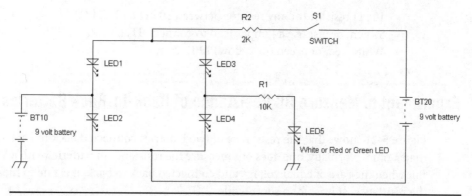

FIGURE 5-19   Electronic switching circuit using four LEDs instead of diodes.

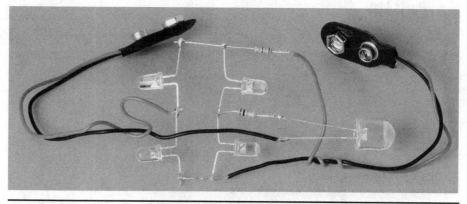

FIGURE 5-20   Three-dimensional (3D) prototype of LED switching circuit.

B is like a wire. And if A to B is like a wire, then there is connection from the 9-volt battery BT10 to light up LED1 via R1.

Note that in Figure 5-17, CR1–CR4 may be replaced with small-signal diodes such as the 1N914/1N4148. Also, if the reader would like, CR1–CR4 may be replaced with LEDs, as seen in Figure 5-19. Also, the reader is encouraged to try different voltages for BT20, such as 1.5, 3, 4.5, and 6 volts, or different values of R1 (e.g., 2 kΩ, 4.7 kΩ, or 10 kΩ) to see if there is any effect on the brightness on LED1 in Figure 5-17 or LED5 in Figure 5-19. Figure 5-20 shows the completed circuit.

Parts List

- Two 9 volt batteries with connectors
- Four diodes (1N4002, 1N914, or equivalent) or, alternatively, four LEDs (any color)
- 2 kΩ resistor (or any resistor between 1,000 Ω and 3,300 Ω)

- 10 kΩ resistor (or any resistor between 10 kΩ and 22 kΩ)
- Switch, or just use a jumper lead to connect BT20 to R2
- White, red, or green (or yellow) LED

Experiment to Measure DC Resistance of Back-to-Back Batteries

Figure 5-21 shows that the resistance of two batteries connected back to back is less than 1 ohm. Normally, one does not measure the resistance of batteries with a VOM. But when they are of equal voltage and connected back to back, it is safe to measure the resistance, at least for a short while.

Parts List

- VOM (preferably an analog type, but a digital version will work)
- Two fresh batteries connected or taped back to back

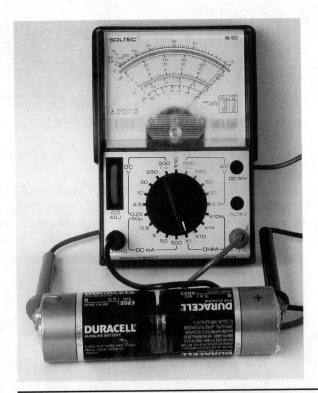

Figure 5-21   Top scale of VOM reads less than 1 Ω.

Diodes as a Gain-Control Circuit

In radios or audio systems, electronic gain control is necessary to prevent overloading amplifiers down the chain. In the next experiment, we will look at how a diode can be used to vary the amplitude of an AC signal. Earlier we found out that both diodes and LEDs have a characteristic of increasing the diode or LED current 10 times when the voltages across them are increased by a small amount, such as 60 mV to 120 mV. What this said is that there is an exponential relationship between the input voltage and output current. This exponential relationship then offers a very useful characteristic in gain control in that the *dynamic or AC resistance* varies inversely in proportion to diode or LED current. For example, in a power diode such as 1N4002, the **measured** dynamic resistance was found to be 405 Ω at 0.10 mA, 47 Ω at 1 mA, and 4.8 Ω at 10 mA. We can show the dynamic resistance effect of a diode or LED by changing the forward-bias current (see Figures 5-22, 5-23, and 5-24). By setting R3 to 100 kΩ, 10 kΩ, 1 kΩ, or 470 Ω, the output signal is attenuated in various degrees.

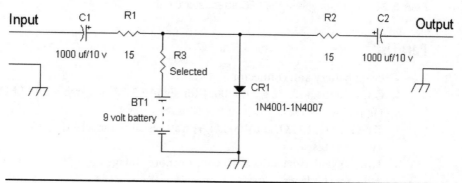

FIGURE 5-22   Diode attenuator circuit.

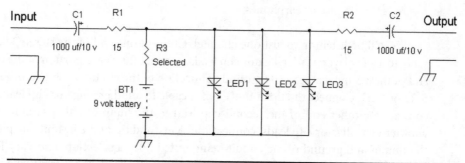

FIGURE 5-23   LED attenuator circuit.

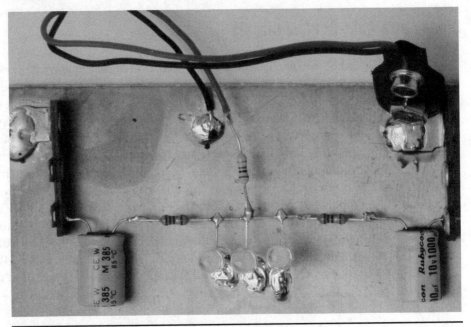

FIGURE 5-24   3D prototype of the LED attenuator circuit.

Part List

- 9 volt battery and connector
- One 1 amp diode (e.g., 1N4002) for Figure 5-22 or three white LEDs for Figure 5-23
- 470 Ω, 1 kΩ, 10 kΩ, and 100 kΩ, ¼ watt resistors for selecting R3
- Two 15 Ω resistors
- Two 1,000 µF capacitors at 10 volts or more voltage rating
- Alligator clip leads
- CD player or radio with a headphone output
- One stereo or mono 3.5 mm cable
- Headphones or earphones

We will be listening to just one channel. Connect one end of the stereo 3.5 mm cable to a CD player. With alligator clip leads, connect the sleeve portion of the other end of the 3.5 mm cable to the ground connection of the circuit, as shown in Figure 5-22 or 5-23. Connect the tip of the 3.5 mm cable to another alligator clip lead, and connect the other end of the second clip lead to the input of either circuit. With another two alligator clip leads, connect the sleeve and tip of the headphone plug to the output and ground of the circuit being tested. Play a selection, and vary R3 to listen for an attenuation effect. *Be sure to remove the headphones from your head before connecting or disconnecting any resistors to avoid hearing very loud clicks.*

Radio-Frequency (RF) Mixing with a Diode

When we think of mixing, there are two types of mixers commonly used in electronics. One is the adding or summing mixer that combines two or more signals in a *linear manner.* When we talk of mixing in a linear manner, we mean that the two or more signals do not interact with each other to cause extra distortion signals. A good example of a linear mixer is an audio mixing board that is able to take multiple audio sources and combine them in an additive way.

The other type of mixer is generally used in radio-frequency (RF) to take two signals and provide an output signal whose frequency is the sum or difference of the frequencies from the two signals. For example, in an AM radio, the incoming RF signal has a frequency of 1.000 MHz. If the AM radio has an RF mixer, the mixer will receive the 1.000 MHz RF signal and a local oscillator signal at 1.455 MHz. The output of the mixer will then provide signals with frequencies of (1.455 MHz + 1.000 MHz) and (1.455 MHz − 1.000 MHz), which are the sum and difference frequencies of 2.455 kHz and 0.455 MHz. See Figure 5-25.

One key to ensuring good RF mixing when using a diode as the mixing element is to provide a larger signal from the oscillator when compared with the incoming radio-frequency (RF) or intermediate-frequency (IF) signal. On the left side of Figure 5-25 we see a linear summing circuit in which the output of the summing circuit is fed to a diode that acts as a half-wave rectifier. Given the inherent turn-on voltages of diodes, the amplitude of the oscillator or beat-frequency oscillator (BFO) should exceed the turn-on voltage of the diode. For example, a silicon diode requires about 0.6 volt to start conducting current. Therefore, the oscillator should have an amplitude level greater than 1 volt peak.

The right side of Figure 5-25 shows the summing circuit as the series connection of two secondary windings for transformers T1 and T2. After summing the incoming RF (or IF) signal and the large-signal oscillator signal, half-wave rectification is applied at the output of the summing circuit, which results in output signals from

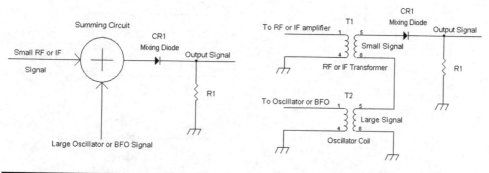

Figure 5-25   A top-level illustration of an RF mixer (*left*). Practical implementation (*right*).

the cathode of the diode CR1 with sum and difference frequencies of the RF signal's frequency and the oscillator signal's frequency.

NOTE A beat-frequency oscillator (BFO) is used for the detection of suppressed carrier signals such as (amateur radio) single-sideband (or double-sideband suppressed-carrier) signals. There will be more discussion of radio circuits and signals in Chapter 10.

We will wrap up the subject of diodes in the next three sections. These include logic circuits with diodes, variable-capacitance circuits using varactor diodes, and voltage regulators using zener diodes. Their uses will become more evident in later chapters when more advanced circuits are introduced.

Logic Circuits Using Diodes

A logic OR circuit consists of two or more inputs that have either a logic high or a logic low level. If any of the inputs are logic high, the output of the OR circuit will be logic high. In order for the output of the OR circuit to be logic low, all inputs must be at a logic low level. In essence, it takes a unanimous agreement for all inputs to a low logic level to create a logic low level at the output of the OR gate. However, any input may receive a veto, such as a single vote to go high to set the OR circuit's output high (see Figure 5-26). For extra inputs, InputN is denoted, and CRN's cathode is connected to OUTPUT1'.

For the AND circuit, we have a somewhat opposite effect. The output provides a logic high level when all its inputs are logic high. Therefore, it takes unanimous agreement to have all inputs set to the high logic level in order to provide a high logic level at the output of the AND circuit. But all it takes is a single low logic level at one

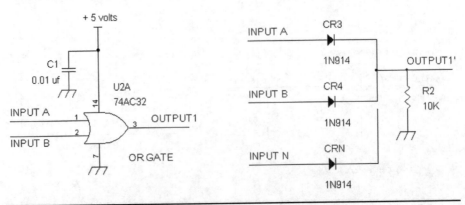

Figure 5-26 OR gate on the left and diode equivalent on the right.

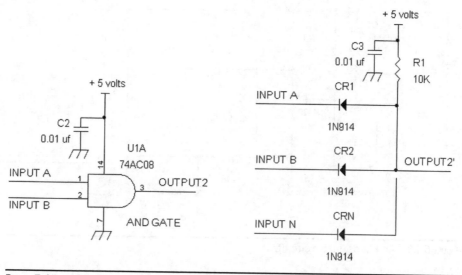

FIGURE 5-27 AND gate on the left and diode equivalent on the right.

of the inputs to veto the output of the AND circuit and provide a logic low output (see Figure 5-27). For extra inputs, CRN is denoted, and its anode is connected to OUTPUT2'.

Variable-Capacitance Diodes

Traditional variable capacitors have required a mechanical adjustment to vary the capacitance. Usually, a shaft is turned to rotate the plates of the variable capacitor (see Figure 2-46, lower right-hand corner). A variable-capacitance diode (varactor or varicap diode) takes the place of a mechanical variable capacitor and has the advantage of varying the diode capacitance electronically via a control signal, which allows faster speeds. Uses for the varactor include tunable RF amplifiers and oscillator circuits that are used in radios. The varactor diode can be used in an FM (frequency-modulation) transmitter to instantaneously change the FM carrier frequency by impressing an audio signal into the varactor. Figure 5-28 shows a low-power FM oscillator using a varactor diode (MV209) to modulate the FM carrier.

Note that the identification of the anode and cathode of Varactor1 (MV209), which comes in a two-lead TO92 plastic case, is as follows: when viewing the flat side in front with the two leads protruding down, the anode lead is on the left, and the cathode lead is on the right. The varactor diode works in reverse-bias mode, and the capacitance is maximum at about 0 volt between the cathode and the anode, and it has decreasing capacitance as the positive reverse-bias voltage is increased at the cathode. For example, with its anode grounded, the cathode of the varactor is biased

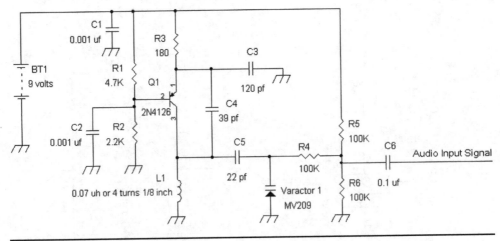

Figure 5-28   An FM oscillator circuit.

via a large-value resistor of at least 100 kΩ (R4) to a positive voltage (via R5 and R6). The higher the positive voltage, the smaller is the capacitance delivered across the cathode and anode of the varactor diode. This advanced-level circuit can be built by the reader, but care must be taken, and the leads of all components around Q1 must be as short as possible. The oscillator's frequency works within an FM broadcast band. The audio input signal may be taken from the output of a CD player or transistor radio via the headphone output. Tune an FM radio close to coil L1, and listen for the audio from the CD player or transistor radio.

NOTE   Other types of diodes can be used as a varactor, such as a power diode (e.g., 1N5401) or a 12 volt zener diode.

Zener Diodes for Voltage Regulation

Another diode that is reverse biased is the zener diode, where the cathode is fed to a positive source and the anode is connected to a negative source. We do not hook up voltage sources (e.g., a battery or power supply) directly to a zener diode because that will cause overheating and destruction. Instead, the zener diode usually is driven with a current-limiting resistor or a *current source*. With a limited current flowing into the zener diode, the reverse biasing causes the diode to break down in a nondestructive manner. The breakdown voltage due to the reverse biasing is called the *zener voltage*.

We generally hook up via a resistor to the zener diode with a higher voltage than the zener voltage. For example, suppose that we have a raw 15 volt DC voltage that includes 60 Hz ripple and want to provide a clean 5 volt DC voltage without the ripple? One way is to connect a series resistor to a 5.1 volt zener diode's cathode with its anode grounded and take the final voltage from the cathode of the zener diode to provide a regulated 5.1 volt supply that is free of 60 Hz ripple. If the 110 volt AC line goes up or down by 10 percent, the raw 15 volt DC supply will vary by ±1.5 volts or have a range of 13.5 volts to 16.5 volts DC. However, the output voltage at the zener diode still will provide a steady or solid 5.1 volts DC.

Zener diodes have generally two important characteristics—zener voltage and power rating. For example, a 1 watt, 12 volt zener diode allows a maximum current of about 80 mA flowing into it. Recall, by Ohm's law, that P = IV, which leads to I = P/V. For this example, V = 12 volts, and P = 1 watt; therefore, I = 1/12 amp, or one-twelfth of an amp = 83.3 mA.

For a rule of thumb, we generally limit the zener diode current to about 60 percent its maximum rating. In the preceding example, we would set the zener diode current to 0.6 × 83 mA ≈ 50 mA. Note that the zener diode current is also the maximum current available for providing the regulated voltage. If the load's current drain exceeds the zener biasing current, the voltage across the zener diode will drop below the zener voltage and will become unregulated.

We now close out this chapter with a power supply for providing ±17 volts unregulated and also ±12 volts regulated at approximately 50 mA (Figures 5-29 and 5-30).

Note that the zener diodes and 100 Ω resistors will get pretty warm. Any small component will get warm to at least mildly hot when dissipating at ¼ watt or more. So be careful when you touch the components.

From here on out in this book, the parts list will include the reference designations from the schematic. The reason is that with more complex circuits that include

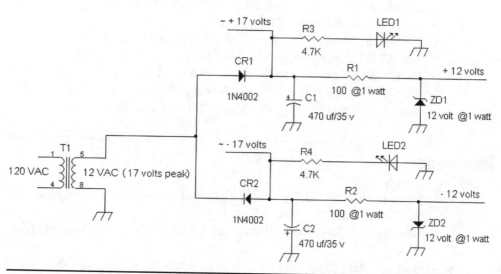

Figure 5-29 A ±12 volt regulated power supply.

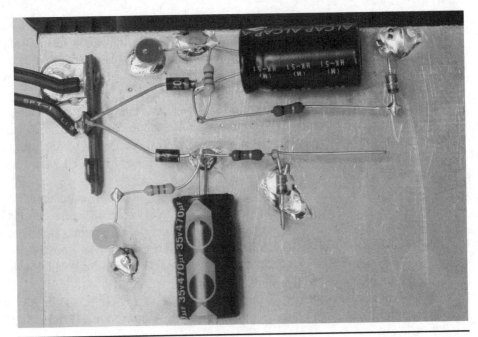

Figure 5-30   Prototype of ±12 volt regulated power supply.

a variety of parts, having the reference designations (e.g., R1, C1, ZD1, LED1, etc.) and their values helps as a double-check against the schematic for accuracy.

Parts List

- 12 volt AC wall adapter transformer of at least 300 mA for T1
- Two IN4002 diodes for CR1 and CR2
- Two 470 μF or more capacitors at 35 volts or more for C1 and C2
- Two 4.7 kΩ, ¼ watt resistors for R3 and R4
- Two 100 Ω resistors at 1 watt or more for R1 and R2
- Two 12 volt zener diodes (e.g., 1N4742) at 1 watt or more for ZD1 and ZD2
- Two of any kind of LEDs for indicators LED1 and LED2

References

1. *Motorola Small-Signal Transistors, FETs, and Diodes Device Data.* Motorola, Inc., Schaumburg, IL, 1993.
2. *Motorola TVS/Zener Device Data, 2nd. Ed.* Motorola, Inc., Schaumburg, IL, 1994.
3. *Motorola Zener Diode Manual.* Motorola, Inc., Schaumburg, IL, 1980.

CHAPTER 6

Transistors, FETs, and Vacuum Tubes

Chapters 4 and 5 introduced light-emitting diodes (LEDs) and diodes, which we found can be used as switches or attenuators to control the strength of a signal. What neither LEDs nor diodes could do was to make the original signal larger. When a signal is switched, the signal is either on or off. And when a signal is already attenuated (i.e., reduced in strength) via a resistive divider circuit, further attenuation using LEDs or diodes as variable resistors does not provide for amplification. Therefore, in this chapter we will explore *amplifying* devices, which allow a signal to become larger in terms of either voltage or current or both.

Before delving into the concept of amplification, we will explore the concept of current sources. We can all relate to voltage sources such as batteries and power supplies, but we only had one introduction to current sources via Chapter 4 when photodiodes were presented as light to current-source devices. All active devices such as transistors, FETs (field effect transistors), and vacuum tubes can be characterized as some type of constant current source or variable current source that can output different amounts of electric current based on voltage across their input terminals.

We will also look briefly into operational amplifiers (op amps). In closing out this chapter, we will show briefly how an FET and a transistor can be used as a variable resistor.

Current Sources and Voltage-Controlled Current-Source Devices

The concept of a *constant current source* is that current flows into a resistor such that the resistor current is the same no matter what the resistance value is. For example, a constant current source that provides 1 mA (0.001 A) will cause 1 mA to flow into 100 Ω resistor R1 or 10,000 Ω resistor R2. That is, when a 100 Ω resistor is connected to a 1-mA source, the voltage across the resistor is (0.001 A × 100 Ω =) 0.100 volt. If

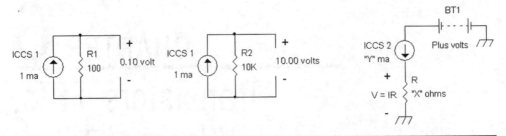

FIGURE 6-1   (*From left to right*) Current sources depicted by different load resistors R1 and R2 and a "practical" current source using a battery BT1.

the 10,000 Ω resistor is connected instead, the voltage across that resistor is (0.001 A × 10,000 Ω =) 10.000 volts. Figure 6-1 shows various circuits.

In general, a current source needs a power supply to enable it for providing current. See the right-hand side of Figure 6-1. A simple way create a constant current source is to take a voltage source such as a battery and connect it in series with a resistor. See Figure 6-2. At first glance, this looks like a "lossy" voltage source, which, in a way, a current source is.

We know if we connect an external (load) resistor R2 across points A and B, the current into R2 will be BT1/(R1 + R2). *Note:* M1, the ammeter, has zero resistance. For example, if BT1 = 1.5 volts and R1 = 100 Ω, we know if a 100 Ω resistor is connected to points A and B, the current through R2 is 1.5 volts/(100 Ω + 100 Ω) = 0.0075 amp. If we change the value of R2 to 10 kΩ, the current is 1.5 volts/(100 Ω + 10 kΩ) = 0.000148 A. At this point, we do not have a constant current source because the current is not constant when R2 has a different value. What we want is if R2 is changed in value, the current through it stays essentially the same.

But what if we make the battery voltage BT1 really large, say, 100 volts, and the resistor R1 a large value such as 10 mega ohms (10M Ω)? See Figure 6-3.

Repeating the values of R2 = 100 Ω and 10 kΩ, we get R2 resistor currents of 0.009999 mA (9.999 μA) and 0.009990 mA (9.990 μA), respectively. Note that the two resistor currents are essentially the same. As you can see, such a circuit with a high-voltage supply and a large-value resistor approximates a constant current source generator. However, we can make current sources with transistors and FETs operating at lower voltages such as 3 volts to 15 volts.

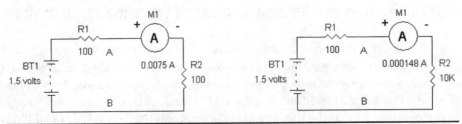

FIGURE 6-2   Two examples of lossy voltage sources with R1 = 100 Ω.

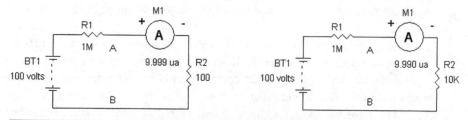

FIGURE 6-3 A lossy "high" voltage source that more closely resembles a constant current source.

Constant Current "Diodes" Using JFETs

One of the simplest ways to make a constant current source is to use a field-effect transistor (FET). As noted earlier, a FET has three terminals: a gate, a source, and a drain. Also known as a current source or current regulator diode, this circuit is nothing more than a JFET (junction field effect transistor) where the gate is tied to its source. Examples of current source diodes are the E500–E507. Before we look into this simple circuit, let's take a look at JFETs. See Figure 6-4.

As seen on the left side in Figure 6-4, VR1 forms a voltage divider circuit with BT1 to apply a negative to zero voltage to gate G of Q1 with respect to Q1's source S. In Figure 6-1, we showed a practical current source via ICCS2, which is shown again in the middle circuit of Figure 6-4, denoted by terminals A and B. By simply connecting the gate to the source of an N-channel JFET of Q2 (e.g., Siliconix FET, Part Numbers E105–E114 and E204), a constant current source circuit is provided, as seen on the far right in Figure 6-4. The battery voltage BT1 is large enough such that the voltage across the drain and source of Q2 is at least 1 volt.

JFETs in general are "depletion mode" FETs, which means that it takes a negative voltage from the gate to source for proper operation for an N-channel device. That is, for an N-channel FET, the gate and source terminals are in reverse bias. A zero

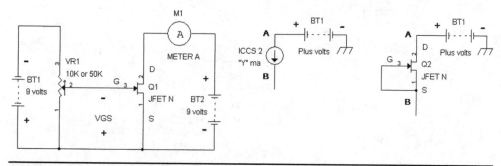

FIGURE 6-4 (*From left to right*) A FET biasing circuit, a constant current source ICCS2, and a JFET implementation of ICC2.

volt potential across the gate and source of a depletion-mode N-channel FET is also allowable.

In a depletion-mode N-channel FET, zero volt across the gate and source denotes the maximum current for the FET, while a negative voltage across the gate and source that causes the FET's drain current via M1 to go toward zero (e.g., a microamp) causes the FET to cut off current or to shut off. The "pinch-off" voltage in an FET is then the voltage across the gate and source such that no drain current flows. See the circuit on the left in Figure 6-4, where VR1 can be adjusted to a high negative voltage to cut off drain current from Q1. Typical pinch-off or cut-off voltages across the gate and source for N-channel JFETs vary from −1 volt to −10 volts.

When the gate to source voltage is reverse or zero bias, there is essentially no current flowing into the gate. But normally applying a positive voltage across the gate and source is not recommended because a forward bias condition occurs between the gate and source.

The circuit on the left side of Figure 6-5 shows the gate tied to the source to provide a fixed constant current to load resistor R1. By inserting a fixed or adjustable resistor in series with the source, the current can be varied (see Figure 6-5, *right*).

Figure 6-5 shows an "improvement" over a constant current source by inserting a variable resistor VR1 (as shown) connected to Q3. The current source nodes are A' and B'. The reader is encouraged to build either circuit in Figure 6-5, with the JFET being any N-channel JFET such as a 2N3819, J110, E108, etc. with R1 = 100 Ω and a 1 kΩ or 10 kΩ potentiometer for VR2 and measure the voltage at point B or B' referenced to ground while varying the battery voltage of BT3 or BT4 from 3 volts to 9 volts. What should happen is the voltage at B or B' will measure the same regardless of varying the battery voltage from 3 volts to 9 volts. The current is adjusted by varying VR2, which forms a voltage due to the current from the source terminal of Q3

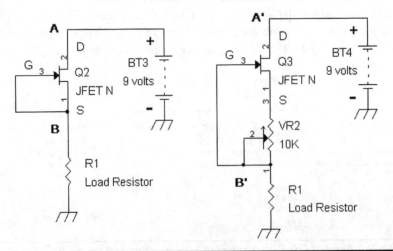

Figure 6-5   A constant current source (*left*). An adjustable current source (*right*).

flowing into VR2. The voltage is *positive* from the source S to the gate G of Q3. That is, connect the positive lead of a VOM to the source and the negative lead to the gate to read a positive voltage. Then connect the positive lead of the VOM to the gate and negative lead to the source of Q3, and you will read a negative voltage—which makes sense because the test probe leads of the VOM are reversed. This means the voltage from the gate G to the source S is negative, as required for depletion-mode FETs.

And in essence, VR2 is sometimes called a *self-biasing resistor* that does not require a negative voltage supply for biasing an N-channel depletion-mode FET (Q3 in Figure 6-5).

NOTE   Some JFETs are symmetrical in terms of their drain and source connections. That is, swapping the drain and source leads is acceptable and generally provides the same result. In general, though, it is better to stick to connecting to the leads as labeled. Also note that the pin-out connections of various JFETs vary. So, accordingly, the reader should look at the labeling on the schematic for D, G, and S and not necessarily rely on the pin-out numbers 1, 2, and 3 on the drawing.

We will return to looking into FETs later, but for now, let's take a look at bipolar transistors.

Bipolar Transistors as Current Sources

A bipolar transistor has three terminals: the emitter, base, and collector. When a transistor operates as a current source, a small amount of current flows into the base by applying a forward bias voltage across the base-emitter junction. The collector-to-base junction is generally reverse biased, while the collector provides the output current.

To operate an NPN transistor as current source, the voltage across the base to emitter is at its turn-on voltage of about 0.6 volt. Similarly, the *emitter-to-base* voltage of a PNP transistor has to have about 0.6 volt. NPN transistors are "opposite" in polarity to PNP transistors. And current sources can be made with both types; however, the current flow is different and opposite between NPN and PNP transistors. For now, we will investigate making current sources out of both.

In an FET, the source, gate, and drain terminals are analogous to the emitter, base, and collector terminals of a bipolar transistor. Unlike the FET, which operates without current flowing into the gate, the bipolar transistor requires base current. See Figure 6-6.

NOTE   If the circuits from Figures 6-6 and 6-7 are constructed, a copper-clad prototype board is recommended to avoid the transistors

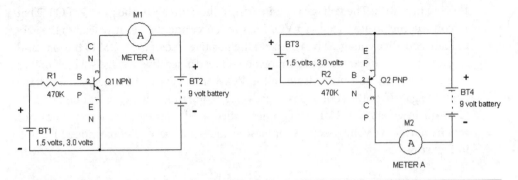

Figure 6-6   Biasing NPN and PNP transistors via R1 and R2 to provide NPN and PNP current sources.

from oscillating. It was found that constructing these circuits on superstrip prototype boards resulted in unstable reading due to oscillations in the circuit. Thus, it is also highly recommended that a 1 μF capacitor with short leads is connected to the base-emitter and another 1 μF capacitor is connected to the collector-emitter terminals to each of the transistors. The capacitors will prevent the transistors from oscillating. Also, these circuits are suitable for testing transistors for current gain β.

There is a relationship between the base current and corresponding collector and emitter currents when the bipolar transistor is in an amplifying or current-source mode. This relationship is as follows: The collector current, IC = β (IB) and the emitter current, IE = (β + 1)IB, where β is the current gain, which varies from 10 to 1,000 depending on the transistor, and IB is the base current. For β or the current gain at 10 or greater, which is typical of most transistors, β ≈ (β + 1), which leads to IC = IE that says that for all practical purposes, the collector and emitter currents are equal. For example, if we take the worse case, where β = 10, then (β + 1) = 10 + 1 = 11. Thus, 10 is approximately 11 if we say that anything within 10 percent is roughly equal. Of course, if you are counting money, that's another story when talking $10 and $11, and if you give someone $10 for an $11 payment, someone is going to be out by a buck!

Figure 6-6 shows a way of measuring collector current when driving the base with current via a base resistor of R1 = 470 kΩ. For example, on the left-side circuit in Figure 6-6, with a 2N3904 or 2N4124 NPN transistor for Q1, the following measurements were tabulated in Table 6-1 for R1 = 470 kΩ and 47 kΩ. Voltage measurements also were taken across the base-emitter junction of Q1. Base currents IB were calculated via IB = (BT1 − VBE)/R1 = (BT1 − VBE)/470 kΩ and IB = (BT1 − VBE)/R1 = (BT1 − VBE)/47 kΩ.

TABLE 6-1   Tabulation of Measurements for VB and IC with Calculated Values for IB and β

BT1 Voltage	VBE	IB	IC	β = IC/IB	R1
1.5 volts	0.623 volt	1.86 μA	187 μA	100	470 kΩ
3.0 volts	0.655 volt	4.99 μA	625 μA	125	470 kΩ
1.5 volts	0.686 volt	17.32 μA	2.98 mA	172	47 kΩ
3.0 volts	0.709 volt	48.74 μA	8.80 mA	180	47 kΩ

From Table 6-1, we see that the current gain β is not a true constant and that it changes value depending on the collector current. Also, if the reader measures the collector current with a 2N3904 or 2N4124 transistor, chances are that there will be different numbers from Table 6-1 because transistors in general have a wide range of current gain.

NOTE   For measuring VBE with R1 = 470 kΩ, an inexpensive digital VOM has only a 1 MΩ input resistance, which would cause the VBE measurement to be in error by being slightly lower. Usually, one characteristic of an inexpensive digital VOM is that the resistance range only measures a maximum of 2 MΩ. For a more accurate reading, a midpriced digital VOM (around $30 or more) usually has a 10 MΩ input resistance, which is preferable for this experiment. The middle- to higher-performance digital VOMs usually have a resistance measurement range of at least 20 MΩ and sometimes to 2,000 MΩ. Also, most autoranging VOMs have 10 MΩ or more input resistance.

Concept of Measuring Collector Currents and Voltage Gain Due to Changes in Input Signals

In the next experiment, we replace the amp meter M1 with a load resistor RL1 and measure the voltage across RL1. See Figure 6-7.

When the load resistor RL1 is increased from 1 kΩ to 10 kΩ, we should expect to see more voltage across it and, more important, a greater change in voltage across RL1 as BT1's voltage is changed when RL1 is increased in value. See Table 6-2.

TABLE 6-2   Measured Voltages Across RL1 via M3 and Calculated VCE from BT2 and M3

BT1 Voltage	RL1	M3 Voltage	BT2	VCE = (BT2 − M3 Voltage)
1.5 volts	1 kΩ	0.16 volt	9.05 volts	8.89 volts
3.0 volts	1 kΩ	0.52 volt	9.05 volts	8.53 volts
1.5 volts	10 kΩ	1.63 volts	9.05 volts	7.42 volts
3.0 volts	10 kΩ	5.32 volts	9.05 volts	3.73 volts

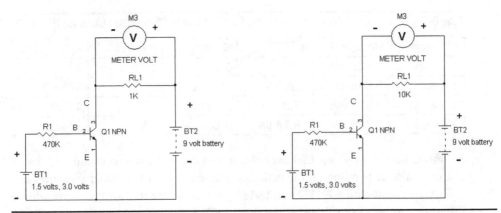

Figure 6-7   Measuring changes in collector current for different values of collector load resistors for RL1.

An extra column is added in Table 6-2, which calculated the collector to emitter voltage VCE based on BT2's battery voltage minus the voltage drop across RL1 that is measured by M3. To measure VCE directly, place the positive lead of the VOM at the collector and the negative lead at the emitter of the transistor. Note that the direction of the changes in voltages for M3 is opposite for the changes in VCE. As the battery voltage in BT1 is increased from 1.5 volts to 3.0 volts, there is an increase in collector current, which is reflected in the voltage across RL1 = IC × RL1. However, the voltage at the collector of the transistor is VCE, and that is equivalent to (BT2 – IC × RL1) = VCE. *Note:* The M3 voltage = IC × RL1. So, as the voltage at the input via BT1 increases, the collector current increases, which also increases the voltage across RL1, but the voltage between the collector and emitter (VCE) decreases. When talking about AC signals, we say that the voltage measured at the collector (and measured not across RL1) referenced to the emitter is *out of phase* with respect to the input signal (i.e., BT1 changing from 1.5 volts to 3.0 volts). The collector current and voltage across RL1 as measured by M3 is *in phase* with the input signal since an increase in BT1 results in an increase in collector current and thus an increase in the voltage across RL1. That is, we can think of the voltage across RL1 as an equivalent battery voltage, V = IC × RL, where RL = RL1, as seen in Figure 6-8.

As can be seen in Figure 6-8, the "Equivalent Battery 1" is connected back to back with the positive terminal of BT2. Thus, the resulting voltage at the collector of Q1 is (BT2 – V) where V = IC × RL. Therefore, the voltage at the collector of Q1 with respect to ground is VCE = (BT2 – IC × RL).

In terms of thinking about the changes in input battery voltages of BT1 from 1.5 volts to 3.0 volts, we can represent these two voltages as an AC signal that has a DC level-shifting voltage with the negative peak at 1.5 volts and the positive peak at 3.0 volts. Then we can also take a look at VCE as AC voltages with their DC level-shifting voltages. Figure 6-9 shows BT1 and VCE represented as sine waves for RL1 = 1 kΩ and 10 kΩ.

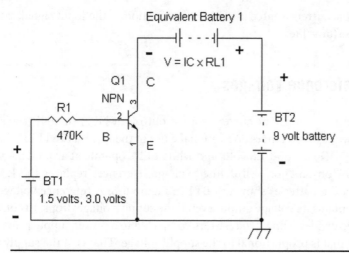

FIGURE 6-8   An equivalent battery voltage in place of the voltage across RL1.

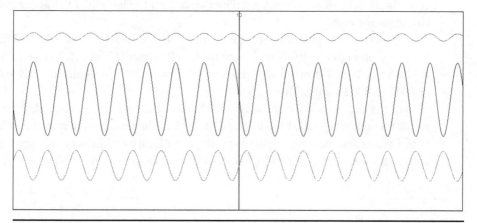

FIGURE 6-9   Three sine waves from bottom to top representing voltages of BT1, VCE @ RL1 = 10 kΩ and VCE @ RL1 = 1 kΩ. The vertical axis is the amplitude from 0 volt to 10 volts, and the horizontal axis is time.

From Figure 6-9, the bottom waveform is a sine wave from 1.5 volts to 3.0 volts, which represents the BT1 voltage range in Figure 6-7. BT1 can be thought of as the input voltage. The middle waveform, which is opposite in polarity or phase from the bottom waveform, represents the voltage range (7.42 volts to 3.73 volts) across the collector emitter (VCE) of Q1 in Figure 6-7 when BT1 changes from 1.5 volts to 3.0 volts for RL1 = 10 kΩ. Note that the VCE sine wave signal at the "output" via the collector of Q1 is larger than the "input" waveform, which is then amplification. Finally, the top waveform shows the VCE voltage swing of 8.89 volts to 8.52 volts when RL1 = 1 kΩ, which is smaller than the "input" waveform. Note that the "gain" of the amplifier is determined by the value of the load resistor RL1. Therefore, a transistor used as a variable current source provides amplification to the input signal, given a certain load resistor value.

Using LEDs as Reference Voltages

Now let's construct a constant current source out of an NPN or PNP transistor using LEDs as voltage references. We can make use of the fact that red LEDs have a turn-on voltage (VF) around 1.8 volts and white LEDs operate at around 3.0 volts. Using LEDs as "constant" or well-defined voltage references replaces the 1.5 volt and 3.0 volt biasing batteries shown as BT1 in Figure 6-6. A reference voltage is a device that maintains its voltage output even if the supply voltage drops. In contrast, a simple voltage divider using two resistors connected to a power supply provides a divided output that is proportional to the supply voltage. That is, if the supply voltage drops by 50 percent, the output of the voltage divider circuit drops by 50 percent Figure 6-10 shows a voltage divider circuit on the left and a voltage reference circuit on the right.

The output voltage from a voltage divider in general is BT1 × [R2/(R1 + R2)]. For the circuit in Figure 6-10, Vout_Unregulated = 9 volts × [1.8 kΩ/(6.8 kΩ + 1.8 kΩ)] = 1.88 volts. If BT1 drops from 9 volts to 4.5 volts, Vout_Unregulated = 4.5 volts × [1.8 kΩ/(6.8 kΩ + 1.8 kΩ)] = 0.94 volt.

For the circuit on the right using LED1 as a voltage reference, Vout_Regulated is about 1.8 volts whether BT1 is 9 volts or 4.5 volts. The reader is encouraged to build both circuits and measure the output voltages with different battery voltages from 3 volts to 9 volts. Here is what was found from my experiment via Table 6-3.

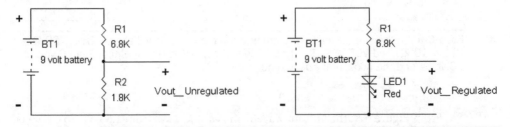

Figure 6-10 (*From left to right*) A voltage divider circuit and a voltage reference circuit using a LED.

Table 6-3 Measured Voltages for Voltage Divider and LED Voltage Reference Circuits with Different Battery Voltages

BT1 Voltage	Vout_Unregulated	Vout_Regulated
3.32 volts (two AA cells)	0.675 volt	1.74 volts
9.05 volts (9-volt battery)	1.90 volts	1.80 volts

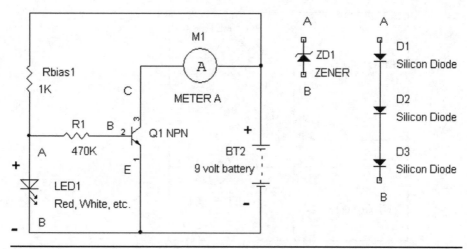

FIGURE 6-11   An NPN current source using a LED (*left*). Alternate voltage reference diodes (*right*).

As can be seen in Table 6-3, the voltage output from the LED circuit is much more stable than the voltage divider circuit with varying battery or supply voltages.

Now let's look at an example of using an LED and diodes for voltage references. See Figure 6-11.

Figure 6-11 shows LED1 to bias transistor Q1. Note that a red LED will provide about 1.8 volts across points A and B, whereas a white LED will give out about 3.0 volts. Alternatively, LED1 can be replaced with a zener diode or a series-connected string of diodes, each having a voltage drop of about 0.6 volt. Zener diodes come in voltages ranging from about 1.2 volts to over 100 volts. In this figure, one has to remember that the voltage reference device (LED, zener diode, or silicon diode) must have a voltage across A and B that is less than the battery voltage BT1. Put another way, the battery or supply voltage must be greater than the reference voltage; otherwise, the reference device will not be turned on. Note that the supply voltage should not be connected directly across the voltage reference diodes.

With the three diodes in series as shown (D1–D3) and biased via resistor Rbias1 (1 kΩ), the voltage across points (nodes) A and B will be about 1.8 volts, given that each diode has a forward bias voltage of about 0.6 volt. If we look at Figures 6-6 and 6-11, each transistor is driven with a base resistor to set up a collector current. This collector current is directly dependent on the current gain β of the transistor. However, the current gain of transistors even of the same part number or manufacturing lot will vary easily by 2:1 or even 3:1. For example, for the 2N3904, the range for β is 100 to 300. Thus, biasing a transistor with a single base resistor (e.g., R1 in Figure 6-11) is generally not done in circuit design unless the transistor has been selected for a range of current gains. This type of transistor current gain selection was and still is used today for certain inexpensive consumer products such as toys.

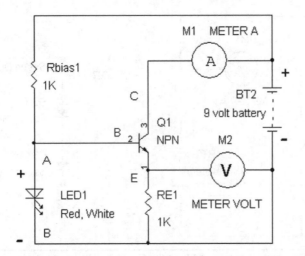

FIGURE 6-12   A stabilized collector current circuit using emitter resistor RE1.

For example, if Q1 in Figure 6-11 is replaced with various NPN transistors (e.g., 2N3904, 2N2222, and 2N5089) or even different 2N3904 or 2N4124 transistors, the current meter M1 will show a range of collector currents that can be twofold in difference.

By adding an emitter resistor and connecting a reference voltage at the base, a transistor circuit can be made to provide a known collector current essentially independent of the current gain as long as the transistor's β is greater than 20. Figure 6-12 shows the addition of an emitter resistor to provide a collector current that is more stable and not as dependent on the current gain, β.

In Figure 6-12, a second meter M2 is used for monitoring the voltage across emitter resistor RE1. If we recall, for transistor current gains β of at least 10, the collector current through ammeter M1 is essentially equal to the emitter current. The emitter current is just the voltage across RE1, which we can define as VRE1 divided by the resistance of RE1, which is 1 kΩ in this example. That is, Q1's emitter current IE = VRE1/RE1.

For the next experiment, which can be made on a superstripe prototype board because we will be changing Q1 from a 2N3904 (β typically 100–300) to a 2N5089 (β typically 400–1,200), we will be comparing the circuits in Figures 6-11 and 6-12 in terms of sensitivity to current gain. Table 6-4 provides the measured results. If the reader tries this experiment, different collector currents may be measured, but in terms of sensitivity to β, the current gain, we will come to the same conclusion. That is, the circuit in Figure 6-12 will be superior in providing a more stable collector current IC with the two different transistors. See Table 6-4, where an LED1 is a red LED.

From Table 6-4, we see that indeed the circuit in Figure 6-12 is better in providing a stable collector current with transistors of different current gains. Note that VRE1 = LED1's turn-on voltage minus VBE, Q1's base-emitter turn-on voltage.

TABLE 6-4 Comparing Two Circuits for Collector Currents and Sensitivity to Current Gain

Q1 Transistor	Figure 6-11 Collector Current	Figure 6-12 Collector Current	VRE1 (Figure 6-12 Only)	Emitter Current VRE1/1 kΩ (Figure 6-11 Only)
2N3904	0.29 mA	1.26 mA	1.25 volts	1.25 mA
2N5089	1.84 mA	1.32 mA	1.33 volts	1.33 mA

Given that the circuit in Figure 6-12 is stable in collector current "regardless" of the current gain of the transistor, as long as the current gain is greater than 10, and that the regulator diodes provide a "stable" voltage even when the power supply drops in voltage, we now have the makings of a constant current source that can be applied elsewhere. Table 6-5 shows the stability of the collector current when BT2 is changed in voltage for the circuit in Figure 6-12.

TABLE 6-5 Collector Current Stability with Variation in Voltage for BT2

Q1 Transistor	BT2	Figure 6-12 Collector Current
2N5089	4.80 volts	1.24 mA
2N5089	6.50 volts	1.28 mA
2N5089	9.05 volts	1.32 mA

Note that the circuit in Figure 6-11 also would yield a stable collector current with a variation in BT2's voltage, which is characteristic of constant current sources. That is, the current output is relatively independent of the voltage across the current-source circuit.

Using a Constant Current Source to Provide Low Ripple at the Power Supply's Output

In Chapter 5, we saw how to make a regulated power supply using 100 Ω resistors to bias 12 volt zener diodes. Although the regulation is good at 12 volts, this circuit, as seen in Figure 6-13, can be improved further.

The voltage at the (+) side of C1, for example, has a ripple voltage at 60 Hz or 50 Hz depending on what part of the world T1 is plugged into. If we can supply ZD1 with a constant current source, the +12 volt output will have further lowered the ripple voltage. See Figure 6-14, where just the +12 volt supply is modified with a current source. Here the 100 Ω resistor R1 in Figure 6-13 is replaced by a constant current source using LED2 (red LED), Q2 (TIP32 or other power transistor), RE2 (22 Ω), and LED2 biasing resistor Rbias2 (3.3 kΩ). The collector current is determined by the voltage across RE2, which is the 1.8 volt LED2 voltage drop minus the 0.7 volt

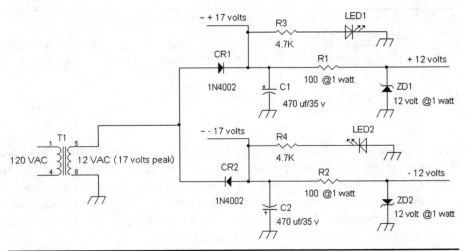

Figure 6-13 Regulated power supply using resistors R1 and R2 to bias zener diodes ZD1 and ZD2.

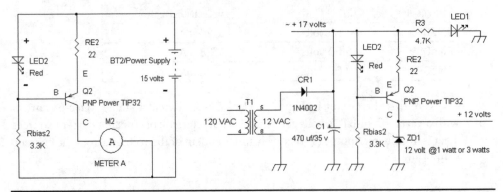

Figure 6-14 Left-side circuit shows a constant current source Q2, and right side shows this circuit integrated into a +12 volt power supply.

VEB turn-on voltage of Q2. Therefore, the voltage across RE2 is 1.8 volts – 0.7 volt = 1.1 volts. The current flowing through RE2 is then (1.1/22) amp ≈ 0.050 amp or 50 mA, which is also the collector current. To lower the collector current, a 24 Ω or 27 Ω resistor can be used for RE2 instead.

Note that a power transistor (TO-220 case) is used instead of a small transistor. That is, the raw supply's voltage is 17 volts, and there is about 1 volt across RE2, so the voltage at the emitter of Q2 is 17 volts – 1 volt = 16 volts. Given that the zener diode ZD1 establishes about 12 volts at the collector of Q2, the voltage across the emitter and collector of Q2 is (the voltage at the emitter *minus* the voltage at the collector) = VEC = 16 volts – 12 volts = 4 volts.

With about 50 mA of collector current and about 4 volts across the emitter-collector junction, the power is IC × VEC (emitter-collector voltage) = 50 mA × 4 volts = 0.2 watt. Although a 2N3906 or 2N2907 may work, either would get really hot. It is safer, then, to use a 40 watt power transistor such as the TIP32 or, better yet, a 75 watt MJE2955.

In determining the current flow through Rbias2, we want to provide sufficient base current from transistor Q2 to generate about 40 mA of collector current. The voltage across Rbias2 is about 17 volts from the raw supply minus 1.8 volts from the LED2 voltage drop, which yields about 15 volts. Thus, the current flowing through Rbias2 is approximately (15 volts/3.3 kΩ) $\approx$ 4.5 mA. Because the current gain β of the TIP32 power transistor is at least 50 for a collector current of 50 mA, the required base current is 50 mA/50 = 1 mA. Thus, we have 4.5 mA available, which is more than enough for the base current requirement of 1 mA for providing 50 mA of regulated current.

If Rbias2 were made larger in value, such as 33 kΩ, the maximum base current would have been 0.45 mA, and with a β of 50, and the maximum collector current would have been 40 $\times$ 0.45 mA = 18 mA, which is short of the 40 mA we need. The bottom line is that if the collector current is not "regulating" to the desired current, decrease the base biasing resistor (Rbias2) to increase the current capability into the base of Q2. But just be aware of the power dissipated into Rbias2. The power into Rbias2 is about (15 volt)^2/Rbias2. For a 3.3 kΩ, ¼ watt resistor, the power into it is (225/3,300) watt $\approx$ 0.068 watt, which is safely less than ¼ watt.

For a ±12 volt regulated power supply for 50 mA maximum current, Figure 6-15 provides a schematic of the circuit, and Figure 6-16 shows a prototype. Note that R1, R2, LED3, and LED4, which were the indicator lamps for the raw supply, can be removed. The reference LEDs, LED1 and LED2, can now serve for indicating voltage at the raw supplies (+17 volts and −17 volts).

Parts List

- Two 470 µF, 35 volt electrolytic capacitors for C1 and C1
- Two 1N4002 or 1N4003–1N4007 diodes for CR1 and CR2
- Two red LEDs for LED1 and LED2
- Two any visible light LEDs for LED3 and LED4 (optional; may be omitted)
- Two 3.3 kΩ, ¼ watt resistors for Rbias1 and Rbias2
- Two 22 Ω, ¼ watt resistors for RE1 and RE2
- Two 4.7 kΩ, ¼ watt resistors for R1 and R2 (optional; may be omitted with LED3 and LED4)
- One NPN TO-220 power transistor such as a TIP31, TIP41, TIP3055, or MJE3055
- One PNP TO-220 power transistor such as a TIP32, TIP42, TIP2955, or MJE2955

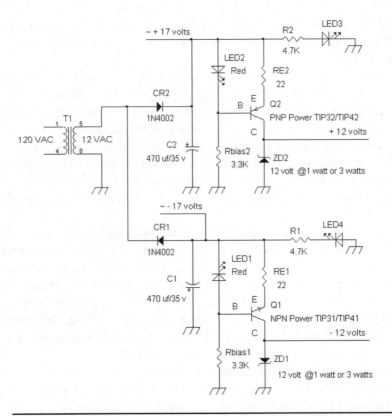

~ + 17 volts

R2
4.7K

LED3

LED2
Red

RE2
22

CR2
1N4002

B
E
Q2
C
PNP Power TIP32/TIP42
+ 12 volts

T1
1   5
120 VAC   12 VAC
4   8

C2
470 uf/35 v

Rbias2
3.3K

ZD2
12 volt @1 watt or 3 watts

~ - 17 volts

R1
4.7K

LED4

CR1
1N4002

LED1
Red

RE1
22

C1
470 uf/35 v

B
E
Q1
C
NPN Power TIP31/TIP41
- 12 volts

Rbias1
3.3K

ZD1
12 volt @1 watt or 3 watts

FIGURE 6-15   Regulated +12 and −12 volt power supply at approximately 50 mA.

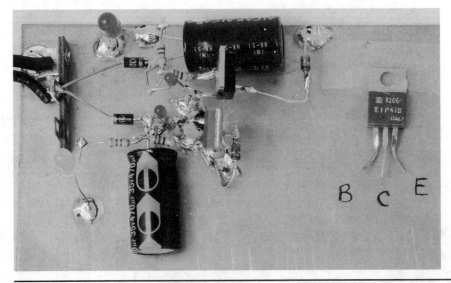

FIGURE 6-16   Voltage regulator circuit (*left*). Pin-outs B, C, and E to power transistors (*right*).

- One AC adapter wall transformer with 120 VAC primary and 12 VAC secondary at 300 mA or more (for countries running on 220 volts AC, use an AC adapter with 220 VAC primary and 12 VAC secondary)
- Two 12 volt zener diodes at 1 watt (1N4742A) or preferably zener diodes at 3 watts (1N5913B or 3EZ12D5)

JFETs, MOSFETs, and Vacuum Tubes Used as Current Sources

JFETs, MOSFETs, and pentode vacuum tubes have current-source characteristics similar to those of bipolar transistors with at least one distinction. The gates of FETs and the grids of vacuum tubes are analogous to the base of the bipolar transistor, and the distinction between gates and grids compared with bases is that they do not draw current. In a bipolar transistor, base current is essential to producing an output collector current, whereas in an FET or a vacuum tube, only the voltage across the gate and source or a voltage across the grid and cathode is necessary to provide an output current. In this section, we will show similar experiments on biasing the devices. The main object is to show the reader some basic characteristics of these devices.

The reader at this point may be thinking that if all of these devices, that is, bipolar transistors, JFETs, MOSFETs, and tubes, can be used for constant current sources, is that the only use? The answer is no, and all these devices are also voltage-controlled variable current sources as well that lend themselves to providing for amplification.

For N and P channel biasing at the gates of JFETs, see Figure 6-17. It should be noted that JFETs normally are used for low-noise and RF (radio-frequency) applications. They are small-signal devices and are not available to handle higher power.

Examples of N-channel JFETs are the J110, 2SK152, MPF102, and 2N3819, and examples of P-channel JFETs are the 2N5460–2N5462 and J174– J177. As shown in Figure 6-17, the left-side circuit, a negative supply, BT1 is fed to VR1, which becomes an adjustable-voltage divider. The slider of VR1 is connected to the gate of Q1,

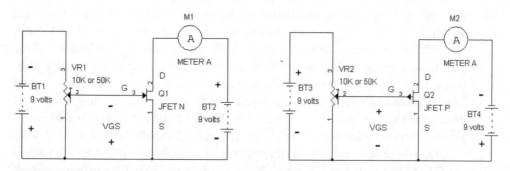

Figure 6-17   Reverse biasing of the gates of N- and P-channel JFET depletion-mode devices.

applying an adjustable negative voltage with respect to the source S. That is, we are measuring the gate-to-source voltage VGS.

The output drain current for an N-channel JFET is measured by ammeter M1. Maximum drain current is specified when the gate-to-source voltage is zero. In terms of the cutoff voltage that causes a near-zero drain current (e.g., 3 nA = 0.003 µA), also known as the *pinch-off voltage*, this varies from –0.5 volt to –10.0 volts in most cases. Even within the same part number, the pinch-off voltage may vary 9:1. For example, an E107 N-channel JFET has a cutoff voltage from –0.5 volt to –4.5 volts for VGS. The question then arises with the wide range of pinch-off voltages how can we use them when manufacturing a product? One answer is to select the JFETs, and another is to provide an adjustment for the negative bias voltage. There is another solution as well, such as using a *negative-feedback* circuit to "servo" the drain current to a particular value by sensing the drain current or voltage and applying a voltage back to the gate of the FET.

On the right side of Figure 6-17 is the connection for a P-channel depletion-mode JFET; also note that everything is the same for the N-channel circuit except that the batteries are reversed in polarity. For just a feel of how JFETs behave, see Table 6-6.

TABLE 6-6   Applying VGS and Measuring Drain Current for Two N-Channel JFETs

JFET (N-Channel)	VGS (Gate-to-Source Voltage)	ID (Drain Current)
MPF102	0.00 volt	5.57 mA
MPF102	–1.50 volts	0.73 mA
MPF102	–2.30 volts	1.7 µA
J110	0.00 volt	92 mA
J110	–1.50 volts	7.36 mA
J110	–3.0 volts	1.7 µA

We now turn to MOSFETs, which are insulated-gate FETs. In a JFET, the gate-to-source terminals are reversed biased, and if the gate-to-source voltage is forward biased with +0.5 volt or more at the gate with respect to the source, there will be gate currents very much like forward biasing a diode. In MOSFETs, the gate is insulated, which means that any forward bias voltage does not incur gate current. Most MOSFETs are manufactured to be enhancement-mode devices. This means that the gate-to-source voltage is positive for N-channel devices. Despite the positive *VGS*, there is no current flowing into the gate. Enhancement-mode MOSFET devices come both in small-signal and large-signal power devices. For example, N-channel enhancement-mode MOSFETs are the VN2222L, VN10K, and IRF610, and examples of P-channel enhancement-mode devices are the VP0550, VP0610, and IRF9630.

However, it is possible to obtain depletion-mode N-channel MOSFETs such as the DN3545N3-G or LND150N3-G. Figure 6-18 shows the biasing of enhancement

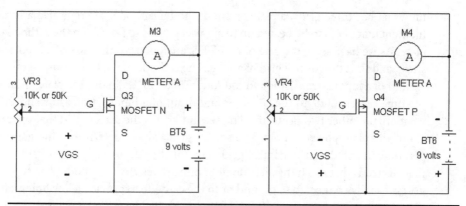

Figure 6-18 Biasing enhancement-mode MOSFETs for N- and P-channel devices.

devices. Note that for the P-channel device, the polaritiy of the battery is reversed when compared with an N-channel transistor.

For the next experiment, both medium- and high-power N-channel enhancement-mode devices will be measured for drain current as VGS is varied. It should be noted that enhancement-mode MOSFETs have turn-on voltages (aka *threshold voltages*) just like bipolar transistor. However, whereas the bipolar transistor has a well-known turn-on voltage of approximately 0.6 volt for silicon, enhancement-mode devices have turn-on voltages that vary quite a bit. For example, an N-channel enhancement device, such as the VN2222L, has a VGS turn-on voltage that varies from +0.6 volt to +2.5 volts. Tests of two N-channel MOSFETs were done using the circuit in Figure 6-18. If the VGS voltage is sufficiently high, greater than 2 volts, the MOSFET will act more like a switch. So VR3 had to be turned down to 0.00 volt first and slowly turned up. It is advisable not to exceed 100 mA of drain current or damage will be incurred in the MOSFETs or the ammeter. Table 6-7 provides some measurements.

We now turn our attention to "hollow-state" devices, which are vacuum tubes. After all, why should solid-state devices such as bipolar transistors and FETs get all the attention? For this chapter, we will concentrate on the pentode, a five-element

Table 6-7 Measurements of Drain Current on N-Channel MOSFETs with Various Gate-to-Source Voltages

MOSFET (N-Channel)	VGS (Gate-to-Source Voltage)	ID (Drain Current)
VN2222L	0.00 volt	0.00 µA
VN2222L	+1.00 volt	2.40 µA
VN2222L	+2.00 volts	69.00 mA
VN10K	0.00 volt	0.00 µA
VN10K	+1.00 volt	3.30 µA
VN10K	+1.64 volts	66.00 mA

tube that has three grids, a plate or anode, and a cathode. There is also a heater or filament that needs to be heated up to allow electrons to flow from the cathode to the plate. As we have seen, the gate of an FET is analogous to the base of a bipolar transistor, and so is the control grid of a tube. The cathode of the tube is analogous to the emitter of a bipolar transistor and the source of an FET. Finally, the plate or anode of the tube corresponds to the collector and drain of a bipolar transistor and FET. So what are the other two grids for? The second grid is the screen grid that operates at about the plate power supply. The screen-grid voltage determines the gain of the pentode, but so does the control grid. The third grid, the suppressor grid, ensures that electrons that reach the plate do not bounce back to the cathode. The suppressor grid is usually connected to ground or to the cathode to provide a low potential such that any electrons bouncing off the plate get repelled back to the plate.

A pentode may be "converted" to a triode by connecting the screen grid to its plate. This results in a less than ideal current generator but generally provides for lower harmonic distortion and allows for lower plate voltages. For example, a 6AG5 or 6AU6 in pentode mode requires about 100 volts for operation, but if the screen grid is tied to the plate for triode mode, a plate voltage as low as 10 volts can be used.

Figure 6-19 shows a triode V1 with G1 as the control grid and a pentode V2 with G1 as the control grid, G2 as the screen grid, and G3 as the suppressor grid. Also, a pentode has its screen grid G2 connected to its plate to form a triode, and note that G3 is connected to its cathode, K. H1 and H2 are the heater or filament connections. Generally, the first single- or double-digit number for a tube denotes the filament voltage. For example, a 1R5 has a 1.4 volt filament voltage, whereas a 6AU6 and 35W4 have 6.3 volt and 35 volt filaments, respectively. It should be noted that most triodes (e.g., 6AT6, 6CG7, 12AX7, 12AU7, 12AT7, etc.) will work down to a plate voltage of about 10 volts.

In the 1950s, prior to the widespread use of transistors, specially made "space charge" vacuum tubes allowed pentodes, pentagrid converters, tetrodes, and triodes to operate "normally" at a plate voltage as low as 10 volts. Pentodes such as the 12DZ6, 12EA6, and 12AC6 were used commonly for car radio circuits requiring 12 volts DC. Thus, 12 volts supplied the 12 volt heater/filament and also ran the plate power supply. Figure 6-20 shows a bias experiment using one of these tubes. Note that although a 12DZ6 is shown in the figure, one can use a 12EA6 or 12AC6 for the

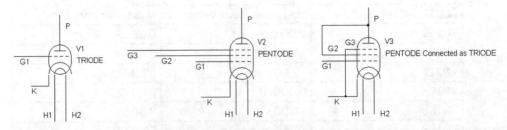

Figure 6-19   A triode, pentode, and pentode configured for triode mode.

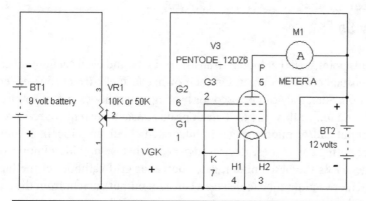

FIGURE 6-20 Measuring plate current as the grid-to-cathode voltage is varied.

experiment since the pin-outs are the same. Note that pin-out connections are not the same for all triodes or all pentodes. Be sure to go to the Web or check a tube manual to identify the plate, cathode, grid(s), and heater pin-out assignments for various tubes.

NOTE The grid-to-cathode voltage ranges from 0 volt to a negative voltage, just like an N-channel JFET having a reverse-bias voltage across its gate and source. Also note that Figure 6-20 is very similar to the N-channel depletion-mode JFET measuring circuit in Figure 6-17 (*left*).

Tubes, like depletion-mode FETs, have cut-off voltages as well, and for a particular tube, the cut-off voltage is much better defined than for a JFET. Note that tubes are only N-type devices. There are no P-channel equivalents in vacuum tubes.

For the next experiment, we will include a space-charge low-plate-voltage tube (12EA6) and run some measurements. Also, a high-voltage pentode such as the 12BA6 will be connected in triode mode by connecting the plate's pin 5 to the screen grid's pin 6. Measurements will be made with just 12 volts for the plate voltage. See Table 6-8.

TABLE 6-8 Plate Current Measurements for Vacuum Tubes While the Grid-to-Cathode Voltage Is Varied

Tube	VGK (Grid-to-Cathode Voltage)	IP (Plate Current)
12EA6 (pentode mode)	0.00 volt	7.78 mA
12EA6 (pentode mode)	−1.64 volts	2.05 mA
12EA6 (pentode mode)	−3.30 volts	0.68 mA
12BA6 (triode connected)	0.00 volt	3.38 mA
12BA6 (triode connected)	−1.64 volts	0.23 mA
12BA6 (triode connected)	−3.30 volts	44.00 μA

Brief Summary So Far

We have at this point seen that bipolar transistors, FETs, and even vacuum tubes can be used as constant current sources. On their own, constant current sources serve a limited function. Sure, they can be used for biasing zener diodes and other circuits, which you found out in this chapter (sneak peak: constant current sources are also used in the input and/or middle stages in integrated circuit amplifiers). A constant current source using any of these devices just means that we provide a known (constant) voltage across the base-emitter, gate-source, or grid-cathode of the bipolar transistor, FET, or vacuum tube to deliver a constant output current from the collector, drain, or plate.

In the experiments shown so far, we also varied the voltages at the input, which caused a change in output current. In a transistor, we can measure the change in input voltage across the base-emitter junction, or alternatively, we can determine the change in base current. Either way, a change in voltage or current at a bipolar transistor's base terminal causes a change in collector current.

To provide amplification, we just can connect a load resistor to the collector of the transistor and determine the voltage swing across the load resistor or measure the collector voltage swing. In general, increasing the value of the load resistor increases the voltage gain. For example, see Figure 6-7 and Table 6-2. Similarly, in the experiments concerning FETs and vacuum tubes, we find that a change in voltage at the input to the gate-to-source junction of an FET or the grid-to-cathode connections of a vacuum tube will lead to a change in drain or plate current, respectively.

Just For Fun: A Low-Power Nightlight Using Transistor Amplifiers and LEDs

Here's a circuit that turns on at night and shuts off in daylight or room light. The reader is encouraged to try out other photoreceiving devices than the LED used for sensing light. For example, a photodiode will have much more sensitivity to light (see Figure 6-21).

This circuit makes use of the current amplification characteristic of bipolar transistors. Generally, these transistors are also considered more to be voltage-to-current converters, but they are also current-to-current converters. So the photocurrent from LED1, which acts as a photodiode, is connected to the base of transistor Q1 to multiply the photocurrent by a current gain $\beta 1$, and this multiplied current exits at its collector of Q1. The collector of Q1 sends the amplified current to Q2's base, which amplifies again by a factor of $\beta 2$ the current gain of the Q2. The collector current of Q2 then is the photocurrent from LED1 $\times \beta 1 \times \beta 2$. When sufficient collector

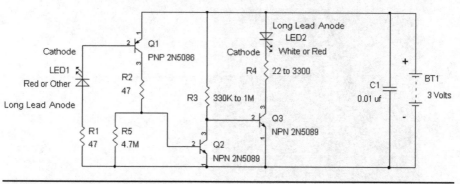

FIGURE 6-21 An automatic night light.

current is pulling from Q2, R3 will develop a voltage drop that causes the collector voltage at Q2 to approach 0 volt and shut off Q3 to turn off LED1. Thus, when there is sufficient light falling into LED1, LED2 shuts off. When LED1 receives very little light, there is insufficient photocurrent to be amplified, and the collector current of Q2 is too small to develop a sufficient voltage drop across R3. This allows current to flow from the power supply BT1 via R3 into the base of Q3 such that Q3 behaves as a current amplifier or a switch that turns on LED2.

Parts List

- Two 47 Ω resistors for R1 and R2
- One 330 kΩ to 1 MΩ resistor for R3 (try 330 kΩ first)
- One 22 Ω to 1 kΩ resistor for R4 to adjust brightness
- One 4.7 MΩ resistor for R5 for setting the threshold of the amount of light to turn off LED2 (start with 4.7 MΩ, but you can also try values from 1 MΩ to 3.9 MΩ)
- Two AA batteries with two-cell holder and connector for providing 3 volts DC
- One red LED or a photodiode for LED1
- One white LED (or other color) for LED2
- Two 2N5089 (or 2N5088) transistors for Q2 and Q3
- One 2N5086 (or 2N5087) transistor for Q1
- One 0.01 μF capacitor for C1 (any capacitor from 0.01 μF to 1 μF, film or ceramic, will do)

Note that the 47 Ω resistors for R1 and R2 are for protecting transistors Q1 and Q2 should there be excessive current due to too much photocurrent or should LED1 be inadvertently connected in reverse. Figure 6-22 shows the prototype.

Figure 6-22   LED night-light prototype circuit.

One can also use this circuit to test transistors. By substituting another PNP transistor for Q1 or an NPN for Q2, one can determine whether the transistor is working or not or has low current gain. In doing so, it is recommended that a photodiode (e.g., Osram SFH 229 from www.Mouser.com) should be used instead for LED1 so that sufficient current is generated by the light (R5 may have to be lowered to less than 1 MΩ). If an LED is used as the photodiode, you may have to place LED1 close by a lamp or shine a bright flashlight into it to turn off LED2. If a transistor is broken, it will either stay on all the time or stay off all the time regardless of how much light LED1 receives.

A Quick Look at Operational Amplifiers (Op Amps)

Integrated circuit (IC) operational amplifiers (op amps) are very common today in lieu of using discrete transistor amplifier designs. By nature, they have very large gains, typically greater than 10,000 (> 10,000). For example, if there is a DC voltage of 0.1 mV across the inputs of an op amp with a gain of 10,000, the output will be 1 volt. Let's take a quick look at these devices for now. See Figure 6-23.

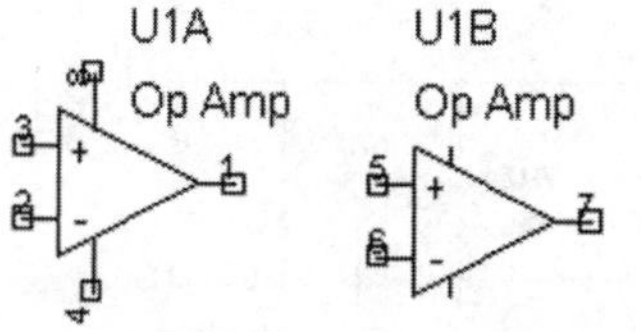

U1A
Op Amp

U1B
Op Amp

Figure 6-23   Example of a dual op amp.

As seen in Figure 6-23, an op amp has two terminals for the power supply, two terminals for the inputs labeled "+" and "−", and an output terminal at the tip of the triangle. The two power-supply terminals for an op amp provide flexibility whether a plus and minus supply is used or either a single plus or minus supply is connected. In the single-power-supply example, one of the power-supply terminals is connected to common or ground. See Figure 6-24. When plus and minus supplies are connected, this allows the op amp to behave exactly like an ideal amplifier in terms of accepting AC signals and outputting amplified AC signal without having to use capacitors for coupling or level shifting.

Op amps come with one or two op amps packaged in an 8 pin IC (integrated circuit), and four op amps in a 14-pin package. One should note that when a single supply is used, coupling capacitors are generally used to allow the transistors inside the op amp to have the proper DC biasing voltages.

When powering an op amp, the polarity of the plus supply terminal V+ and minus terminal V− must be obeyed accordingly. For example, most dual op amps have the V+ terminal as pin 8 and the V− terminal as pin 4. For proper operation, there must be a positive voltage measured from pin 8 to pin 4. That is, if one places the positive lead of a VOM on pin 8 and the negative lead on pin 4, a positive voltage should be read. An incorrect polarity into these terminals will generally result in permanent damage to the op amp. Also note that op amps require power-supply decoupling capacitors (C1–C4) on the order of at least 0.01 μF and preferably 0.1 μF or larger.

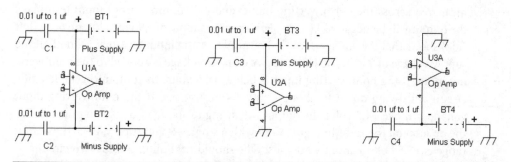

Figure 6-24   Op amps powered by plus and minus supplies, a single positive supply, and a single negative supply.

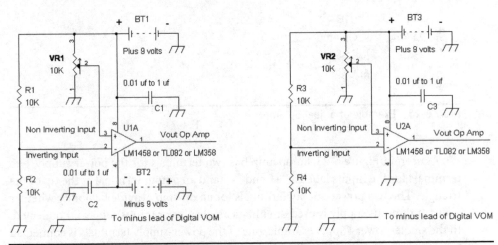

Figure 6-25   (*Left to right*) Op amps with high gain can be used as comparators when powered by dual and single power supplies.

We now look at the input terminals of the op amp. See Figure 6-25. An op amp has a gain, which is called the DC *open-loop gain*. The DC open-loop gain can be thought of as the ratio of the output voltage to the input voltage across the (+) and (–) inputs. For example, if we put 1 microvolt (1 μV) of DC across the (+) and (–) inputs and get 10 volts out, the DC open-loop gain is then (10 volts)/(1 μV) = 10 million = open-loop gain for a DC signal.

Suppose that our power supply were limited to BT1 = V+ = +12 volts and BT2 = V– = –12 volts, and the op amp (on the left side of Figure 6-25) could output reliably from +11 volts to –11 volts, but we put 10 μV across the input. If the power supply were "unlimited" in voltage and the op amp could accept the unlimited power-supply voltage, the output would be 10 μV × 10 million = 100 volts. However, with the +12 volt and –12 volt power supplies, the output of the op amp will be limited to 11 volts for a 10 μV input. Depending on the polarity of the voltage at the input, the output will be positive or negative 11 volts. For example, if +10 μV is measured across the (+) and (–) input terminals, the output will be +11 volts, and if +10 μV is measured across the (–) (inverting input) and (+) (noninverting) input terminals, the output will be negative 11 volts. For a power supply of ±9 volts, Figure 6-25 (left side) shows that the voltage at the (–) or noninverting input terminal is one-half the supply voltage of 9 volts, or +4.5 volts. A variable voltage is applied to the noninverting input. If the noninverting input is below +4.5 volts, the output on the left side circuit will swing down to about –7.5 volts in practice. If the noninverting input voltage is above +4.5 volts, the output will swing to about +7.5 volts in practice. For the op amp on the right in Figure 6-25 with a single +9 volt supply, BT3, the output range is +7.5 volt to about +1.5 volts. One should try building these two circuits to get a feel for the output range of op amps. Why are these two circuits called comparators? They are comparing two voltages from VR1 or VR2 against the +4.5 volts via R1 and R2 or via R3 and R4. Any voltage from the variable resistors higher than +4.5 volts is a "logic" high state, and for any voltage below +4.5 volts, the output of the op amp swings to a low logic state.

Is it possible to try to set the input voltage very carefully to have a set output voltage such as 0 volt for the circuit on the left that has plus and minus supplies? The answer is no, not normally, unless some type of feedback mechanism is included. Any slight variation in power-supply voltage or any changes in resistance to R1, R2, or VR1 will cause the output to tip toward the plus or minus supply voltage.

Therefore, op amps normally use some type of feedback, where there is a DC path from the output to the inverting input terminal. Figure 6-26 shows a variable resistor VR1 connected from the output terminal to the inverting input. The noninverting input has a DC voltage of approximately 5 percent of +9 volts, or about +0.45 volt. By varying VR1, the output voltage will vary from about +0.45 volt to about +5 volts. If one measures the DC voltage at the noninverting and inverting input terminals, one will find almost the same voltage. Or put another way, if the voltage is measured across the noninverting and inverting input terminals at pin 3 and pin 2, the voltage will be close to 0 regardless of the setting on VR1. One can try this measurement with a digital VOM to confirm that indeed when an op amp has negative feedback as shown in Figure 6-26, putting the positive lead on pin 3 and the negative lead on pin 2 results in a VOM reading of nearly 0 volt.

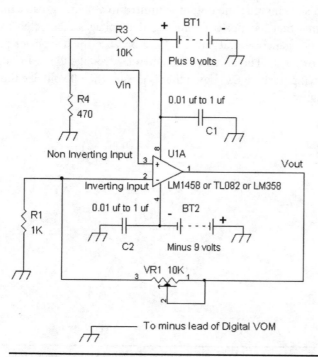

Figure 6-26   Negative-feedback configuration for an op amp.

The 0 volt across the noninverting and inverting inputs is sometimes called a *virtual short circuit* across the input. But, in reality, there is no short circuit here! The 0 volt condition is really a consequence of the open-loop gain being very high. Recall that for high-open loop gains, all it takes is on the order of a few millivolts or less at the input to cause the output to swing toward the power-supply voltages. Therefore, as long as the output of the op amp is less than its maximum output voltage, the input voltage, or, more appropriately, the voltage across the (+) and (−) inputs will be on the order of less than a few millivolts. For example, for an op amp with an open-loop gain of 3,000, a few millivolts across the (+) and (−) inputs will yield 3,000 × 3 mV = 9 volts at the output. For higher open-loop gains, the voltage across the (+) and (−) inputs will be smaller—hence coming to the "conclusion" that the (+) and (−) inputs approach 0 volt and thus become a "virtual" short circuit. (I suppose that the (+) and (−) inputs under a negative-feedback condition "play" a short circuit on TV, but in reality, the two inputs are not really a short circuit—a takeoff on the old joke about an actor saying that he or she is really not a doctor but only plays one on TV.)

We now turn to how to correctly use op amps with a single supply. By the way, all op amps advertised as single supply op amps will work with plus and minus supplies, and all op amps that generally are shown to work with plus and minus supplies will work on a single supply *if biased correctly and if there is sufficient voltage (e.g., usually at least 6 volts) to power the op amp.*

On the left-side circuit of Figure 6-27, the op amp is powered with a single positive power supply. This means that the output is limited to voltages greater than 0 volt. With an AC signal that has positive and negative voltages connected to the input, the output cannot generate a negative voltage because the V− power pin is connected to ground or 0 volt. Thus, we need to add a DC voltage to the AC signal such that we get a varying positive DC signal that is greater than 0 volt. See the circuit on the right in Figure 6-27.

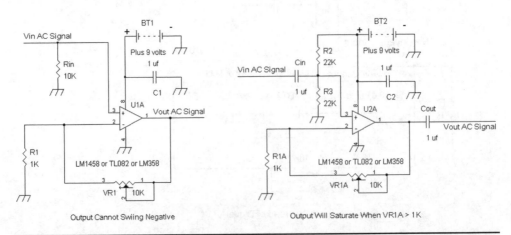

FIGURE 6-27 Two op amp circuits connected to single supplies, but they require more work for proper operation.

   In the circuit on the right in Figure 6-27, the input signal "Vin AC Signal" is connected to a coupling capacitor Cin, which allows R2 and R3 to level shift the AC signal to average at one-half the plus-supply voltage. This results in having +4.5 volts at the (+) input terminal, which is a good start to solving the problem. If VR1A is set to 0 ohms, the output is connected to the inverting input. By virtue of the fact that the voltage at the (–) input is the same or very close (within a few millivolts) to the (+) input, the output will then also be +4.5 volts because the (+) input terminal has +4.5 volts. Pretty cool, eh? The op amp servos the output voltage to the same voltage as the (+) input for VR1A = 0 Ω.

   Now let's see what happens if VR1A is set to 1 kΩ. Because R1A is also 1 kΩ, the voltage at pin 2, the (–) input, is half the output signal at pin 1 because a divide by two voltage divider circuits consists of two equal-valued resistors. The (+) input is +4.5 volts, so the op amp will want to make the output 9 volts because VR1A = 1 kΩ and R1A = 1 kΩ will divide the 9 volts from the output in half to +4.5 volt. Unfortunately, a real op amp such as an LM1458 will deliver about +7.5 volts maximum at its output, the result is that the output has now "clipped" or saturated and is no longer working as an amplifier. Let's take another example by setting VR1A to 9 kΩ. The voltage division formed by a 9 kΩ (VR1A) and a 1 kΩ resistor resulted in a division of voltage of 10. Because there is +4.5 volts at the (+) input terminal, in order to get +4.5 volts at the (–) input terminal, the output would have to be +45 volts, which is impossible given that BT2 is a 9 volt supply. That is, the maximum output voltage of the op amp is less than or equal to the power-supply voltage. It would be a great trick if we could get 45 volts out of an op amp powered with just 9 volts—but sorry, this will not work.

   So how do we solve the problem? Ideally, we want the output to be sitting at about one-half the power-supply voltage so that its voltage swings are about equal going toward +9 volts and down toward 0 volt. It looks like an extra capacitor (C3) has come to the rescue to solve our problem (see Figure 6-28).

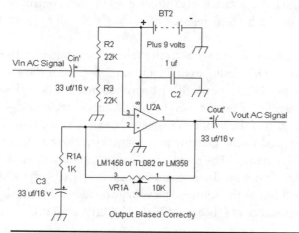

Figure 6-28   Single-supply op amp biased correctly.

In Figure 6-28, let's take a look at our "new" DC voltage divider circuit VR1A, R1A, and C3. Instead of having the bottom lead of R1A going to ground, we put a "large" capacitor in series with it before connecting to ground via the negative terminal of C3. If we follow the DC voltage from VR1A to R1A, initially, C3 will act like a short circuit to ground, but because the DC voltage is being provided by pin 1, via VR1 and R1A, C3 is being charged up, and eventually, there will be no more current flowing into C3 after a warm-up period of less than a second. Note that it does not matter what value of resistance VR1A has, eventually, C3 will be charged up. If there is no DC current flowing into C3, there is also no DC current flowing through VR1A and R1A. Therefore, the DC voltage across VR1A is 0 volt because by Ohms law, $V = IR$, and because $I = 0$, $V = 0R = 0$. We can then see that for DC purposes, the DC voltage at pin 1 is the same DC voltage as at pin 2, the inverting input. But the voltage at pin 3 is +4.5 volts, and since the voltages at pins 3 and 2 are the "same," the output, pin 1, has to be also +4.5 volts.

Another intuitive way to look at this is that for DC purposes, C3 is an open circuit, so take C3 out for finding out the DC voltage at the output. This leaves a voltage-divider circuit consisting of only VR1A because the bottom of R1A is now connected to an open circuit or to "nowhere." A voltage divider consisting of just one resistor provides no voltage division at all, and all the voltage going into the resistor (VR1A to pin 1, the output terminal) also comes out with the same voltage at the other end (VR1A to pin 2, the (–) input). Therefore, we have the +4.5 volts at pin 3, the (+) input, to servo the output to +4.5 volts so that the (–) input is also +4.5 volts.

JFETs and Bipolar Transistor as Voltage-Controlled Resistors

Finally, we turn our attention to using FETs and bipolar transistors as voltage-controlled resistors. When used as part of a voltage-divider circuit, that is, the resistor that goes to ground but is a variable resistor, we get to use a control voltage to adjust the resistance of the bottom resistor of a voltage-divider circuit (see Figure 6-29).

One of the differences that you will see with the voltage-controlled resistors in Figure 6-29 is that there is no DC biasing to drain and no collector for the FET Q1 and for the bipolar transistor Q2, respectively. It turns out that JFETs make very good voltage-controlled resistors. They are used in low-distortion Wein bridge sine wave oscillators to control gain of the oscillator's system. For an N-channel device, a negative bias voltage is applied to the gate-to-source junction. The resistance is a function of the negative bias voltage. The more negative the gate-to-source voltage is, the higher is the resistance from the drain to source. Less negative bias provides a lower resistance from the drain to the source. The voltage divider is formed by R1 and the drain-to-source resistance of Q1 (e.g., MPF102, 2N3819, or J110).

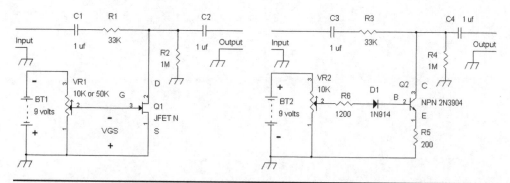

FIGURE 6-29   JFET Q1 variable resistor and a bipolar transistor Q2 voltage-controlled resistor.

A voltage-controlled resistor can be made from a bipolar transistor. The method is tricky and not obvious. The circuit you see on the right in Figure 6-29 is a practical circuit used in a Superscope CD-302A cassette tape recorder for its automatic level control amplifier. What is shown is just the variable-resistance portion of the amplifier, and it is not the complete automatic level control amplifier. A forward bias voltage from BT2 and VR2 turns on Q2 via diode D1. Applying a voltage to R6 causes the collector-to-emitter resistance to decrease. Thus, the collector-to-ground resistance changes and provides more attenuation. The initial value for VR2 is 10 kΩ, but a lower-resistance potentiometer may work better, such as 1 kΩ. Typically, the input AC signals levels are less than 1 volt peak to peak, and the outputs are followed with ×100 or more gain amplifier.

References

1.   *Power Products Data Book*. Dallas, TX: Texas Instruments, 1985.
2.   *FET Data Book*, Santa Clara, CA: Siliconix, Inc., July 1977.
3.   *Transistors Small Signal Field Effect Power*. Santa Clara, CA: National Semiconductor, May 1974.
4.   *RCA Receiving Tube Manual RC-29*. Harrison, NJ: RCA Corporation, 1973.

PART TWO

Intermediate-Level Electronics

CHAPTER 7

Amplifiers and Feedback

We will cover some basic concepts of negative feedback first before showing some of the amplifier circuits employing negative feedback. For most amplifiers, negative feedback is used for a variety of purposes that include setting the gain of the amplifier with a couple of resistors, reducing distortion, and increasing frequency response, just to name a few. Voltage-controlled amplifiers are often used with negative-feedback systems such as automatic gain control circuits, and an automatic level control amplifier for audio signals will be presented. Finally, we will present a couple of examples where positive feedback in an amplifier is not desirable.

What Are Negative-Feedback Systems?

Generally, a negative-feedback system is where a reference signal is at the input of the system, and the output of the system is following the reference signal in some manner. If the output strays too low or too high, the output is compared with the input reference signal, and a correction signal in the *opposite* direction is given to maintain the desired output. For example, probably one of the oldest negative-feedback systems is the biological systems in mammals by which a body temperature is maintained. For example, for humans, the body temperature is regulated to about 98.6 degrees Fahrenheit.

In terms of a mechanical system, a motor with weights (aka a *governor*) attached to arms that will fly open when the speed is too high and collapse if the speed is too slow is a negative-feedback system. (Think of an ice skater who uses the extension of the arms to control spin speed.) In electric motors for tape recorders built in the mid-1960s, two weights were attached to the contacts of two switches that shorted out two resistors in series with the motor's winding. When the speed was too high, the weights flew out to open the switch contacts. The motor was reduced in current because the two series resistors were now enabled to reduce the current into the

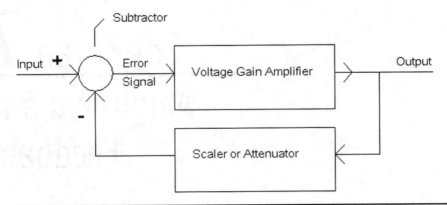

FIGURE 7-1   A negative-feedback system.

motor, which then reduced the speed of the motor. When the motor slowed down sufficiently, the weights collapsed such that the switches were turned back on to short out the series resistors to deliver more current to the motor and speed it up. There is then a constant back and forth of the motor's speed going slow and fast due to the governor's arms extending and collapsing, which results in a long-term constant speed. If the motor with its governor is designed correctly and is coupled with a flywheel to average out the short-term speed variations of the motor, the tape speed is constant during the record and playback modes. A model of a negative-feedback system is shown in Figure 7-1.

A reference voltage is fed to the input, and the output of the system is feedback and is scaled or attenuated or reduced in strength and then subtracted from the reference input signal. A difference signal then provides an error signal to the voltage gain amplifier.

At first, the block diagram in Figure 7-1 may seem abstract. For instance, what does it mean to have an attenuated output subtracted from the reference input? So let's take a look at a concrete example, a heating system with a thermostat and furnace (see Figure 7-2).

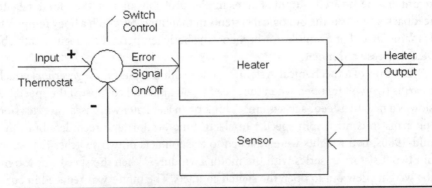

FIGURE 7-2   A furnace heating system.

A thermostat sets the desired room temperature (e.g., 70 degrees Fahrenheit). The furnace turns on, and some of its heat is transferred to the sensor. When the room is heated to 70 degrees or slightly higher, the furnace turns off. As the room cools below 70 degrees, the sensor develops a voltage to turn the furnace back on.

Now let's take a look at whether the sensor is insulated or put in the colder corner of the room. What will happen is that the furnace will stay on until that part of the corner of the room is heated to 70 degrees. But this will cause the rest of the room to be hotter than 70 degrees. Alternatively, do not move the sensor but wrap some heating insulation around it. This will cause the sensor to receive attenuated heat, and thus the room will be heated to beyond 70 degrees because the sensor has to reach 70 degrees after the heat goes through the wrapped insulation.

In both of these cases, if the output heat is attenuated by distance to the sensor or by insulating heat from the sensor, the result will be a higher output from the heater, which leads to higher than 70 degrees in the room. The negative-feedback system is merely compensating to have the sensor reach 70 degrees whether it has been put in a normal area of the room, a remote location, or the sensor has been blocked from the heat by having insulation around it.

A generalized electronic feedback system consists of a differencing circuit that has the input reference signal subtracted by a return, or attenuated, output signal. For a specific feedback amplifier, see the right block diagram of Figure 7-3 that contains a *differential* input amplifier that mimics the subtractor circuit with (+) and (−) inputs.

Negative-feedback systems include a feedback network. In the furnace heater system, the sensor acts as the feedback network. Positioning the sensor determines the received heat and thus determines whether the room is hotter than the set temperature at the input via the thermostat. In an amplifier, the feedback network usually attenuates the signal via a resistor voltage divider such that the output signal is of higher amplitude than the input signal. In a voltage follower configuration, the feedback network does not attenuate the output signal to the inverting input of the amplifier. The gain of this specific negative-feedback system is unity or one.

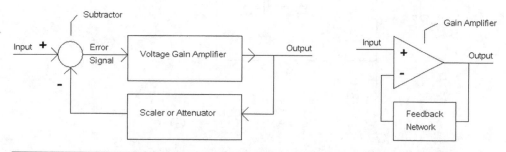

Figure 7-3   (*From left to right*) A generalized feedback system and a specific feedback amplifier system.

Another Look at the Non-Inverting-Gain Operation Amplifier (Op Amp)

Chapter 6 introduced the non-inverting-gain amplifier. We will look at this configuration with a slightly different view for this chapter. See Figure 7-4. Intuitively, we observe that the input voltage is applied to the (+) input terminal, so we should expect that the output signal in *in phase*. That is, a positive voltage into the input yields a positive output voltage. Note: For the input notations of amplifiers, the "(+)" input is equivalent to the "+" input, and the "(−)" input is equivalent to the "−" input.

For an example, first look at the open-loop gain of the system by removing RF and setting R1 = 0 Ω. The open loop gain A_0 refers to the raw gain of the amplifier prior to adding a feedback network around it.

$$Vout = A_0 \times [V(+) - V(-)] \tag{7-1}$$

where V(+) is the voltage at the (+) input terminal, and V(−) is the voltage at the (−) input terminal. This says that the output voltage of the op amp is equal to the voltage across the (+) and (−) inputs multiplied by the open-loop gain A_0. This relationship holds whether the op amp has no feedback or has negative feedback.

When the open-loop gain is very large, the voltage difference across V(+) and V(−) is very small. Put another way, via algebraic manipulation of dividing A_0 on both sides of Equation (7-1), the output voltage divided by the open-loop gain is just the voltage across the (+) and (−) inputs, which says

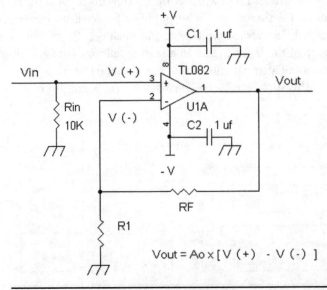

Figure 7-4 A non-inverting-gain amplifier using negative feedback with a voltage divider circuit, RF and R1.

$$Vout/A_0 = [V(+) - V(-)] \tag{7-2}$$

For example, if the open-loop gain $A_0 = 100,000$, a common figure for many op amps such as the LM741, TL082, or NE5532, and the output is +10 volts DC, the voltage across the (+) and (–) inputs is +10 volts/100,000, or 0.10 mV, which is one-tenth of a thousandth of a volt. See Equation (7-3).

$$+10 \text{ volts}/100,000 = [V(+) - V(-)] = 0.0001 \text{ volt} = 0.10 \text{ mV} \tag{7-3}$$

Thus, for all practical purposes, we say that the voltage at the (+) and (–) inputs is essentially the same, within a few millivolts give or take. See Equation (7-4). A consequence of having a very high open-loop gain results in that the voltages at the input terminals are essentially the same potential.

$$[V(+) - V(-)] = 0.1 \text{ mV} \approx 0 \text{ volt} \rightarrow V(+) \approx V(-) \tag{7-4}$$

Now let's look at a couple of examples. Let RF = R1. We know with equal-valued resistors, a divide-by-two voltage divider is formed from the output of the op amp to the (–) input. Therefore, if we connect +1 volt at the (+) input, we should also get +1 volt at the (–) input by virtue of the fact that the two input voltages have the same potential *when negative feedback is applied and the negative-feedback system is working properly with a large open-loop gain A_0; that is, it is not oscillating, or the output is not latched or stuck to one of the power-supply voltages.*

We then ask the question, what voltage divided by two is +1 volt? The answer is +2 volts at the output of the op amp—at least, intuitively, one might guess that.

So what is really happening here? The op amp via negative feedback is "servoing" the output to provide a voltage when attenuated by a factor of two via the voltage divider to drive the voltage at the (–) input to match the voltage at the (+) input. Again, some books will just state that there is a virtual short circuit between the (+) and (–) inputs of an op amp when negative feedback is applied. But this virtual short circuit is really not a short circuit at all. It is the feedback system trying to do its best to change the output of the amplifier to match the (–) input voltage to the (+) input voltage after going through the feedback network. If the output voltage, for example, overshoots and exceeds +2 volts, there will be a bit more than +1 volt at the negative input terminal of the op amp, which is larger in magnitude that the input voltage at the (+) terminal of exactly +1 volt.

By nature of the (–) input voltage being larger than the (+) input voltage, the output voltage then has to move down from slightly greater than +2 volts to something lower. Similarly, if the output voltage is a little less than +2 volts, yielding slightly less than +1 volt at the (–) input, the input voltage at the (+) input terminal at exactly +1 volt would be greater than the voltage at the (–) input and thus would cause the output voltage to increase. Thus, there is a self-correcting nature about negative-feedback systems or amplifiers. We then get an equilibrium state such that the two input voltages are the same.

What if we change RF to 9 kΩ and keep R1 = 1 kΩ? We end up with a divide by 10 voltage divider circuit where the voltage divider formula from Vout to the (–) input terminal is [R1/(R1 + RF)] = voltage divider formula. This leads to V(–) = Vout[R1/(R1 + RF)] = Vout[1 kΩ/(1 kΩ + 9 kΩ)] = Vout[1 kΩ/(1 + 9) kΩ] = Vout(1 kΩ/10 kΩ) = Vout (1/10). Therefore, if we have +1 volt at the (+) input, what voltage at the output when divided by 10 gives +1 volt at the (–) input? Again, just by guessing, the answer would be 10 volts.

Therefore, the gain of the two examples is defined as the ratio of the output voltage to the input voltage. For RF = R1 = 1 kΩ, the gain Vout/Vin = +2 and for RF = 9 kΩ and R1 = 1 kΩ, and thus, Vout/Vin = +10. If we look more closely, we find that the gain is just the reciprocal of the voltage divider formula [R1/(R1 + RF)] → Vout/Vin = (R1 + RF)/R1, which sometimes is expressed as Vout/Vin = (RF + R1)/R1 and sometimes an alternate expression, is Vout/Vin = [1 + (RF/R1)].

Let's look at another way of looking at the equation V(–) = Vout[R1/(R1 + RF)]. We know via negative feedback that V(+) = V(–) and that Vin is connected to the (+) input of the op amp in Figure 7-4. So Vin = V(+) = V(–) via negative feedback. Therefore,

V(–) = Vin = Vout[R1/(R1 + RF)]   or   Vin = Vout[R1/(R1 + RF)]

which leads to when we divide by Vout on both sides:

Vin/Vout = [R1/(R1 + RF)]

and if we take the reciprocal of both sides, we get the gain as:

Vout/Vin = (R1 + RF)/R1 = [1 + (RF/R1)]   (7-5)

Inverting-Gain Configuration

The inverting op amp configuration is not used as often as the noninverting amplifier circuit. Noninverting op amp circuits allow for high input resistance so as not to load the input signal and also lend themselves to low-noise amplifier configurations such as microphone and phono-cartridge preamps or even moving-coil pre-preamps.

When one uses the inverting-gain configuration, there is a higher likelihood of loading down the input signal source and generating more noise. However, the inverting-gain amplifier is often used as photodiode or transresistance amplifier, where the input signal is a current and the output signal is a voltage. The gain of a transresistance amplifier is Vout/Iin, and if we recall Ohm's law, we know that something that has voltage divided by current will yield a resistance. Thus, Vout/Iin is measured in resistance. We will discuss the transresistance amplifier later. For now, Figure 7-5 shows an inverting-gain voltage amplifier.

Intuitively, this configuration is somewhat more mysterious or difficult to understand. We can draw one good guess, that if the voltage is positive at Vin, the output

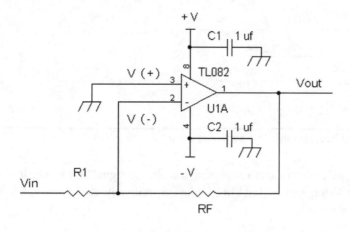

$$Vout = Ao \times [\, V\, (+)\ - V\, (-)\ ]$$

FIGURE 7-5   An inverting-gain op amp circuit.

Vout has to be a negative voltage because R1 is connected to the (–) input. Thus, we can also surmise that the output is out of phase with the input, so if the input voltage is negative, the output has to be positive.

However, we can still take the one feature we know about an op amp when negative feedback is applied, and that is that the voltages at the (+) and (–) inputs are essentially the same. This means that the voltage difference between the (+) and (–) inputs is zero. Given just this information, let's see what happens when RF = R1 = 1 kΩ and Vin at R1 = +1 volt. We see that the (–) input terminal is grounded or is at 0 volt. Therefore, the op amp will generate an output to "servo" the voltage at the (–) input terminal to 0 volt. The input and output voltages can be represented by two voltage sources, Vin and Vout, as seen in Figure 7-6, along with RF and R1.

Given that Vin = +1 volt DC and that the two resistors are equal in resistance, what voltage must Vout be in order to have 0 volt where the two equal resistors are connected? Our first guess would be –1 volt DC = Vout because we can redraw the circuit from Figure 7-6 as a summing circuit, as shown in Figure 7-7.

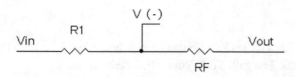

FIGURE 7-6   Voltage sources Vin and Vout with resistors RF and R1.

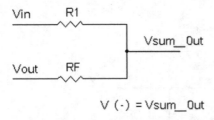

$$V(-) = Vsum\_Out$$

FIGURE 7-7   Vin and Vout redrawn as a summing circuit.

If we have equal resistors and a summing circuit, the signal at the output of the resistor network Vsum_out = (½)(Vin + Vout). Specifically, if Vin = +1 volt and Vout = –1 volt, then

$$Vsum\_out = V(-) = (½)(+1 \text{ volt} + -1 \text{ volt}) = 0 \text{ volt}$$

Now let's see what happens when we change one of the resistors RF to 10 kΩ and keep R1 = 1 kΩ. We no longer have a simple summer because the resistor values are unequal. However, we do know this: if there is a 1 volt drop across R1, we will get 0 volt at the junction of the two resistors. Put another way, because R1 = 1 kΩ, we need 1 mA flowing through it to develop 1 volt across R1. But the total resistance of the divider circuit is R1 + RF = 1 kΩ + 10 kΩ = 11 kΩ.

We have +1 volt at Vin, and we need a total of 11 volts across R1 and RF via voltage sources Vin and Vout. The voltage across the series resistances of R1 and RF, which total 11 kΩ, is just the potential difference between Vin and Vout, which is (Vin – Vout). We know that Vout should be a negative voltage such that Vin – Vout = +11 volts. One of the concepts we learned about subtracting a negative number is that a positive number results, at least in this case. So we have the following: +1 volt minus a negative voltage Vout should equal +11 volts. At this point, if you guessed –10 volts, you are correct. Therefore, Vout = –10 volts.

What has happened here, then? The op amp via negative feedback "servoed" the output voltage to provide a negative voltage such that at the (–) input terminal there is 0 volt via RF and R1, and Vin = +1 volt. In this feedback amplifier, as shown in Figure 7-5, there are actually two reference voltages going on here. The first one is the (+) input being connected to 0 volt or ground. This anchors what the voltage at the (–) input terminal will be. The second reference voltage is Vin, which determines as a whole what the output signal is going to be. There is much more complicated math that can be shown, but for now, with the examples shown, we so far know that for an inverting amplifier as shown in Figure 7-5:

$$Vout = -(RF/R1)Vin \tag{7-6}$$

With Vin = +1 volt, when RF = R1 = 1 kΩ, Vout is –1 volt, and when RF= 10 kΩ with R1 = 1 kΩ, Vout = –10 volts. The gain Vout/Vin is then deduced as:

$$(Vout/Vin) = -(RF/R1) \tag{7-7}$$

Alternatively, divide Equation (7-6) by Vin on both sides to get Equation (7-7). Note that if RF < R1 or, put another way, R1 > RF, the magnitude of inverting gain is less than 1. For example, make R1 = 100 kΩ and RF = 1 kΩ, and we get a gain of –1/100 = –0.01. This sets up a unique feature. With a gain of less than 1, the input voltage can exceed the power-supply voltage (e.g., typically ±12 volts) of the op amp. For example, with a gain of –0.01, an input signal of +200 volts DC as Vin into R1 will only yield an output voltage of –2 volts DC.

> **NOTE** The maximum input signal voltage into the (+) input of a non-inverting-gain op amp configuration (e.g., Figure 7-4) is generally less than the power-supply voltage.

We know that volts divided by amps equals a resistance. So we can also rewrite Equation (7-6) in terms of an input current; that is, Iin = Vin/R1. Then we can reinterpret Equation (7-6) as the following:

$$\text{Vout} = -(\text{RF}/\text{R1})\text{Vin} = -\text{RF}(\text{Vin}/\text{R1}) = -\text{RF}(\text{Iin}) \quad \text{or} \quad \text{Vout} = -\text{RF} \times \text{Iin} \qquad (7\text{-}8)$$

Note that Iin is the "positive" current flowing into the (–) input terminal via R1. In a photodiode circuit with the anode grounded and the cathode connected to the (–) terminal, the photocurrent Iphoto = –Iin because the photocurrent is flowing down and away in the opposite direction from the (–) input terminal. See Figure 7-8.

In Figure 7-8, the photodiode preamp is also known as a *transresistance amplifier* because the input signal is a current and the output signal is a voltage. The gain or transfer function that characterizes this amplifier is Vout/Iin, which has a unit of resistance. But because the input is a signal current, we call Vout/Iin a *transresistance transfer function* and we describe the gain magnitude in terms more of volts per amp

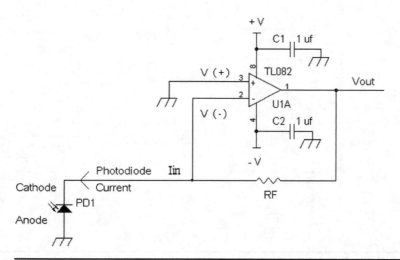

FIGURE 7-8   A photodiode amplifier using an inverting op amp configuration.

or milliamp. For example, with RF = 1 kΩ, the transfer function for the transresistance amplifier is 1 volt per 1 mA. That is, if 1 mA is inputted, the output will have a voltage magnitude of 1 mA × 1 kΩ = 1 volt. Therefore, we get 1 volt (output) per 1 mA of input. Also, note that (1 volt per mA) = 1 volt/mA = 1 kΩ = RF.

Two Projects: A Photodiode Sensor and an Automatic Level Control Amplifier

Photodiode Sensor

Speaking of photodiodes, let's build a photodiode sensor using an op amp and comparator to sense whether a white or light piece of paper is present. The circuit consists of a white light-emitting diode (LED), a photodiode (a real one, not using an LED this time), and a dual op amp. The first section of the dual op amp will amplify the photocurrent reflected from LED1, and the second section of the op amp will be used as a comparator. The comparator is set to turn on another LED, LED2, to indicate the presence of the white paper (see Figures 7-9 and 7-10).

LED1 is positioned below photodiode PD1 so that stray light will not leak into the photodiode. Thus, the photodiode will only pick up reflected light from the white paper that is illuminated by LED1.

Parts List

- One LM358 or TLC272 op amp for U1 with an **8 pin IC** socket
- One 22 pF capacitor for C1 (C1 is optional)

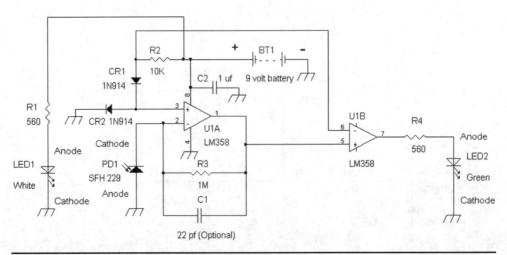

Figure 7-9 Schematic diagram of a white paper sensor.

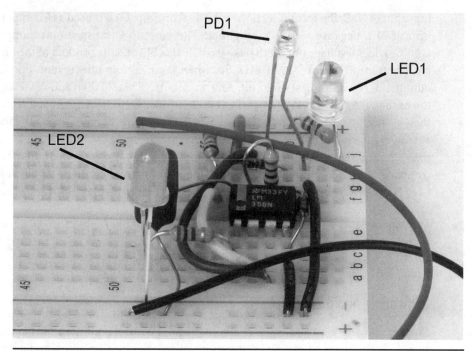

Figure 7-10   Prototype circuit of the white paper sensor.

- One 1μF capacitor for C2
- One 9 volt battery and connector for BT1
- Two 1N914 or 1N4148 diodes for CR1 and CR2
- One SFH229 or other photodiode for PD1
- One 24,000 mcd (millicandela) white LED for LED1
- One green or red LED for LED2
- Two 560 Ω resistors for R1 and R4 (other values between 330 Ω and 1 kΩ may be used)
- One 10 kΩ resistor for R2
- One 1 MΩ resistor for resistor R3
- One makeshift barrier made of a generally opaque material such as cardboard (optional)

Here are a few notes to add on the photodiode sensor circuit. First, the cathode of photodiode PD1 is biased up to about +0.6 volt via CR2 so that if a phototransistor is substituted, there is sufficient DC bias on the phototransistor. Also, biasing the cathode of the photodiode reduces its inherent reverse-bias capacitance across the cathode and anode. A +1.2 volt reference voltage source is established by the series connection of CR1 and CR2 with R2, providing DC forward bias for these two diodes. At the anode of CR1, the 1.2 volt reference voltage is connected to the inverting input of U1B, the second section of the dual op amp; U1B is used as a comparator instead of a negative-feedback amplifier. The output of transresistance amplifier U1A provides a voltage output that is +0.6 volt via CR2 *and* the product of the photocurrent from PD1 and R3 = 1 MΩ. So, when there is no photocurrent, pin 1, the output of U1A, is sitting at +0.6 volt. Any photocurrent via R3 then adds a voltage on top of the +0.6 volt. The output of the transresistance amplifier is then connected to the (+) input of the comparator at pin 5 of U1B. When the photocurrent exceeds about 0.6 μA via the light reflected from the paper from LED1 to PD1, the transresistance amplifier will output a voltage of more than +1.2 volts, which then causes comparator U1B to output a high-level output of nearly the power-supply voltage (~+7.5 volts DC) at pin 7, which then supplies current to turn on LED2.

 **NOTE** The output of the transresistance amplifier at pin 1 = (photocurrent x 1MΩ) + 0.6 volt.

Controlling Audio Signal Levels

The next project is an audio signal automatic level control amplifier. This amplifier combines circuits such as a DC restoration, voltage doubler, op amps, and a bipolar transistor voltage-controlled attenuator.

An automatic level control (ALC) amplifier will allow normal operation of amplifying until a certain amplitude level is exceeded. Then the gain of the amplifier is decreased to limit the amplitude of the AC signal in a nondistorting manner. ALC amplifiers are very useful when recording or amplifying a live event signal from a microphone so that the signal is limited in the dynamic range so as to prevent distortion in the recording or amplifying device (see Figures 7-11 and 7-12).

This amplifier includes an inverting op amp input amplifier U1A, which has a gain of $-R2/R1 = -10$ kΩ$/47$ kΩ $= -0.212$. The output of U1A is fed to a 30 kΩ series resistor that forms a voltage divider circuit via the collector-to-ground variable resistance of Q1. This variable resistance is varied by a control voltage from R9. The output of the variable voltage divider is AC coupled via C5 to a non-inverting-gain amplifier U1B, whose gain is $[1 + (R7/R6)] = [1 + (220$ kΩ$/3.3$ kΩ$)] \approx 67$. The AC signal output is coupled via output resistor R11 to Vout to be connected to an amplifier or recorder. However, the AC signal output of U1B is also converted to a DC voltage by a DC restoration circuit, C8 and CR1. At the cathode of CR1, the audio or AC signal has the negative peak clamped or DC restored to nearly 0 volt via germanium diode CR1. The peak of the clamped AC signal at the cathode of CR1 is transferred by rectifier CR2 to a hold capacitor C9. The voltage from C9 is drained slowly via R12, a 1 MΩ resistor. Thus, R12 controls the release time of this peak limiter amplifier circuit.

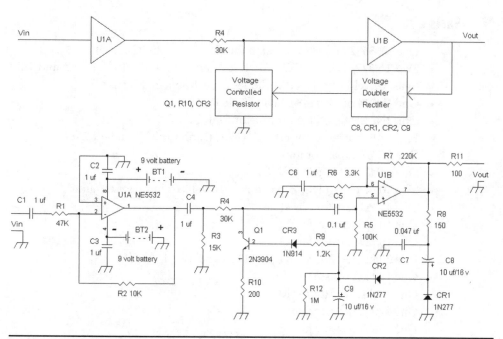

FIGURE 7-11 Block diagram and schematic of the ALC amplifier.

FIGURE 7-12 Prototype circuit of the audio ALC (automatic level control) amplifier.

Parts List

- One NE5532 or LM833 op amp for U1 with 8-pin IC socket
- Two 1N277, 1N270, 1N34, or 1N60 germanium diodes for CR1 and CR2 (alternatively, two Schottky diodes or rectifiers may be used)
- One 1N914 or 1N4148 silicon diode for CR3
- One 2N3904 or 2N4124 transistor for Q1
- Five 1 μF capacitors for C1, C2, C3, C4, and C6 (C6 can be 2.2 μF)
- One 0.1 μF capacitor for C5
- One 0.047 μF capacitor for C7
- Two 10 μF electrolytic capacitors at 16 volts or more for C8 and C9
- One 47 kΩ resistor for R1
- One 10 kΩ resistor for R2
- One 15 kΩ resistor for R3
- One 30 kΩ resistor for R4 (33 kΩ can be used instead)
- One 100 kΩ resistor for R5
- One 3.3 kΩ resistor for R6
- One 220 kΩ resistor for R7
- One 150 Ω resistor for R8
- One 1.2 kΩ resistor for R9
- One 200 Ω resistor for R10 (either 180 Ω or 220 Ω can be used instead)
- One 100 Ω resistor for R11 (any value from 47 Ω to 1 kΩ can be used instead)
- One 1 MΩ resistor for R12

The ALC circuit was found to limit at around 1.4 volts peak-to-peak output for input signals from about 210 mV to 7,000 mV peak to peak. CR3 is in series with the base of Q1 to allow more negative value voltages at the collector of Q1 without forward biasing its base-collector junction. Also, the series combination of CR3, a silicon diode with a 0.7 volt turn-on voltage, and the 0.7 volt turn-on voltage of Q1 for its base-emitter junction starts the ALC action at about 1.4 volts. The voltage doubler and DC restoration circuit consisting of CR1 and CR2 will take the peak-to-peak AC voltage from the output and transform that AC voltage to a DC voltage. For example, if the AC output voltage is 1 volt peak to peak (e.g., a sine wave with a positive cycle of +0.5 volt and a negative cycle of –0.5 volt), the output of CR2 is then about a +1.0 volt DC signal. Thus, the turn-on voltages from CR3 and Q1, at +0.7 volt and +0.7 volt = 1.4 volts, determines that the peak-to-peak output AC signal is limited to 1.4 volts.

Another Voltage-Controlled Amplifier Circuit

Previously, we saw voltage-controlled attenuators such as the ones shown in Chapters 5 and 6 that used diodes including LEDs, bipolar transistors, and FETs as voltage-controlled resistors. We can also control the DC bias of bipolar transistors, FETs, and even vacuum tubes to vary the amplifier's gain. A simple voltage-controlled amplifier is a bipolar transistor common emitter or grounded emitter amplifier where the collector current is varied and the base of the transistor is driven with a low-resistance voltage source. For the next experiment, we will use a combination of op amps and a transistor amplifier to show how gain is varied by varying the collector current of the transistor.

By varying the collector current, we will then uncover a characteristic of the bipolar transistor and learn something extra, which is the concept of *transconductance*. The concept of *transresistance* was briefly discussed as a current input signal with a voltage output signal. For understanding transconductance, we first need to determine what conductance is. *Conductance* is that ability to allow electrons to flow through a material such as silver, copper, aluminum, lead, glass, and so on. Intuitively, we know that silver is the best conductor, followed by copper, and one of the worst conductors is glass. Thus, we would say that silver has higher conductivity than copper. Put another way, silver has very high conductivity that has less resistance than copper. Glass, with very poor conductivity (close to zero conductivity), has very high resistance. Therefore, conductance is merely the reciprocal of resistance. That is, conductance, whose symbol is G, has $G = 1/R = 1/resistance$. From Ohm's law, we know that $V/I = R$; thus, $G = I/V$.

Transconductance, which is the ratio of output current to input voltage, is then measured or expressed in *mhos* (*ohms* spelled backwards), or in the more modern era, transconductance is often measured in *siemens,* whose abbreviation is S. That is, $1 S = 1$ mho. Transconductance, whose symbol is g_m, is characterized in amps per volts (mhos), or it is more commonly expressed in milliamp(s) per volt (mmhos).

Voltage-Controlled Common Emitter Amplifier

This experiment will measure an AC signal from an AC adapter, which will provide a 60-Hz sine-wave signal. Figure 7-13 shows the schematic. By adjusting the DC voltage to a base resistor R3 and measuring the voltage across an emitter resistor RE1, we will verify that by increasing the DC emitter current, which leads to almost exactly the same increase in DC collector current, the gain of the amplifier will increase proportionally. An autoranging digital volt-ohmmeter (VOM, e.g., Extech MN26 Series or Extech MiniTec 26 Series) or a digital VOM that can measure full-scale 2 volts AC is required. The cheaper digital VOMs start out with 200 volts AC full scale and are not suitable for this experiment since the AC signals from this amplifier will be typically less than 1 volt.

Potentiometer VR1 is adjusted for 500 mV AC (0.500 volt AC) at its slider terminal 2 or at the input of R1 Vin. The AC signal voltage at VB is 1/100 of Vin via resistive divider R1 and R2. To measure transconductance, we measure the AC collector current and then divide by the AC signal voltage at the base of Q1 VB. The resulting

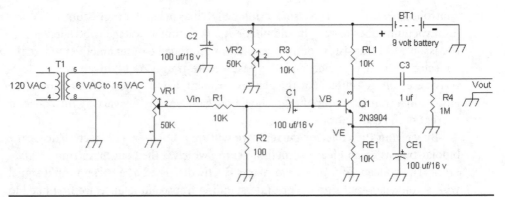

Figure 7-13   Voltage-controlled common emitter amplifier and schematic.

voltage across the base-emitter junction of Q1 is then about 5 mV, small enough to ensure accurate readings without gross distortion of the sine wave at the output Vout. This circuit is relatively easy to build, and the reader is encouraged to do so and take measurements. Table 7-1 summarizes various measurements for different DC collector currents.

Table 7-1   Measurements on Voltage-Controlled Common Emitter Amplifier

VB AC	Vout AC (mV)	VE (volts DC)	IC = DC Collector Current (µA)	Iout = Vout/10 kΩ AC	Transcondconductance (g$_m$ = Iout/VB)	Magnitude of Gain = IVout/VBI
5.0 mV	60.0	0.33	33	6.0 µA	1.2 mS (mmho)	12.0
5.0 mV	125.0	0.67	67	12.5 µA	2.5 mS (mmho)	25.0
5.0 mV	189.0	1.01	101	18.9 µA	3.78 mS (mmho)	37.8
5.0 mV	280.0	1.49	149	28.0 µA	5.60 mS (mmho)	56.0
5.0 mV	383.0	2.04	204	38.3 µA	7.66 mS (mmho)	76.0

What we can see from this table is that increasing the DC collector current linearly increases the gain and transconductance gm for this common emitter amplifier.

Negative Feedback to Stabilize or Self-Bias Collector Currents

In Figure 7-13, an emitter resistor RE1 and a DC voltage to the base of Q1 via RB1 established the collector current by establishing a voltage across RE1. The question then arises whether we can self-bias a transistor for setting their collector current. The answer is yes, and we have actually seen this previously in Chapter 3, Figure 3-22. See Figure 7-14.

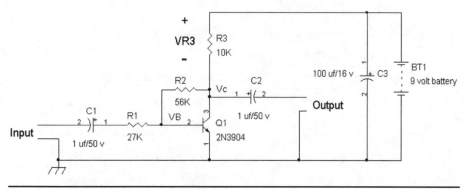

Figure 7-14 One-transistor amplifier from Chapter 3.

In this amplifier, the collector current is set by "defining" the collector voltage. First, let's make an approximation that the base current of the transistor is low such that the voltage across R2 is approximately 0 volt because the current gain β of transistor Q1 is high such that β > 100. Recall that a high current gain of a transistor results in low base currents. Note that there is a way to find a more exact voltage drop (Vc – Vb) via a set of equations and "lots" of algebra, which we will not show here. For now, and in practice, chasing the exact answer is usually not an efficient use of one's time. Also, if one is concerned about base currents, just replace Q1 with a high-β transistor such as a 2N5089, where β > 500, and then forget about the base currents that will cause a voltage drop across R2, the 56 kΩ feedback resistor.

We know that the transistor turns on at about +0.6 volt for VBE, and because the emitter is grounded, VB = VBE = +0.6 volt. With about 0 volt across R2, we have VB = VC, so VC = +0.6 volt. Now all we need to know is the battery voltage for BT1, which is 9 volts. Therefore, the voltage across R3 = 10 kΩ, the collector load resistor, is (9 volts – 0.6 volt) = 8.4 volts = VR3 (see Figure 7-14). The collector current then is IC = VR3/R3 = 8.4 volts/10 kΩ = 0.84 mA = IC.

One question is, why did this circuit bias the collector voltage to 0.6 volt in the first place? The answer is that R2 is a feedback resistor, and the transistor amplifier Q1 can be modeled or thought of as an op amp configured as an inverting amplifier with the (+) input terminal *not* connected to ground but connected to a VBE voltage source of +0.6 volt. See Figure 7-15.

For the simple one-transistor amplifier, the common emitter amplifier is an inverting input amplifier. That is, the more positive the voltage applied to the input at the base, the more collector current is provided, which then forms a larger voltage drop across R3, which is the collector current times the resistance of R3, or IC × R3. But the collector voltage VC is (BT1 – VR3) = (BT1 – IC × R3) = VC. Therefore, VC's voltage goes down as the base voltage VB goes up, which then characterizes an inverting amplifier. Note that we have negative feedback via R2 from the collector to the base of Q1, and the AC gain ≈ –R2/R1 ≈ –56 kΩ/27 kΩ ≈ –2.

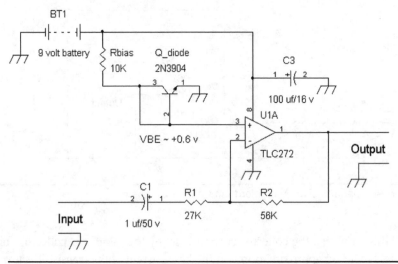

FIGURE 7-15 Inverting op amp representation of the one-transistor amplifier.

For one more perspective on why this one-transistor amplifier is a negative-feedback system, we note that if the collector voltage VC rises, it will send an increased voltage via R2 to the base of Q1. The increased voltage into the base of Q1 will then increase collector current IC, which will then *decrease* the collector voltage at VC. Recall that VC = BT1 − IC × R3, so if IC increases, VC has to get smaller because the product IC × R3 or voltage drop across R3 subtracts *more* from the battery voltage BT1. Thus we have a system that is self-adjusting in terms of collector current and collector voltage.

The next question is, can we adjust VC to some other voltage? The answer is yes, by adding a resistor R4 from the base of Q1 to ground. See Figure 7-16.

But why change the collector voltage in the first place? For the circuit in Figure 7-14, it is actually optimized for approximately symmetrical AC voltage swings of

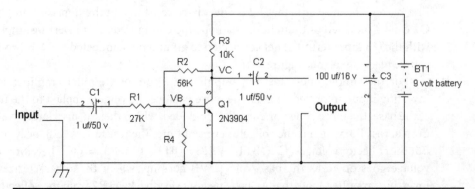

FIGURE 7-16 Modified one-transistor amplifier with R4 to raise the DC collector bias voltage.

±0.6 volt, which would imply that BT1 → 1.5 volts, approximately. However, if BT1 is a higher voltage (e.g., 9 volts), the amplifier is limiting itself to going only 0.6 volt in the negative direction and about 8 volts in the positive direction. If we can bias VC to about half of BT1, or (½) × 9 volts = 4.5 volts, a larger symmetrical voltage swing would be provided.

So how do we calculate R4? We know in a voltage divider with a supply voltage VC that the output of the voltage divider to the base of Q1 is

$$VB = 0.6 \text{ volt} = VC \times R4/(R2 + R4)$$

In this case, we know approximately what VB is, but we do not know what VC is. If we multiply both sides of the equation by (R2 + R4)/R4, we get

$$0.6 \text{ volt} \times (R2 + R4)/R4 = VC \tag{7-9}$$

Let's try a couple of examples. Suppose that R4 = R2 = 56 kΩ; then VC = VB × (56 kΩ + 56 kΩ)/56 kΩ = 0.6 volt × 2 = 1.2 volts = VC. Suppose that we want to set VC to a specific voltage. How do we find the resistors? We start with Equation (7-9) and divide by VB = 0.6 volt on both sides to get

$$(R2 + R4)/R4 = VC/VB \rightarrow (R2/R4 + 1) = VC/0.6 \text{ volt}$$

by subtracting 1 from both sides, that is,

$$R2/R4 = [(VC/0.6 \text{ volt}) - 1]$$

and taking the reciprocal of both sides, we get

$$R4/R2 = 1/[(VC/0.6 \text{ volt}) - 1]$$

and multiplying by R2 on both sides, we *finally* get

$$R4 = R2/[\mathbf{(VC/0.6 \text{ volt}) - 1}] \tag{7-10}$$

If we have BT1 = 9 volts, how do we set VC to 4.5 volts when R2 = 56 kΩ? Using Equation (7-10), we get

$$R4 = 56 \text{ kΩ}/[(4.5 \text{ volts}/0.6 \text{ volt}) - 1] = 56 \text{ kΩ}/(7.5 - 1) = 56 \text{ kΩ}/6.5 = 8.6 \text{ kΩ} = R4$$

After grinding out Equation (7-10), I would just try different values for R4. I know that 4.5 volts for VC is about 7 times 0.6 volt. And I know that the ratio R2/R4 is roughly the gain of the system DC-wise. That is, R2/R4 ≈ (R2/R4 +1) if R2/R4 is a lot larger than 1. Thus, R4 will be approximately 56 kΩ/7 ≈ 8 kΩ ≈ R4, which is not too bad of an approximation.

When we look at the DC output at VC, it has a voltage that is the result of multiplying VBE. That is, VC = VBE × (R2 + R4)/R4, where the multiplying factor is (R2 + R4)/R4. A circuit consisting of only Q1, R2, and R4 is sometimes called a *VBE (voltage) multiplier circuit*, and it has uses in power output stages in amplifiers such as stereo high-fidelity receivers, headphone amplifiers, and many op amps.

Power Output Stages and Using the VBE Multiplier Circuit

In the following schematics, we will show a progression of circuits on how to improve the one-transistor amplifier into a simple power amplifier that can drive a headphone or small loudspeaker. Figure 7-16 as it stands will work as a preamplifier that can be fed to the auxiliary input or CD input of a stereo receiver. However, it does have limited output current drive. One improvement to this circuit is to add an emitter follower amplifier Q2 (see Figure 7-17).

With the addition of Q2 and R5, this amplifier has a lower output resistance at the emitter of Q2, which is useful when driving long cables that have sizable capacitances. The emitter follower amplifier inherently has a voltage gain of around unity or 1 and has an advantage of high input resistance (base terminal of Q2) while having low output resistance and providing improved output current capability at its emitter output terminal. The addition of R6, a series 100 Ω resistor, is to ensure that the emitter follower Q2 does not directly load to the capacitance of an external cable. We can further improve on lowering distortion and frequency response by rerouting R2 so that it applies negative feedback not from the output of Q1 but from the output of Q2. See Figure 7-18.

With R2 "sensing" the output of Q2, at least two things are happening. One is that the output resistance at Q2's emitter is further lowered from Figure 7-18, and second, there should be a wider frequency response when the signal is taken from the emitter of Q2. A disadvantage of using an emitter follower consisting of Q2 and R5 is that although signals from the positive cycle of an AC waveform have good current capability, the negative cycle or negative flowing current of an AC signal is

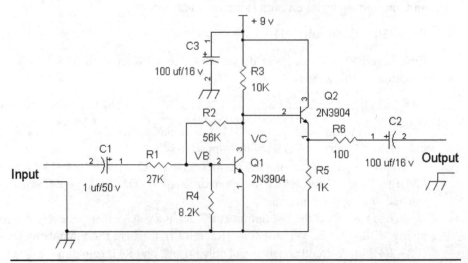

FIGURE 7-17   One-transistor amplifier with a second transistor added for increased output current capability.

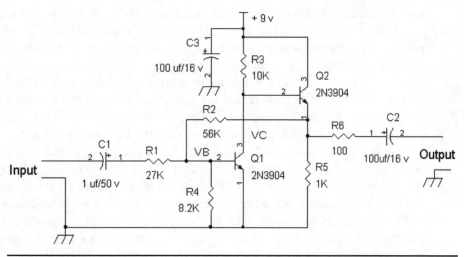

Figure 7-18 Negative-feedback resistor R2 rerouted to apply feedback at the emitter of Q2.

limited by R5. R5 can be lowered in value (e.g., in Figure 7-18, R5 → 470 Ω) to increase the standing current of Q2. Recall that the emitter current of Q2 is just the DC emitter voltage of Q2 divided by R5. But such lowering of the resistance value of R5 would consume too much power on standby. What would be desirable is to design an output stage that idles low current on standby (e.g., no signal at the input) while having the capability of good current drive for both positive and negative cycles of AC signal. See Figure 7-19.

In this circuit, a third transistor Q3 that is PNP allows the capability of increased current in the negative direction. This is needed for the negative cycle of the AC waveform. Q2's emitter handles the current for the positive cycle of the AC signal,

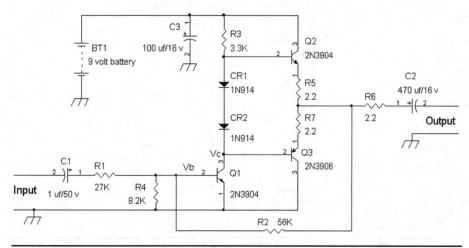

Figure 7-19 One-transistor amplifier with Q2 and Q3 as a push-pull output stage.

and Q3's emitter handles the current for the negative cycle. Diodes CR1 and CR2 are used to bias emitter currents of Q2 and Q3. However, because CR1 and CR2 often do not have the same turn-on voltage characteristic as the VBE and VEB turn-on voltages of Q2 and Q3, respectively, there is a good chance that the emitter currents of Q2 and Q3 may be too low, which leads to crossover distortion (i.e., dead-band distortion). If the emitter currents of Q2 and Q3 are too low, one can add a third diode in series, but that may cause the emitter currents of Q2 and Q3 to be too high. And, if CR1 and CR2 initially cause too high emitter currents for Q2 and Q3, one can use one diode instead and, for instance, short out CR2, but that may lead to crossover distortion due to insufficient emitter bias currents in Q2 and Q3. So what can be done to adjust the emitter or collector currents of Q2 and Q3? What would be desirable is a variable diode voltage reference generator. But we have already seen that in Figure 7-14, where VCE, the DC bias collector-to-emitter voltage of Q1, is just VBE(1 + R2/R4) = VCE. We can use another transistor as a VBE voltage multiplier to replace CR1 and CR2. See Figure 7-20.

In Figure 7-20, the collector-to-emitter voltage of Q4 is VBE(1 + R8/R9). R9 is nominally 10 kΩ to set the VCE voltage to about 2 diode voltage drops or 2 × VBE = 2 × 0.6 volt = 1.2 volts = VCE. However, the asterisk in R9 denotes that it can have a selected resistance value, or R9 can be a variable resistor (e.g., 20 kΩ) to have a range from 20 kΩ to about 5 kΩ. The emitter current of Q2 or Q3 is just the voltage across R5 or R7 divided by 2.2 Ω. Setting up the emitter current of Q2 and Q3 requires that the AC input signal be disconnected or turned off.

Note in Figure 7-20 that R3 is lowered from 10 kΩ to 3.3 kΩ to bias Q1 to a higher collector current so that more current is available to the bases of Q2 and Q3 for higher output current drive. Also, R5 and R7 are set to 2.2 Ω, and C2 is increased from 1 μF to 470 μF for better low-frequency bass response when using this amplifier to drive a headphone or small loudspeaker.

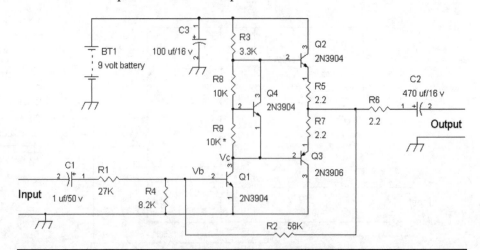

Figure 7-20 A fourth transistor Q4 replaces CR1 and CR2 for biasing the output transistors.

Misadventures in Positive Feedback

There is the old saying about amplifiers turning into oscillators and oscillators turning into amplifiers. In this section, we will explore just a couple of examples where positive feedback gives an undesirable result.

One of the misconceptions in using op amps is that it does not matter how the output is connected to one of the inputs. For an op amp to work properly as an amplifying device, there must always be negative feedback. Figure 7-21 provides one example.

I have seen a hobbyist magazine and a ham radio book show a circuit like the one in Figure 7-21. Unfortunately, this circuit will latch to one of the power-supply voltages, or it will oscillate. The problem is that there is no servo system to drive the output voltage going to the (+) input of the op amp to match the input voltage at the (–) input. When a positive voltage is connected to the (–) input of the op amp, the output has to produce a negative-valued voltage. However, we see that the input terminals at pin 2 and pin 3 of the op amp are taking the arithmetic difference of two voltages (input and output voltages) that are opposite in polarity. This results in a large output voltage. This large output voltage builds on itself to eventually saturate the output voltage, or the op amp becomes an oscillator.

In other words, the voltages are in opposite polarities to each other at the (+) and (–) inputs, so it is incorrect to think that the input signal at the (+) terminal can servo the output that is connected to the (–) input to the same voltages. For example in Figure 7-21, if the input = +1 volt, then the output equals some negative voltage, and one cannot reconcile that a positive voltage equals a negative voltage.

In a negative-feedback configuration, the input terminals are taking the arithmetic difference of two like polarity voltages, which results in a smaller output voltage. And that smaller output voltage is under control. The two identical-polarity signals at the input under negative feedback can be driven to the same voltage. Thus, if someone tells you that there is such a thing as a negative or inverting voltage follower, something is wrong. A proper voltage follower is shown in Figure 7-22 with RF = 0 Ω and R1 taken out.

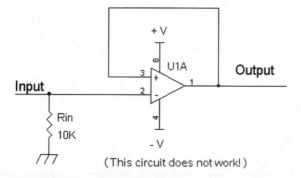

(This circuit does not work!)

Figure 7-21   An erroneous inverting voltage follower that does *not* work.

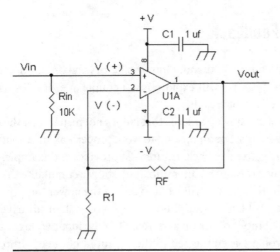

FIGURE 7-22 A proper voltage follower is implemented when RF = 0 Ω and R1 is removed.

Another problem arises when an op amp is directly connected to a capacitive load. See Figure 7-23. An op amp is a system that consists of at least two phase-shift stages. A third phase-shift stage can cause the op amp circuit in Figure 7-19 to oscillate (see Figure 7-24).

In Figure 7-24, an op amp is represented by a differential amplifier A1, a voltage-gain amplifier A2, and a unity gain of 1 in the output stage (e.g., emitter follower gain of 1). Equivalent internal low-pass filters are shown via R10, C10, R20, and C20. An internal output resistor of the op amp is shown as Rout. External biasing resistor Rin and external feedback resistor RF and R1 are shown outside the box along with the capacitive load C_Load. Each low-pass filter delivers up to 90 degrees of phase shift. With two stages shown inside the box, less than 180 degrees of phase shift is provided,

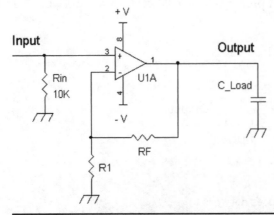

FIGURE 7-23 Pure capacitive loading can lead to oscillation.

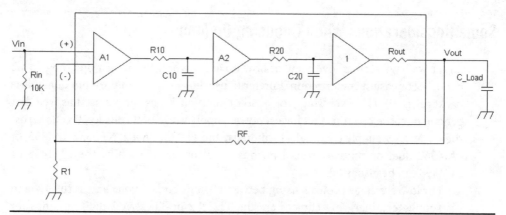

FIGURE 7-24 An op amp representation with multiple voltage gain stages and low-pass filters.

but the extra phase shift via Rout and C_Load may have just enough added phase shift to push the amplifier into oscillation. The reason is that when there is a net 180 degrees of phase shift eventually going into the inverting input of the amplifier via A1, the total phase shift is 360 degrees, which looks like 0 degree. Or, in other words, the 180 degrees of phase shift from R10, C10, R20, C20, Rout, and C_Load essentially flips the polarity of the (–) input into a (+) input, and positive feedback is provided in a "bad" way that would result in oscillation. Also note that oscillation will occur even if the input signal is 0 (e.g., Vin = 0 in Figure 7-24). A solution to ensure that there is no oscillation is to just add a series resistor as shown in Figure 7-25.

Generally, a series resistor Rseries of 22 Ω or larger will suffice in allowing stability without oscillation when connecting to a capacitive load such as a long cable.

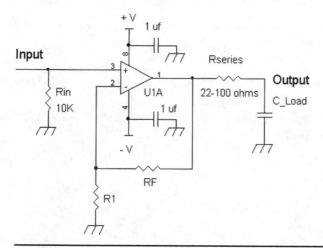

FIGURE 7-25 An added series resistor isolates the capacitive load to provide freedom from oscillation.

Some Considerations When Choosing Op Amps

So far we have seen an overview of how op amps work. But which op amps do we choose? For easiest construction, I prefer to use dual op amps in the 8 pin dual inline package (DIP). However, single or quad op amp packages can be used as well. For general purpose op amps that have a power supply range of 9 volts to 30 volts across the +V and −V supply terminals, I would use the TL072, TL082, NE5532, LM833, or LM4562 dual op amps. General purpose dual op amps such as the LM1458 or RC4558 can be also used.

For lower voltage projects using between 3 volts and 15 volts across the +V and −V supply terminals, my choices are the TLC272 or TLC27M2 dual op amps for working with frequencies below 20 kHz.

The LM358 dual op amp works from a supply of 3 volts to 30 volts across the +V and −V supply terminals. However, this op amp has a *unique* output stage that exhibits crossover distortion when loading into less than 10 kΩ. One can "fix" this by adding a pull down resistor of about 3.3 kΩ from the output terminal to the −V supply terminal. Alternatively, a 3.3 kΩ resistor from the output terminal to the +V supply terminal will also work in many situations.

If low voltage capability and "rail to rail" performance is desired from 3 volts to 30 volts across the +V and −V supply terminals, then you may want to consider the surface mount SO-8 (small-outline 8 pin) dual op amps, ISL28208, ISL28217, or ISL28218. However, you will need an adapter board (see Figure 11-26) or very careful soldering techniques to work with these op amps since they are not through-hole devices.

Finally for low noise applications, you may consider the AD797, an 8 pin DIP single op amp.

References

1. Paul R. Gray and Robert G. Meyer, *Analysis and Design of Analog Integrated Circuits, 3rd ed.* New York: Wiley, 1993.
2. Thomas M Frederiksen, "Intuitive IC Op Amps." Santa Clara, CA: National Semiconductor, 1984.

CHAPTER 8

Audio Signals and Circuits

Chapters 5 through 7 covered diodes, light-emitting diodes (LEDs), amplifying devices, and operational amplifiers. In particular, we saw on a top level how to use negative feedback with op amps. We will build on some of the circuits and integrate them into more complex circuits. For example, we will show another situation where a constant current source is used for setting the emitter current of a differential transistor amplifier.

A main objective of this chapter is to discuss different types of preamplifiers for microphones and phonograph (phono) cartridges. Note that this chapter will not cover audio power amplifiers, which people have written entire books on.

Signal Levels for Microphones, Phono Cartridges, Line Inputs, and Loudspeakers

Audio signals are measured in terms of root-mean-squared (RMS) voltage or current. A sine-wave signal has a peak voltage, and the peak-to-peak voltage is twice the peak voltage (Figure 8-1).

A sine-wave voltage is not a constant direct-current (DC) voltage. It fluctuates between a positive and a negative voltage, and at times, its voltage is zero. When the power is calculated from the peak sine-wave signal, we actually get half the power. For example, a 10 volt peak sine-wave voltage V_p across a 1 Ω resistor yields a power $P = (\frac{1}{2})(V_p)^2/1\ \Omega = (\frac{1}{2}) \times 10^2$ watts = 50 watts. In contrast, a 10 volt DC voltage across a 1 Ω resistor yields $(10\ \text{volts})^2/1\ \Omega = 100$ watts.

By Ohm's law, power is proportional to the square of the voltage or current. So how do we express an equivalent voltage V_{RMS} related to the peak sine-wave voltage Vp that matches the DC voltage in terms of power? That is, can we find an equivalent alternating-current (AC) voltage without the $\frac{1}{2}$ factor that includes $(V_p)^2$ for the

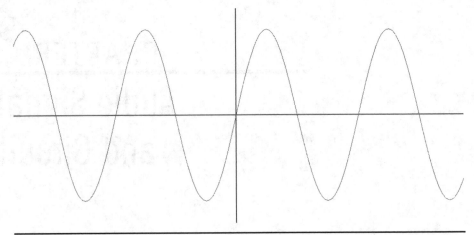

FIGURE 8-1   Sine-wave signal at 1 volt peak, where the vertical axis is amplitude and the horizontal axis is time.

power formula $P = (\frac{1}{2})(V_p)^2/R$? Let's come up with a new voltage called V_{RMS} to represent an AC voltage that has the same power as a DC voltage that does not require the $\frac{1}{2}$ factor. Now let's equate the two power formulas this way:

$$(V_{RMS})^2/R = P = (\frac{1}{2})(V_p)^2/R$$

We can multiply by R on both sides to "solve" for what V_{RMS} is in terms of V_p, which yields

$$(V_{RMS})^2 = (\frac{1}{2})(V_p)^2$$

Now let's take the square root of both sides:

$$V_{RMS} = \sqrt{V_p} \quad \text{or} \quad V_{RMS} = 1/\sqrt{V_p}$$

Equivalently, we can find V_p in terms of V_{RMS} by multiplying by the $\sqrt{2}$ on both sides, which yields

$$V_{RMS}(\sqrt{2}) = V_p \rightarrow V_p = (\sqrt{2})V_{RMS}$$

In an example for a 1 volt peak sine wave, the equivalent RMS voltage:

$$V_{RMS} = (1/\sqrt{2})\ 1\ \text{volt} = 0.707\ \text{volt RMS}$$

In power supplies, the secondary transformer winding voltage may be $V_{RMS} = 12$ volts AC, which leads to a peak sine-wave voltage of $V_p = 12(\sqrt{2})$ volts peak = 16.9 volts peak.

   Audio signals can be measured in terms of peak, peak-to-peak, and RMS voltages (or currents). The most common measurement used in meters is the RMS voltage. Table 8-1 lists various audio signal levels.

TABLE 8-1   Various Audio Levels for Audio Devices in RMS AC Voltages

Device	Voltage Level V_{RMS}
Dynamic microphone	~1 mV output
Electret microphone	~10 mV output
Phono-cartridge moving magnet	~5 mV output
Phono-cartridge moving coil	~0.5 mV output
Line level (aux input)	70 mV to 2,000 mV output
Headphone (32 Ω)	~0.18 volt for 1 mW level into the headphone
Loudspeaker	0.5 volt to >10 volts into the loudspeaker

The microphone output levels shown in Table 8-1 are for normal speaking levels. Obviously, placing them in front of musical instruments can result in output levels of over 100 times nominal levels. Moving-magnet phono cartridges, which are the most commonly used types, also can generate output signals on the order of 50 mV to 100 mV depending on the recording. As for line-level audio signals, there are the "old school" devices such as phono preamps, FM tuners, and so on that can generate as low as 70 mV, but today's CD, DVD, and Blu-Ray players will deliver audio signals on the order of 1 volt to 2 volts.

Balanced or Differential-Mode Audio Signals Used in Broadcast or Recording Studios

Audio signals used in consumer equipment are unbalanced, which means there is a hot and a ground lead. Over long distances, the ground lead, which is supposed to shield again hum or other noise pickup, can itself be induced with noisy signals. To further reduce induced noise pickup over wires, a balanced or differential-mode audio signal can be used.

When audio signals are delivered over long lines such as telephone lines or microphone cables, a balanced or differential-mode signal is outputted, and this signal is received with a balanced or differential amplifier. Noise will be induced or added in a common-mode manner. The balanced audio line has two leads that have equal and opposite audio voltages with respect to ground. Noise is induced equally on both lines but with the phase or polarity as a common-mode noise signal.

To further understand this concept, we can try the following experiment, as shown in Figure 8-2, where we apply a hum signal injected to the common-ground signal while applying a balanced line audio signal. Both the common-mode hum signal and the balanced-mode audio signal are connected to a common amplifier (U1A) that amplifies signals referenced to ground, whereas the balanced-mode amplifier (U1B) only amplifies the signal across the balanced line and rejects the common-mode hum signal that is referenced to ground.

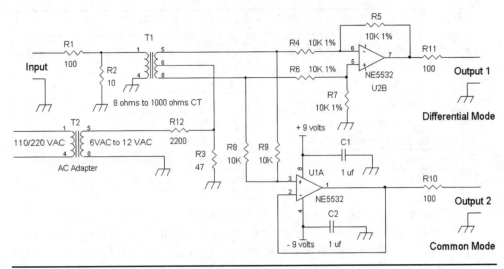

Figure 8-2 An experiment with common-mode and balanced audio signals.

This experiment involves connecting a CD or radio headphone output signal to transformer T1 to provide music (or voice) signals. The secondary winding of T1 then has music across pins 5 and 8, the outer terminals. At the center tap of T1, we add a hum signal via T2, an AC adapter, to provide a hum signal on both pins 5 and 8. At the receiving end, there are two different types of amplifiers, a differential-mode (aka balanced-mode) amplifier via U1B and a common-mode (aka unbalanced-mode) amplifier via U1A. The balanced mode only "reads" voltages that are the difference between the two outer terminals of T1. The music signal has a balanced output at T1, so U1B will receive the music signal fine while subtracting out the hum signal that is the same on both terminals at the secondary winding of T1. This subtracting effect accounts for the noise-reducing advantage of using balanced-output and balanced-input amplifiers.

The common-mode amplifier, by summing resistors R8 and R9, will add the hum signal that is applied at the center tap of T1 to be amplified while nulling out the differential-mode music signal. The reason why the differential mode signal is nulled is that the music signal is equal and opposite at pins 5 and 8, the output terminals of T1 referenced to ground. Note that better nulling of the balanced-mode signal can be achieved by replacing R8 and R9 with a 20 kΩ or 50 kΩ potentiometer (pot) where the middle (slider) terminal of the pot is connected to pin 3 of U1A and the outer leads of the pot are connected to pins 5 and 8 of T1. Note that T1, Catalog/Part Number 273-1380, is available at RadioShack.

A common-mode amplifier is not normally used, and this experiment just shows how it can be used. Nevertheless, there have been systems that have both differential- and common-mode signals sent across a balanced pair of wires plus the ground connection. For example, when I worked at a radio station, the audio signal from the studio to the transmitter room was delivered by a pair of balanced audio signal wires. To turn on and off the transmitter from the studio, a common-mode DC signal was supplied to both the balanced audio signal wires (e.g., +24 volts DC) by a center-tap connection of a transformer and ground. The transmitter received the AC audio signal fine via transformers, and the control circuitry sensed whether there was a DC voltage referenced to ground.

Balanced lines are commonly used in broadcast and recording studios. Long runs of wire are prone to pick up noise such as hum. However, most of the noise is common-mode noise, and thus, an audio system with balanced outputs and balanced inputs will reduce this type of noise substantially.

Originally, the balanced outputs and inputs were implemented with transformers, and although they are not used as often today, transformers can be used. It's more of a question of space and cost. High-quality audio transformers are relatively expensive when compared with a balanced-output or balanced-input amplifier with integrated circuits (ICs).

Typical balanced-output and balanced-input levels for 0 VU on an audio meter vary from 0 dBm to +8 dBm. The level of 0 dBm is 1 mW across a 600 Ω resistor or 0.775 volt RMS. See the VU meter in Figure 8-3. Note that most analog meters include a set screw to adjust the nominal position at zero signal level. This meter is adjusted deliberately slightly to the left side to show more clearly the –20 dB and 0 percent markings.

We can find out what +8 dBm is in terms of voltage by the following formula. If we call 0 dBm a reference-voltage level at 0.775 volts RMS, then +8 dBm is a multiplying factor |Vout/Vin| of the 0 dBm level. We can express the multiplying factor as follows:

$$|Vout/Vin| = 10^{\text{number of dB/20 dB}} \tag{8-1}$$

FIGURE 8-3   An analog VU meter.

Thus +8 dB translates to $10^{8\,dB/20\,dB} = 10^{8/20} = 2.51$ times over 0 dBm. And thus +8 dBm = 2.51×0 dBm = 2.51×0.775 volt = 1.945 volts RMS.

Therefore, the range of voltages in a broadcast or recording studio is 0.775 volt to 1.945 volts referenced to 0 VU. By the way, VU stands for "volume unit." By looking at the VU meter's marking in decibels (dB) and percentage, we see that –1 dB is about 90 percent, –2 dB is about 80 percent, –3 dB is about 70 percent (it's actually 70.7 percent), –6 dB is half or 50 percent, and so on. Equation (8-1) gives the exact reading when converting dB (decibels) into a number. For example, +6 dB is 2 or 200 percent. The VU meter shows a relative audio level, and generally, when recording an audio signal, it is useful to keep the meter's reading above –6 dB (~50 percent) and below +3 dB (~141 percent). There are other standards such as dBV, which has 1 volt RMS at 0 VU instead of 0.775 volt. Using a VU meter does not limit the audio signal to balanced audio signals, and unbalanced signals are often used with equipment having VU meters.

Now let's take a look at how balanced audio signals are provided and received. Figure 8-4 shows balanced output and input amplifier circuits.

For generating a balanced output signal via the Balanced Output Amplifier in Figure 8-4, connect a unity-gain inverting op amp (U1B) as shown, where R20 = R30, and use a voltage-follower amplifier (U1A). The output voltage across U1A and U1B via Output1 and Output 2 is twice the voltage from the single-ended outputs of either U1A or U1B. Although originally the output resistors for U1A and U1B were set to 300 Ω each to provide a total balanced-output resistance of 600 Ω, by at least the 1980s, the resistor values (e.g., R40 and R50) were kept as low as possible, typically 10 Ω to 22 Ω, which results in a total balanced-output resistance of 20 Ω to 44 Ω. Because of long cable runs throughout the broadcast or recording studio, the capacitance of the cable was substantial and caused a frequency roll-off at 20 kHz with a balanced-output resistance of 600 Ω.

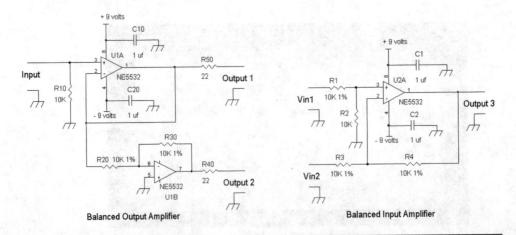

FIGURE 8-4 Balanced audio circuits.

NOTE   Be sure to specify an op amp that has sufficient current drive into a 600 Ω load. Most general purpose op amps (e.g., LM1458, TL082, or LF353) have about a 10-mA to 15-mA maximum output, which is not always sufficient. Op amps such as the LM5532, LM833, or OPA2134 are designed to drive 600 Ω loads without problems.

The Balanced Input Amplifier in Figure 8-4 receives a balanced output signal. Thus, Vin1 = –Vin2. In general, this amplifier can be looked at as a combination inverting-gain amplifier for Vin2 and a non-inverting-gain amplifier for Vin1. Voltage divider circuit resistors R1 and R2 are of the same value, so the voltage at the (+) input of U2A is (½)Vin1. We know that the gain of a non-inverting-gain amplifier using an op amp is (1 + R4/R3) from the (+) input of the op amp U2A to Output 3. Therefore, the output voltage (Output 3) due to Vin1 is

(½)Vin1(1 + R4/R3).

With R4 = R3 and R4/R3 = 1, this results in:

(½)Vin1(1 + R4/R3) = (½)Vin1(1 + 1) = Vin1 (at the output of U2A).

The contribution of Vin2 to the output of U2A is Vin2 × (–R4/R3) = –Vin2 because –R4/R3 = –1, the gain of an inverting-gain op amp.

NOTE   For a balanced signal connected to Vin1 and Vin2 of the balanced-input amplifier in Figure 8-4, the input resistance at Vin1 is R1 + R2 = 20 kΩ. However, the input resistance at Vin2 is 10 kΩ/1.5 = 6.66 kΩ. Therefore, to provide equal input resistances at Vin1 and Vin2, set R1 = R2 = 3.33 kΩ or R1=R2 = R3/3. This results in a 6.66 kΩ input resistance at Vin1, which matches the input resistance at Vin2.

If we now take both input signals Vin1 and Vin2 into account, we find that the output signal is then:

Vin1 + –Vin2 = Vin1 – Vin2 = Output 3

To provide the highest performance in rejecting common-mode signals, R1, R2, R3, and R4 are typically 1 percent or better tolerance resistors.

Microphone Preamplifier Circuits

From Table 8-1, the relative levels from microphones are on the order of millivolts (mV), which are very low signal voltages. One can build an amplifier with a gain of about 1,000 and amplify the microphone's signal to the order of about 1 volt. But the question arises what type of low-noise amplifier we will need to amplify the microphone's signal without adding noise? Generally, a signal-to-noise ratio of >60 dB, which translates to a >1,000:1 (signal-to-noise) ratio, is required to have the amplified signal relatively hiss-free. For 1 mV of signal from the microphone, we need to ensure that the equivalent noise at the input of the microphone preamp is less than 1 mV/1,000 = 1 μV.

But how do we measure the noise? To measure noise for its associated noise voltage, we must always measure noise at the output of a device with a filter. This is because measuring noise without a filter with a specified bandwidth will result in an erroneous figure. Most audio noise measurements use a bandwidth of 20 kHz or less. And it is very common to use an "A" weighting filter, which has about a 10 kHz bandwidth that can be approximated with a 1 kHz high-pass filter and a 11 kHz low-pass filter connected in series. The "A" weighting filter passes signals whose frequencies are generally audible to the human ear. It should be noted that with white noise (e.g., hiss), which is random, the noise voltage is proportional to the square root of the bandwidth, $\sqrt{BW}$.

Many op amps have input noise density voltage specifications that we can use to determine how much noise is equivalent at the input of the op amp. The noise density voltage V_{nd} is specified for a bandwidth of 1 Hz, and to figure out the noise voltage for a specified bandwidth, we have

Total noise = noise density voltage × $\sqrt{BW}$

where BW = bandwidth in hertz (Hz). Or, in other words:

$$\text{Total noise} = V_{nd} \times \sqrt{BW} \quad \text{or} \quad \text{Total noise} = V_{nd} \sqrt{BW} \tag{8-2}$$

Because we choose a 10 kHz bandwidth for an "A" weighting filter:

Total noise = noise density voltage × $\sqrt{10,000}$ =
noise density voltage × 100 = V_{nd} × 100

Low-noise op amps such as the NE5532, LM4562, and AD797 have noise density voltages of about 4.5 nV/$\sqrt{Hz}$, 2.7 nV/$\sqrt{Hz}$, and 1 nV/$\sqrt{Hz}$, respectively. For a 10 kHz bandwidth, the noise voltages are 0.45 μV, 0.27 μV, and 0.1 μV for the NE5532, LM4562, and AD797, which results in signal-to-noise ratios for a 1 mV input signal of about 66 dB (2,000), 71 dB (3,700), and 80 dB (10,000). In practice, we may lose one or two dBs (decibels) in noise performance due to noise voltages generated by the gain-setting feedback resistors or even more due to the noise generated by the microphone itself. A balanced microphone preamplifier is shown in Figure 8-5.

Although we could use a balanced-input amplifier like the Balanced Input Amplifier shown in Figure 8-4, the input resistors have to be of low value, such as having R1 = R3 = 100 Ω and R2 = R4 = 20 kΩ, with U2A = LM4562 or other lower-noise op amp, for a gain of 46 dB or 200. In most cases, this balance-input microphone amplifier will work fine unless the user needs to adjust the input-loading resistor for the particular microphone.

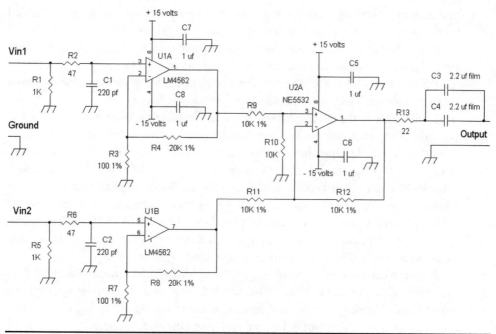

FIGURE 8-5   A balanced microphone preamplifier for XLR connectors.

Generally, if there is a 150 Ω microphone, the load resistance is much higher, such as 800 Ω or so, to avoid overdamping the voice coil in the microphone. Figure 8-5 shows one such balanced microphone preamplifier where the input resistors R1 and R5 that are set nominally to 1 kΩ each can be replaced with resistors with other values for flexibility of input load resistance.

This microphone preamplifier has input load resistors (R1 and R5) that typically allow for a balanced-input resistance of about 2,000 Ω. The preamp consists of two low-noise non-inverting-gain amplifiers U1A and U1B with low-value resistors R3 and R7 to minimize added noise. The two amplifiers U1A and U1B both have a gain of (1 + 20 kΩ/100) or 201. Gain can be adjusted by varying the values of both feedback resistors R4 and R8. Differential (mode) amplifier U2A ensures that common-mode noise is reduced or nulled. Lower-noise op amps can be found that are better in noise density voltage than the LM4562, but as a good compromise in a dual package, this op amp is fine.

Capacitors C1 and C2 provide some RFI (radio-frequency interference) protection, which is generally needed given the high voltage gain of the amplifier. These capacitors roll off the frequency response at RF signals but pass audio signals.

Generally, for a preamplifier such as this, the supply voltages are typically ±12 volts to ±15 volts. This ensures sufficient dynamic range to accommodate the wide range of sound levels that will be encountered. One way to mitigate the large variation in microphone signal levels is to connect the output of this amplifier via C3 and C4 to the peak limiter amplifier that was presented in Figure 7-11.

Microphones for consumer electronics normally do not have balanced-output signal lines and instead have unbalanced outputs including hot and ground leads via the tip and sleeve of the connector. For unbalance microphones, including those with dynamic and electret elements, see Figures 8-6 and 8-7.

In Figure 8-6, the preamplifier has an RFI protection circuit via R2 and C1. Loading of the microphone is set by R1, which can be a resistor with a value from about 500 Ω to 2 kΩ depending on the microphone specifications. This preamplifier is split into two stages, with the first stage having a gain of about 34 via feedback resistors R6 and R7. The second stage is an adjustable-gain amplifier via VR1, which allows setting the second-stage amplifier for a gain of between 2 and about 34. By splitting up the gain distribution over two stages, lower GBWP (Gain Bandwidth Product) op amps may be used if desired, but more important, some control from overload is provided in case the microphone input signal is too high.

For a typical dynamic microphone, a gain of 100 to 1,000 is needed, and for electret types, a gain of about 20 to 100 is needed. Note that the electret microphone requires a DC biasing voltage via a resistor at the input (via R1 in Figure 8-7), and this DC biasing circuit is not to be used with dynamic microphones.

To connect a balanced-output microphone with an unbalanced amplifier, simply ground the (–) terminal to the ground shield and connect in the same manner as an unbalanced dynamic microphone. The electret condenser microphone has a built-in field-effect transistor (FET) amplifier that requires powering via a bias resistor with voltages that vary typically from 1.5 volts to 9 volts. Common biasing voltages are 3 volts or 5 volts. In this particular design, we use a regulated 3.3 volt reference via

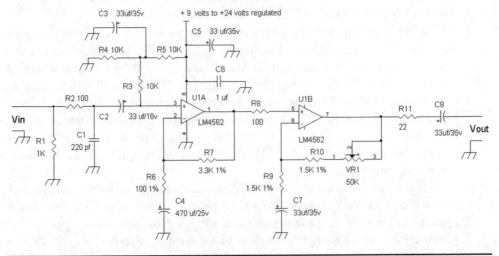

Figure 8-6 Preamplifier for unbalanced-output microphones that have sleeve and tip connectors.

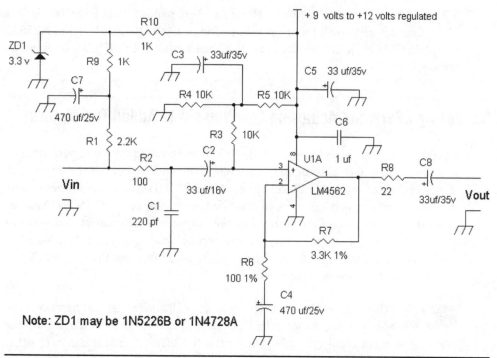

Figure 8-7   Electret condenser microphone preamplifier with biasing circuit at the input Vin.

zener diode ZD1. Because the zener diode is a pretty good white noise generator, low-pass filtering is required to remove the white noise plus any power-supply noise. Thus R9 and C7 form a low-pass filter. The microphone signal is AC coupled and level shifted via C2 to provide the microphone signal that will be amplified by U1A. A gain of 34 is provided via feedback resistors R7 and R6, where the non-inverting-gain op amp gain = [1 + R7/R6] = [1 + 3300/100] = [1 + 33] = 34. Electret microphones provide more signal output, and a gain of about 30 dB is sufficient. Depending on the electret microphone's signal level, R7 may be changed.

When applying resistive negative feedback around an op amp, the frequency response of the amplifier is given by the gain bandwidth product (GBWP) divided by the gain of the system. When referring to bandwidth, we generally are talking about a –3-dB bandwidth, which translates to a drop off in amplitude response of $1/\sqrt{2}$, or 0.707. Most op amp data sheets will list the GBWP. For example, an LM1458 op amp with a 1-MHz GBWP set for a noninverting gain of 34 will have a –3-dB bandwidth out to 1 MHz/34 = 29 kHz. So, at 29 kHz, the gain is dropped from 34 to $34 \times 0.707 = 24$ as opposed to a gain of 34 at a lower frequency such as 1 kHz. With audio signals, the bandwidth or frequency response is often stated as the frequency that causes the signal to drop from 100 percent to 70.7 percent. The op amps were chosen for their high gain bandwidth product (GBWP) such that at maximum gain, the frequency response exceeds 20 kHz by at least threefold. For example, in Figure

8-7, an LM4562 has a GBWP of 55 MHz for a –3-dB bandwidth of 55 MHz/34 = 1.6 MHz. One can also use a lower-priced op amp in Figure 8-7, such as the NE5532 (10-MHz GBWP), which results in a –3-dB bandwidth of 10 MHz/34, or approximately 300 kHz.

Increasing GBWP and Achieving Low Noise with Added Transistors

Of course, there are ways to increase the GBWP and provide low-noise performance in a preamplifier circuit by adding low-noise transistors in front of the op amp, as shown in Figure 8-8. U1A, a TL082 op amp ($18 \text{ nV}/\sqrt{\text{Hz}}$) is about four times noisier than an NE5532, which has an input noise density voltage of $4.5 \text{ nV}/\sqrt{\text{Hz}}$. *Note: In general, noise is generally measured at the output when the input is shorted to ground.* By adding a pair of low-noise transistors as shown by Q1 and Q2 (2N4401s), the "new" equivalent input noise density of the preamp is less than $2.5 \text{ nV}/\sqrt{\text{Hz}}$, which is about half the noise of an NE5532.

NOTE Care must be taken to avoid oscillations that can happen if the feedback resistors R13 and R12 are set for too low of a gain. For example, to avoid oscillation, the gain = [1 + (R13/R12)]≥ 38. If a gain of 40 is desired, then let R12 = 100 Ω and R13 = 3900 Ω.

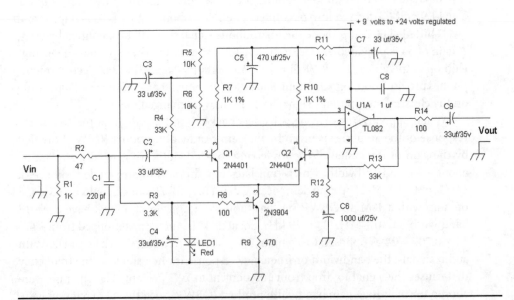

Figure 8-8 Added transistors to increase the overall GBWP of an op amp.

The circuit in Figure 8-8 can be rather tricky to build, and an oscilloscope should monitor the output to ensure that the output terminal Vout is free of oscillations. Basically, low-noise 2N4401 transistors Q1 and Q2 are biased as a differential amplifier. Are there other low-noise transistors that one can try? Yes, there are other bipolar transistors such as the 2SA1316, which has a different pin-out of ECB (Emitter-Collector-Base) instead of the EBC (Emitter-Base-Collector) of the 2N4401. And it's possible to use low-noise JFETs such as the LSK170, but values for R9 = 470 Ω and R11 = 1 kΩ probably will have to be divided by about 5 that results in R9 = 100 Ω and R11 = 200 Ω; and the supply voltage set to +24 volts. Also note that in an FET, the source, gate, and drain correspond in the same order to the emitter, base, and collector of a bipolar transistor. When the circuit in Figure 8-8 is powered on, please wait about 60 seconds for the DC operating points to stabilize.

The collector current of Q3 is determined by the turn-on voltage of LED1 (~1.7 volts DC) and R9. The base-emitter junction turn-on voltage of Q3 is about 0.7 volt, so the voltage across R9 is 1.7 volts – 0.7 volt = 1 volt. Therefore, the emitter current of Q3 is 1 volt/R9 = 1 volt/470 Ω ≈ 2 mA, which is also the collector current of Q3. The op amp via R13 will adjust a voltage at pin 1 of U1A such that equal currents of 1 mA each will flow through R7 and R10. With 1 mA of collector current for Q1 and Q2, the (differential) voltage gain from the bases of Q1 and Q2 to load resistors R7 and R10 is about 38. Basically, increasing the current via R9 will increase the gain proportionally. For example, if R9 is decreased from 470 Ω to 235 Ω, the collector currents of Q1 and Q2 will increase from 1 mA to 2 mA, and the resulting gain will increase from 38 to 76.

The actual gain as defined by the differential output voltage at the collectors of Q1 and Q2 referenced to the input signal at the bases of Q1 and Q2 is:

$$\text{Gain}_{Q1\_Q2} = [(\tfrac{1}{2})\,(\text{IE}_{Q3}) \times R7/0.026\ \text{volt}] = g_m R7$$

where IE_{Q3} is the emitter current of Q3 (e.g., approximately 2 mA for R9 = 470 Ω). For example, if IE_{Q3} = 2 mA and R7 = 1 kΩ:

$$\text{Gain}_{Q1\_Q2} = [(\tfrac{1}{2})\,(2\ \text{mA}) \times 1\ \text{k}\Omega/0.026\ \text{volt}] = 1\ \text{mA}(1\ \text{k}\Omega)/0.026\ \text{volt} = 38.4$$

With 1 mA of collector current for Q1 and Q2, this results in an added gain of 38 in front of the op amp U1A, and the GBWP → 38 × GBWP$_{op\_amp}$. This means that the 4-MHz GBWP TL082 is now extended to a GBWP of 38 × 4 MHz, or 152 MHz. The only down side with this setup with the extra gain is that we add another stage that has a phase shift that can cause oscillations in a negative-feedback amplifier. To avoid oscillations, the gain set by the feedback resistors R13 and R12 always must be set to *higher than* g_mRL, which is 38 in this case. In this example, the gain is about 1,000 because R13/R12 = 1,000 > 38, so we are safe.

Moving-Magnet or Magnetic High-Output Phono-Cartridge Preamps

Although compact discs (CDs) and other digitally recorded media are used commonly today in the twenty-first century, analog vinyl records still are being manufactured and have a niche following. The two most common speeds for playing back records are 33⅓ rpm and 45 rpm (rpm = revolutions per minute). Much older records that play back at 78 rpm (e.g., 10-inch disks recorded up to the 1960s) or 16⅔ rpm (e.g., children's records also from the 1960s) are much less common today.

A stereo magnetic phono cartridge has four leads for the right and left channels. They are right hot, left hot, right ground, and left ground, and these four leads are associated with red, white, green, and blue wires, respectively, in the tone arm. Normally, these cartridges track the records at a stylus force between 1 gram and 3 grams. The stereo output cables of a turntable have a RCA connector plug, where the red-marked cable denotes the right channel, and the left channel is marked in white or black.

Frequency Response and Phase for Recording and Playing Records

A typical vinyl long-playing (LP) record has about 20 to 30 minutes of information on each side. In order to optimize playing time and prevent a cartridge's stylus from mistracking, the record is recorded with an equalization curve that looks like a bass cut and treble boost. That is, the lower-frequency signals at 20 Hz are recorded at a level about 1/10 the level at 1 kHz, and the higher frequency signals at 20 kHz are recorded about 10 times the level at 1 kHz. In general, most music has a frequency spectrum that is attenuated at high frequencies, and the treble boost at the record cutting end is acceptable.

To play back an analog record, a complementary equalization curve must be included such that there is a bass boost and treble cut. To implement the playback equalization curve, there is a roll-off starting at 50 Hz, which levels out at 500 Hz to about 2.12 kHz, where there is a second roll-off at 2.12 kHz. This second roll-off can continue to beyond 20 kHz, but it is not uncommon to have the second roll-off level out at around a frequency of 40 kHz or higher. Figure 8-9 shows the record and playback frequency responses that follow the Recording Industries Association of America (RIAA) curves.

In Figure 8-9, we see 6 dB/octave for the RIAA record equalization curve. This means that as the frequency changes by a factor of 2, the 6 dB/octave slope translates to a twofold increase in amplitude. For example, when measuring the amplitude level from the RIAA recording equalization curve at 5 kHz and 10 kHz, the amplitude level at 10 kHz is twice the level of the amplitude at 5 kHz. For a –6 dB/octave slope seen in the RIAA playback equalization curve, the frequency goes up twice, and the amplitude goes down by a factor of ½ (or the reciprocal of 2). For example, in the RIAA playback curve, the signal level at 8 kHz is half the signal level at 4 kHz.

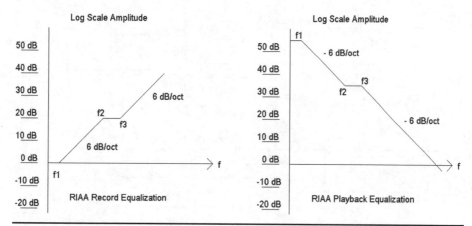

Figure 8-9   RIAA record and playback equalization curves for analog vinyl records at 33⅓ and 45 rpm.

The three frequencies f1, f2, and f3 required to shape the equalization curve can be equivalently stated as time constants. For example, the frequencies f1 = 50 Hz, f2 = 500 Hz, and f3 = 2.12 kHz are equivalently expressed as time constants τ_1 = 3,180 μs, τ_2 = 318 μs, and τ_3 = 75 μs, respectively.

If we choose 50 Hz, for example, we find in an RC low-pass filter that the –3-dB (0.707) cutoff frequency is:

$$f_c = 1/[2\tau] = 1/[2RC]$$

and solving for RC, we will find that τ_1 = RC = 3,180 μs = 3,180 × 10^{-6} second, that is:

$$50\ \text{Hz} = 1/[2\pi(3{,}180\ \mu s)] = 1/[2\pi(3{,}180 \times 10^{-6}\ \text{second})]$$

By *time constant*, we mean that any combination of resistors and capacitors such that the product of their values equal 3,180 μs will provide a cutoff frequency of 50 Hz. For example, combinations of a capacitor of 3,180 pF and a resistor of 1 MΩ or a capacitor of 6,360 pF and resistor of 500 kΩ have the same time constant of 3,180 μs. In this example, doubling the capacitance resulted in halving the resistance.

Three RIAA Equalization Phono Preamplifier Projects

First RIAA Phono Preamp

For the first phono preamplifier project, we will use op amps and active equalization, where the feedback network implements the playback curve entirely (see Figure 8-10).

Parts List

- C1 and C2, 100 pF silver mica or polystyrene capacitors 5% or 10%

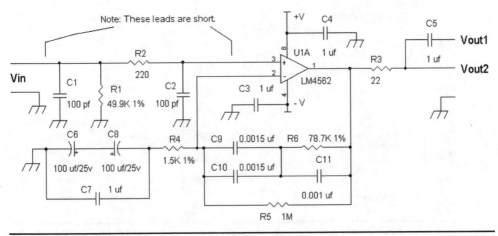

FIGURE 8-10   A phono preamp using a single op amp per channel.

- C3, C4, C5, and C7, 1 μF film capacitors (Mylar or polyester) 5% or 10%
- C6 and C8, 100 μF electrolytic capacitors with at least 25 working volts
- C9 and C10, 0.0015 μF polyester, polystyrene, or silver mica 2% tolerance capacitors or use 5% versions measured with a capacitance meter within 2 percent tolerance
- C11, 0.001 μF polyester, polystyrene, or silver mica 1% or 2% tolerance capacitors or use 5% versions measured with a capacitance meter within 2% tolerance
- All resistors ¼ watt unless specified
- R1, 49.9 kΩ, 1% or alternatively 47.5 kΩ, 1%
- R2, 220 Ω, 5%
- R3, 22 Ω
- R4, 1.5 kΩ, 1%
- R5, 1 MΩ, 1% or 5%
- R6, 78.7 kΩ, 1%
- One 8-pin IC socket
- U1, any of the following op amps: LM4562, NE5532, LM833, TL072, TL082, OPA2134, RC4560, RC4558, JRC4556, LF353, or even LM1458
- Batteries or (+) and (–) power supplies from ±9 volts to ±15 volts for +V and –V

This is a non-inverting-gain amplifier with an active equalization feedback network. The feedback network includes two capacitors, C11 (1,000 pF) and the parallel combination of C9 and C10 (3,000 pF), and two resistors, R5 and R6. The gain at 1 kHz is about 52 or approximately 34 dB. The actual circuit analysis of this network to determine the exact values of the time constants is somewhat tedious because there are interactions between the resistors and capacitors.

However, we can make a rough approximation that at low frequencies, R6 at 78.7 kΩ looks relatively low in resistance when compared with R5 at 1 MΩ, so we can say that for low frequencies, R6 is a short circuit in this context. Thus, R5 is now in parallel to C9 and C10, which forms a time constant of 1 MΩ × 3,000 pF = 3,000 μs ≈ 3,180 μs = τ_1.

It gets a little more difficult to determine the second time constant, τ_2 = 318 μs, because R6 × C9||C10 = 236 μs, so R5 and C11 may be involved with τ_2. Obviously, more calculations would be needed, but not in this chapter. I will spare myself and the reader from the long, drawn-out math.

For determining τ_3 = 75 μs, we can approximate that the 3,000 pf from C9 and C10 acts like a short circuit by 2.12 kHz and that R5 (1 MΩ) is then in parallel with R6 (78.7 kΩ), which then results in the parallel resistance of R5||R6 = 73 kΩ. From this result, the approximate calculated time constant τ_3 = (R5||R6) × C11= 73 kΩ × 1,000 pF = 73 μs ≈ 75 μs.

Well, that's enough calculations for now. Let's take a look at the long view of this phono preamp. It has an RFI filter R2 and C2. The load capacitance to the phono cartridge is about 200 pF due to C1 and C2. The 200 Ω for R2 is low enough to connect C2 in parallel with C1. There are two outputs, AC and DC coupled output terminals for Vout1 and Vout2. Either can be used, and Vout1 should be used for line-level preamps with input resistances of at least 10 kΩ. If less low-frequency roll-off is desired, one can increase the capacitance value of C5 to 2.2 μF (nonpolarized electrolytic or film capacitor). A prototype circuit of the first RIAA preamp circuit is shown in Figure 8-11.

Figure 8-11   Single op amp per channel RIAA prototype phono preamplifier.

Second RIAA Phono Preamp

A more elaborate phono preamplifier can include passive (R5 and C5) and active equalization (R7, R8, and C9) networks, as shown in Figure 8-12.

When I first started designing phono preamps, I used a network similar to the one shown in Figure 8-10. This first preamp was built with a TL072 with an all-in-one active RIAA equalization network; it sounded as good as an audiophile preamp. But then a second high-end preamp was compared with it, and my TL072 preamp lost. So I had to go back to the drawing board. What would I do next to improve the sonic quality (which is always a matter of taste to different listeners)?

Not much time after I designed the TL072 preamp, I thought about implementing the equalization network into two or three stages. By splitting out separate circuits to implement each time constant, possibly a more accurate RIAA playback curve could be achieved. My best estimate was to try to design the new preamplifier in two stages using passive and active RIAA equalization. This was done just before 1980, when I was consulting for an audio cable manufacturer. The challenge at the time was whether I could design a good-sounding do-it-yourself (DIY) preamp for less than $10, not counting the chassis or power supply. I used 9-volt batteries for my tests, which are great temporary "no hum" power supplies.

After building and listening to my first attempt on a passive and active RIAA equalization phono preamp, I was not happy with the sound. But maybe this preamplifier just did not "mate" well with my system but would sound better with another one. I tried out this preamplifier with a different stereo system (different power amplifier and speakers), and everyone liked it because it matched the second high-end preamplifier in sound quality.

The current design shown in Figure 8-12 is actually not the original design from over 30 years ago, but it has many aspects of the original. To ensure buffering and

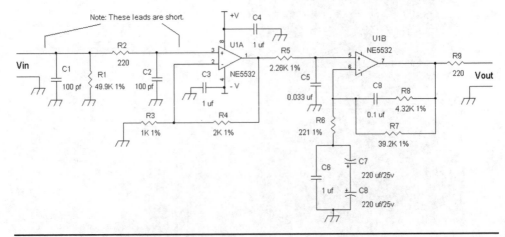

Figure 8-12 A phono preamp with active and passive equalization networks.

some gain, U1A has a gain of 3 that then drives R5 and C5 for the 75-μs time constant low-pass filter. Equalization has the τ_3 75-μs time constant circuit implemented first via R5 and C5, which are 2.26 kΩ and 0.033 μF. Thus, R5 × C5 = 2.26 kΩ × 0.033 μF = 74.6 μs ≈ 75 μs. Also, we can see that a –6 dB/octave slope frequency-response curve is achieved by a simple RC low-pass filter via R5 and C5.

Later on, via U1B, the RIAA equalization with the first two time constants τ_1 and τ_2 was implemented. We have for τ_1 the 3,180-μs time constant provided by R7 (39.2 kΩ) and C9 (0.082 μF) as part of the feedback network in the amplifier with U1B, where the resulting time constant is 3,210 μs ≈ 3,180 μs (within about 1 percent).

The τ_2 318-μs time constant is formed by the parallel combination of R7 (39.2 kΩ) and R8 (4.32 kΩ) and C9 (0.082 μF), which results in 3.89 kΩ × 0.082 μF = 319 μs ≈ 318 μs—pretty close! This circuit using R7, C9, and R8 forms a –6 dB/octave slope starting at 50 Hz, and the –6 dB/octave slopes ends a decade later at 500 Hz.

Beyond 500 Hz, the amplitude response levels off to a gain of:

$$[1 + R8/R6] = [1 + 4.32 \text{ k}\Omega/221 \ \Omega] = 20.5$$

Intuitively, R7 and C9 start the 3,180-μs time constant 50 Hz roll-off, and eventually, at higher frequencies, R9 is not as relevant, and the second time constant 318 μs is "sort of" formed by C9 and R8. But the exact calculation with the math shows that the 318 μs time constant is really implemented by C9 and R8||R9. Eventually, at higher frequencies, C9 becomes an AC short circuit, and we are left with a straight-gain amplifier determined by R8 and R6. The large-valued capacitors C7 and C8 were designed to be essentially an AC short circuit starting from 20 Hz and beyond. Figure 8-13 shows the preamplifier.

Figure 8-13 Prototype circuit of the active and passive RIAA equalization phono preamplifier.

Parts List

- C1 and C2, 100 pF silver mica capacitors 5% or better tolerance
- C3, C4, and C6, 1 μF film capacitors (Mylar or polyester) 10% or better tolerance
- C5, 0.0033 μF film capacitor (Mylar, polyester, or polystyrene) (these should be measured with a capacitance meter to within 2% tolerance)
- C7 and C8, 220 μF electrolytic capacitors with at least 25 working volts rating
- C9, 0.082 μF film capacitor (Mylar, polyester, or polystyrene) (this should be measured with a capacitance meter to within 2% tolerance)
- All resistors ¼ watt unless specified
- R1, 49.9 kΩ or 47.5 kΩ, 1% resistor
- R2, 220 Ω, 5% or 221 Ω, 1% resistor
- R3, 1 kΩ, 1% resistor
- R4, 2 kΩ or 2.21 Ω, 1% resistor
- R5, 2.26 kΩ or 2.21 kΩ, 1% resistor
- R6, 221 Ω, 1% resistor
- R7, 39.2 kΩ, 1% resistor
- R8, 4.32 kΩ, 1% resistor
- R9, 220 Ω, 5% or 221 Ω 1% resistor
- One eight-pin IC socket
- U1A and U1B, one of any of the following dual op amps: LM4562, NE5532, LM833, OPA2134, TL072, TL082, LF353, RC4556, RC4558, RC4560, JRC 4580, or AD712
- Batteries or (+) and (–) power supplies from ±9 volts to ±15 volts for +V and –V

Third Phono Preamplifier: Less Is More?

Simplicity sometimes works just as well in electronic designs. Op amps contain many more transistors than a discrete amplifier design. For example, a phono preamp can be built with just two or three amplifying devices (e.g., tubes, FETs, and/or transistors). When I was designing phono preamps years ago, I found that some of the simplest designs based on fewer amplifying devices in the signal path had good sonic performance. Also, with one of these designs, I can introduce the reader to how amplifiers are designed on the transistor level. We will look at a two-transistor design that was adapted from vacuum-tube phono preamplifiers and redesign it by having an input bipolar transistor followed by a MOSFET output transistor. This can be powered by an 18 volt power supply or by batteries (Figure 8-14).

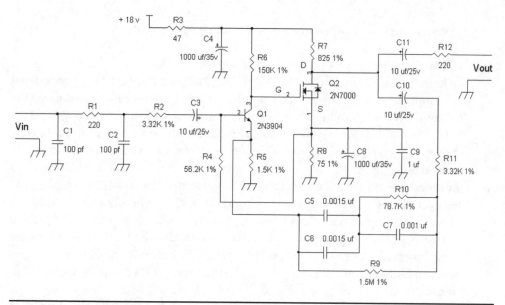

FIGURE 8-14   Two-transistor phono preamplifier circuit.

Parts List

- C1 and C2, 100 pF silver mica capacitors 5% or better tolerance
- C3, C11, and C12, 10 μF electrolytic capacitors with at least 25 working volts
- C4 and C8, 1,000 μF electrolytic capacitors with at least 35 working volts
- C5 and C6, 0.0015 μF film capacitors (Mylar, polyester, or polystyrene) measured for 2% tolerance with a capacitance meter
- C7, 0.001 μF capacitor (Mylar, polyester, or polystyrene) measured for 2% tolerance with a capacitance meter
- C9, 1 μF film capacitor (Mylar or polyester) with 10% or better tolerance
- Q1 2N3904 or 2N4124 NPN transistor
- Q2 2N7000 N channel MOSFET
- All resistors ¼ watt unless specified
- R1 and R12, 220 Ω, 5% resistor
- R2 and R11, 3.32 kΩ, 1% resistor
- R3, 47 Ω, 5% resistor
- R4, 56.2 kΩ, 1% resistor
- R5, 1.5 kΩ, 1% resistor
- R6, 150 kΩ, 1% resistor
- R7, 825 Ω, 1% resistor
- R8, 75 Ω, 1% resistor
- R9, 1.5 MΩ, 1% or 5% resistor

- R10, 78.7 kΩ, 1% resistor
- Batteries or positive voltage power supply for +18 volts

Although we only have a bipolar transistor Q1 and MOSFET Q2, there's a lot going on here. Q1 and Q2 form a two-stage amplifier, and the base of Q1 can be considered a (+) input terminal of a feedback amplifier, while its emitter can be considered the (−) input terminal. Although Q1 is not strictly a differential amplifier like Q1 and Q2 in the microphone preamplifier of Figure 8-8, nevertheless, Q1 in Figure 8-14 constitutes a differential amplifier of some sort. The RIAA equalization curve is implemented by using almost the same RIAA network used in Figure 8-10. All the values are the same except that the 1-MΩ resistor is replaced with a 1.5-MΩ resistor because the open-loop gain or composite gain from Q1 to Q2 is only about 2,000 (66 dB), whereas in an op amp, the low-frequency gain is usually greater than 10,000 (>80 dB). R9 is set to the higher value of 1.5 MΩ to ensure that the 3,180-μs time constant is implemented. Generally, when an RIAA preamplifier is built with discrete components and limited low-frequency gain, a resistor such as R9 has to be increased to ensure that at 50 Hz there is sufficient "boost" relative to 1 kHz.

Note that the DC bias drain current of Q7 is set up by having R4 bias the base of Q1. Because the collector current is very low at Q1, about 100 μA, this also leads to very low base current that flows through R5. Thus, the voltage across R5 << VBE_{Q1} = 0.7 volt. We can estimate that the base voltage reference to ground for Q1 is about 0.7 volt. Thus, the voltage at the source of Q2 is "servoed" to the DC base voltage of Q1 at 0.7 volt. The 0.7 volt DC across the 75 Ω source resistor R8 then sets up a source current of about 0.7 volt/75 Ω ≈ 10 mA. The 10-mA source current sets up a drain current of also 10 mA, and thus 10 mA is flowing through R7. The voltage across R7 is then 825 Ω × 10 mA = 8.25 volts. And thus the voltage at the drain of Q1 is the supply voltage, 18 volts, minus 8.25 volts, which is about 10 volts.

Indeed, there are quite a few things going on with just these two transistors. First, there is a DC bias servo system via R4, and second, there is an AC feedback amplifier that mimics an op amp via the output drain of Q2 and C10 coupled to an RIAA equalization network, which is connected to the (−) input at the emitter of Q1.

The gain of the Q1 first stage is approximately −R6/R5, and the gain of the Q2 second stage is about $-g_{mQ2} \times R7$. Thus the gain of the Q1 and Q2 two-stage amplifier is approximately $(R6/R5)g_{mQ2} \times R7$, where R7 = 825 Ω. *Note:* Two negative numbers multiplied result in a positive number. The transconductance of Q2, g_{mQ2}, is the ratio of the output drain to the input (AC) voltage applied to the gate and source. FETs have varied transconductance figures even among the same part number. For this preamplifier, we just measure the voltage gain of Q2. By direct measurement using a low-frequency signal of about 100 Hz, $g_{mQ2} \times R7$, the gain from the gate to drain of Q2 was found to be about 20. Therefore, the total gain from the base of Q1 to the drain of Q2 is *about* (150 kΩ/1.5 kΩ)20 ≈ 2,000.

This preamplifier also includes a 3.32 kΩ shelving resistor, R11, to level off the roll-off after 20 kHz at about 45 kHz. This causes a slight boost of 20 kHz, which should not matter much in the tonal playback of a vinyl record. A shelving resistor in series with the RIAA equalization network (C5, C6, C7, R9, and R10) allows for improved stability of the feedback amplifier in terms of preventing oscillation by having the feedback amplifier level off to a gain of [1 + (3.32 kΩ/1.5 kΩ)] = 3.2 instead of unity.

There is a very important note for any two- or three-transistor design similar to this. This pertains to the source (or emitter if Q2 has a bipolar transistor instead of an FET) of the second transistor being fed back to the base of the first transistor via a resistor (e.g., R4). The input resistor R2 is usually on the order of 1 kΩ or more to prevent a low-frequency oscillation from occurring when the phono cartridge is connected to the input terminal Vin. Some phono cartridges have DC resistances of less than 100 Ω, and if R2 → 0 Ω, then input capacitor C3 is essentially connected to ground (DC or low-frequency–wise). A motor-boating oscillation will occur due to a low-frequency phase shift provided by C8 and another low-frequency shift by C3 and R4, the input capacitor. By adding a series resistor R2 (3.32 kΩ) with the input capacitor C3, the low-frequency shift by R4 and C3 is reduced or canceled, and the oscillation is stopped. Recall that it takes at least two stages of phase shift to cause an oscillation. By adding series resistor R2, only one stage of phase shift via C8 remains. Figure 8-15 shows a prototype circuit of the two transistor phono preamplifiers.

Figure 8-15 A two-transistor phono preamplifier using a BJT and MOSFET.

Having Fun with Low-Voltage Tube Amplifiers

Before closing out this section on preamplifiers, I want to experiment with using vacuum tubes with just +24 volts for the plates (Figure 8-16). This circuit was built on a Dynaco PAS-3 preamplifier PC-5 circuit board, which is also available on eBay. Also, one can build this circuit on the PC-6 circuit board or just purchase the PC-5 and/or PC-6 fully assembled from eBay and modify the plate, cathode, and feedback resistors. The Dynaco PAS-2 or PAS-3 preamplifier service manual and schematic are available on the Web. One link that has it is www.the-planet.org/dynaco/ Preamplifier/PAS2_3.pdf.

NOTE   We will not be using the original high-voltage supply power from the original chassis. A suggested +24 volt power supply and filament supply will be shown later in this chapter.

If you are really into modifying tube Dynaco preamplifiers or power amplifiers, see the following links:

- http://curcioaudio.com/index.htm
- http://curcioaudio.com/pasdes_3.htm

Now let's take a look at working with low-voltage tube line amplifiers. Although the tube specified in the schematic shows a nine-pin 12AU7, other tubes such as the 12BH7, 12U7, and 12FQ7/12CG7 may be used. The gain is set for 11 via resistors R6 and R2. However, because the open-loop gain is not large, the actual gain is about 7 for |Vout/Vin|. Maximum *peak-to-peak* output voltage is about 5 volts to 8 volts. The 12U7 is a special "space charge" low-voltage tube that works very well. However, it

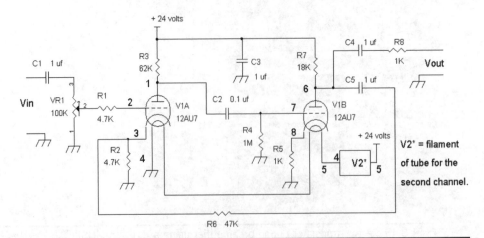

Figure 8-16 A two-stage triode line amplifier.

FIGURE 8-17   Assembled PC-6A phono preamplifier and blank PC-5A line amplifier boards.

was found that the 12BH7 actually slightly outperformed the 12U7 in terms of maximum output swing given the circuit in Figure 8-16. Other 9-pin 12 volt filament twin triodes such as the 12AT7 and 12AX7 were tried but will produce less voltage swing with the fixed-load resistors R3 at 62 kΩ and R7 at 18 kΩ. Other load resistor values may be tried to optimize the gain and output-voltage swing. Figure 8-17 shows two PAS-3 preamplifier boards available on the Web, such as eBay.

Power Supplies for the Preamplifier Circuits

In Figures 5-29 and 6-15, two designs of a ±12 volt regulated supply were shown. An improvement was made in ripple rejection by replacing the resistor in Figure 5-29 with the current sources in Figure 6-15. In this chapter, we will show a further improvement to the regulator design simply by using integrated circuit positive and negative voltage regulators 7812T and 7912T (see Figure 8-18).

The zener diodes and current-source transistors from Figure 6-15 in the previous design have been replaced with voltage-regulator chips. An improvement in the current design (Figure 8-18) has the regulator chips drain less current on standby.

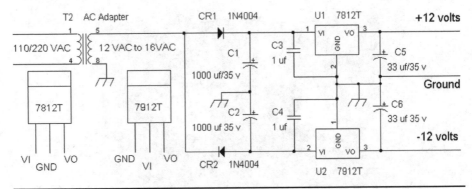

Figure 8-18 A ±12 volt supply using 7812T and 7912T voltage regulator ICs.

Also, the maximum output supply current is good up to 1 amp, provided that the 7812T (U1) and 7912T (U2) chips are mounted on heat sinks.

The previous design with the zener diodes had a maximum supply current of about 40 mA to 50 mA. Note that the pin-outs are different for the negative voltage regulator (7912T pin 1 = ground, pin 2 and tab = input, and pin 3 = output) versus the positive voltage regulator (7812T pin 1 = input, pin 2 and tab = ground, and pin 3 = output). Note also that the middle leads of both chips are connected to the metal tab of the TO220 case. Therefore, although the tab of the 7812 or 7800 Series chips can be mounted directly on a heat sink, the 7900 Series or 7912T voltage regulator cannot be mounted on the same heat sink unless an insulating mounting kit is used to isolate the tab electrically from the heat sink. The power-supply rejection is much better than with the previous designs using the zener diodes. C3 and C4 are positioned very close to the input and ground pins of U1 and U2 so that there is very good input decoupling to avoid instability or oscillations from the voltage regulators.

Note that if the transformer T1 has a 16 volt AC secondary winding, one can replace U1 and U2 with 7815T and 7915T for regulated ±15 volts, or replace U1 and U2 with 7818T and 7918T for a ±18 volt supply. For voltage-regulator chips of this type, the raw DC power-supply voltage connected to the input terminal of the regulator must be at least 2 volts above the regulated output voltage. For example, a 7812T for a +12 volt regulated output should have at least 14 volts at its input terminal. Similarly, a negative voltage regulator such as a 7912T that provides a regulated –12 volts should have at least a negative voltage of –14 volts (e.g., –14 volts DC or less such as –14 volts to –20 volts DC).

In Figure 8-19, one important note to remember is to place the input of U1 near C1 for proper decoupling at the input of the regulator IC. Otherwise, connect a 0.1 μF to 1 μF capacitor close with short leads to the input and ground terminals of U1. This supply can be used to power the two-transistor phono preamplifier.

Figures 8-20 and 8-21 can be used for the low-voltage tube line amplifier experiments pertaining to Figure 8-16. There is 24 volts DC to power the filaments of two tubes in series and the plates. *Note:* A regulated DC source powers the filaments of

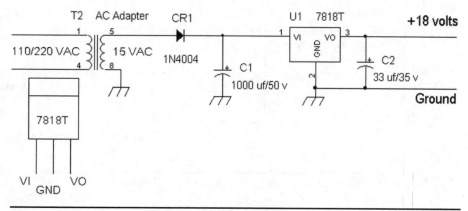

Figure 8-19   A +18 volt power supply for Figure 8-14.

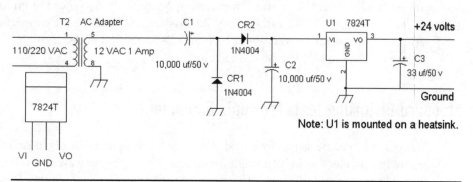

Figure 8-20   A +24 volt supply using a DC restoration voltage doubler circuit.

the tube line amplifier to reduce power supply hum. If AC powered the filaments, then most likely there will be hum at the output of Figure 8-16. Figure 8-20 uses a DC restoration circuit to achieve a voltage doubling effect. C1 and CR1 form a DC restoration circuit to "clamp" the negative peak of the 12 volt AC signal to near zero or ground voltage. The peak voltage from a 12 volt AC RMS source is 12 $\sqrt{2}$ volts = 16.9 volts peak. The peak-to-peak voltage is twice the peak voltage, or about 33.8 volts peak to peak. Thus the waveform at the cathode of CR1 is a varying voltage from about 0 volt to +33.8 volts. The second diode CR2 delivers a peak voltage of +33.8 volts to C2. With its storing capacity, C2 holds the +33.8 volts DC and provides this stored voltage as the raw DC voltage into the 24 volt regulator U1 (7824T). Of course, there are some diode voltage losses in CR1 and CR2 of about 1.5 volts or so, and thus C2 actually would receive about 32 volts instead of 33.8 volts. Again, U1 should be located close to C2, or connect a small capacitor with short leads to the input and ground terminals of U1. Now let's look at Figure 8-21.

A variation of the two half-wave rectifier circuits that provide a plus and minus supply is used here as a voltage-doubler circuit. Originally, the bottom lead of the

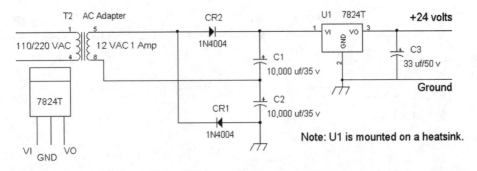

Figure 8-21   An alternate power-supply circuit for +24 volts.

secondary winding, pin 8, was grounded. But because we can float the lead and ground the anode of CR1 instead, we create approximately referenced to ground a +33 volt raw supply at the cathode of CR2. Basically, C1 and C2 are connected in series. Each capacitor delivers about 16.5 volts DC across C1 and C2, and in series, as shown in Figure 8-21, we get the 33 volts.

Standard Distortion Tests for Audio Equipment

Audio amplifiers are generally tested with sine-wave signals. Ideally, a sine-wave generator provides a signal of a single frequency. When the sine-wave signal is fed to an amplifier, usually some form of distortion occurs at its output. For example, when a 1 kHz tone is fed to the amplifier, the output of the amplifier will show an amplified version of the 1 kHz tone but will also have at its output smaller signals at 2 kHz, 3 kHz, and so on. These signals, which are multiples of the test-tone frequency, are *harmonics*. For example, the second harmonic of the fundamental frequency signal at 1 kHz is 2 kHz, and the third harmonic is 3 kHz.

In general, to measure an individual nth harmonic distortion, it is HD_n = amplitude of the nth harmonic signal/amplitude of the fundamental frequency signal. Put in another way the nth harmonic distortion, HD_n, is the ratio of the amplitude of the nth harmonic signal to the amplitude of the fundamental frequency signal.

Total harmonic distortion (THD) is defined as:

$$THD = [(HD_2)^2 + (HD_3)^2 + \ldots + (HD_n)^2]^{1/2}$$

Total harmonic distortion is the result of taking the square root of the sum of the squares of all the harmonic distortions. Fortunately, normally we do not calculate the THD, and there are machines or test equipment to come up with this number. In general, we test the amplifier for THD over a range of 20 Hz to 20 kHz at various output levels. The standard for what constitutes a good number for THD depends on the application. For example, in communication systems such as a voice channel in a two-way radio system, audio THD in the range of less than 3 percent to 10 percent is considered acceptable.

In high-fidelity systems, audio amplifiers generally will have a THD of less than 1 percent, and many will do much better than that—typically less than 0.1 percent. Super-low-distortion amplifiers have distortion levels well below 0.01 percent and sometimes typically less than 0.001 percent, such as the LM4562 op amp when configured for voltage gains of less than 10.

So what causes harmonic distortion? *A root cause of this type of distortion is a change in gain of the amplifier for different operating points of the transistors inside the amplifier.* We found in Chapter 6 that we could make a voltage-controlled amplifier with a simple common emitter amplifier with its emitter AC grounded by changing the collector current. So any amplifier that has a gain variation due to changes in collector or drain current usually causes harmonic distortion. Other causes of harmonic distortion include voltage-controlled capacitances within the amplifier. These voltage-controlled capacitances are always present in all transistors, JFETs, and MOSFETs.

Let's introduce a signal that leads into the next type of standard audio test signal, the two-tone test to measure intermodulation distortion (see Figure 8-22).

For clarity and illustration purposes, the two sine waves are locked in phase. However, normally when this test is done, in reality, the two sine waves are not locked in phase, causing a nebulous blurred look on the oscilloscope for the higher-frequency signal. The object of this test signal is to look for interaction between the two sine-wave signals that will cause other signals, *intermodulation distortion products*, to appear at the output of the amplifier other than harmonic distortion. However, it should be noted that the output of the amplifier will contain a combination of harmonic distortion on either or both test-tone signals plus the intermodulation distortion products. The intermodulation distortion products can be measured in two ways. One is to measure the amplitude of distortion signals centered around the high-frequency signal, and the other is to use an amplitude-modulation demod-

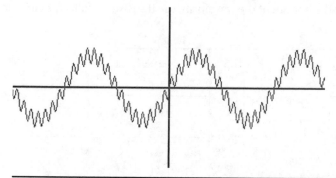

Figure 8-22 A two-tone test signal having a high-amplitude, low-frequency sine wave and a low-amplitude, higher-frequency sine wave. The vertical axis is amplitude, and the horizontal axis is time.

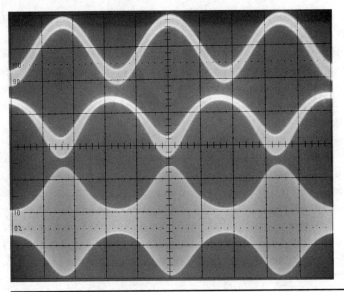

FIGURE 8-23 Test signal connected to the input signal of an amplifier (*top*). Output signal of the amplifier (*center*). High-pass filter applied to the output signal. Vertical axis is amplitude, and horizontal axis is time (*bottom*).

ulator or envelope detector to measure the amount of amplitude modulation on the higher-frequency signal. Figure 8-23 better illustrates what is happening.

By purposely driving a one-transistor common-emitter amplifier as shown in Figure 8-24 with about 40 mV peak to peak at the input to the base of the transistor, we see that at the output (denoted by the center trace in Figure 8-23) it has distortion. The center trace shows two important characteristics. First, the low-frequency signal itself levels off on the top and does not produce a clean sine wave. Second, we can see that the higher-frequency signal's amplitude gets smaller on the top and much larger on the bottom of the waveform. This variation in gain is consistent with what was observed in Chapter 7, which showed higher-voltage gains for higher collector currents. At the bottom of the waveform, the transistor's collector current is at

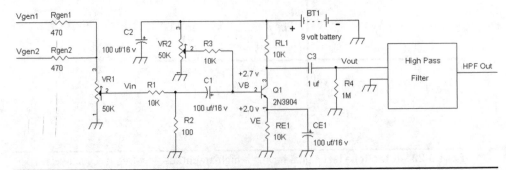

FIGURE 8-24 Circuit setup for IM distortion experiment.

a maximum, whereas on the top of the waveform, the transistor's collector current is at a minimum.

After filtering out the low-frequency signal as shown in the center trace of Figure 8-23, the bottom trace of Figure 8-23 shows an amplitude-modulated signal with the higher-frequency signal as the carrier. Also, the envelope of the amplitude-modulated signal is not a clean sine wave but is a replica of the distorted low-frequency signal inverted in phase.

Figure 8-24 shows the amplifier used for this two-tone distortion experiment. Two signal generators, Vgen1 and Vgen1, are summed with resistors via Rgen1 and Rgen2. The input signal to the base of Q1 is about 40 mV peak to peak, as shown in the top trace of Figure 8-23. The output of the amplifier, shown as Vout, connects to C3 and R4 and has a distorted waveform, as seen in the center trace in Figure 8-23. The output of the amplifier is then connected to a high-pass filter that removes the low-frequency signal, and the HPF Out terminal passes through the higher-frequency signal, which (as seen on the bottom trace of Figure 8-23) is an amplitude-modulated higher-frequency signal. The low-frequency generator Vgen1 is at 60 Hz, and the higher-frequency signal via Vgen2 is at greater than 3 kHz.

Figure 8-24 is adapted from Figure 7-13. VR2 is adjusted to provide 2 volts DC across RE1 to set up a collector current of 200 μA. VR1 is adjusted such that the base of Q1 has a 40 mV composite waveform, as seen in the top trace of Figure 8-23.

In essence, the amplifier in Figure 8-24 acts as a multiplier of sorts by using the low-frequency signal to modulate the amplitude of the higher-frequency signal. We will find later that this circuit is commonly used as an RF mixer for radio-frequency devices.

The SMPTE (Society of Motion Picture and Television Engineers) provided an intermodulation test waveform similar to Figure 8-22 and the top trace of Figure 8-23. The low-frequency signal is at 60 Hz, and the high-frequency signal is set to 7 kHz. This is called the *SMPTE IM distortion test*. The amplitude mix ratio has the 60 Hz signal at 4 times larger than the 7 kHz signal. Distortion signals around the 7 kHz signal have frequencies of 7 kHz ± ($n \times 60$ Hz), where n is a whole number greater than or equal to 1. Distortion is measured via an envelope detector such as a half-wave rectifier, an AM synchronous detector, or analysis of the distortion signals around the 7 kHz signal.

In a perfect amplifier, the high-pass filter output would show a 7 kHz signal with no variation in amplitude. So one question would be, is the amplifier in Figure 8-24 bad? For very low-level signals of less than 2 mV peak to peak, it will provide less than 1 percent distortion. For levels higher than this, the circuit distorts unacceptably for high-fidelity reproduction. However, sometimes distortion is *good*. We can use this circuit to enhance or change the sounds of guitar notes from an electric guitar. Distortion-making devices such as fuzz boxes deliberately distort the input signal to add more overtones to the original guitar string's sound (see Figure 8-25).

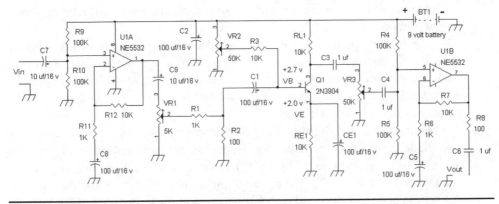

FIGURE 8-25   Experimental fuzz distortion-generator circuit.

The input signal from the electric guitar is amplified without distortion with a voltage gain of about 11 volts via U1A. The gain can be changed by increasing or decreasing the value of R12. U1A's output is fed to VR1, which feeds into R1 the input of the distortion-generating circuit. Resistor R2 may be increased to 1 kΩ if it is desired to generate more distortion. VR1 adjusts the amount of distortion, while VR2 may be set for a particular collector current via a base bias voltage.

Sometimes it may be advantageous to set the base voltage close to 0.7 volts, near the cut-off point of the transistor, to generate even more distortion. If this is done, it may be preferable to increase the signal drive into the base of Q1 via VR1.

U1B is a gain of 11 amplifier, with VR3 setting the final distortion output level. Output capacitor C6 is meant to connect to an amplifier with 10 kΩ or more input resistance. If a low-input-resistance amplifier is connected to C6, simply increase its value to 2.2 µF or larger. A 10 µF or 33 µF capacitor can replace C6, with the positive terminal of the electrolytic capacitor connected to R8. Note that almost any dual op amp will work for U1A and U1B, such as a TL082, LM1458, RC4458, LF353, RC4560, or LM4562 op amp.

References

1. *RCA Receiving Tube Manual*, RC-29. New York: RCA Corporation, 1973.
2. *RCA Solid-State Hobby Circuits Manual*, HM-91. New York: RCA Corporation, 1970.
3. Bob Metzler, *Audio Measurement Handbook*. Beaverton, OR: Audio Precision, Inc., 1993.
4. Paul R. Gray and Robert G. Meyer, *Analysis and Design of Analog Integrated Circuits, 3rd ed.* New York: Wiley, 1993.

CHAPTER 9

Oscillators

To generate an alternating-current (AC) signal, usually we start with an oscillator circuit. Depending on the circuit, oscillators generally provide sine-wave signals or pulsed waveforms. Other types of waveforms can be generated, such as triangle waves, which are generally derived from a square-wave generator circuit (e.g., an "on/off" circuit). This chapter will introduce the reader to some types of oscillators. There are a multitude of oscillator circuits, and some other types will be shown in subsequent chapters.

Therefore, we will explore how oscillation occurs and examine different types of oscillators used in audio and radio-frequency (RF) circuits. Also included will be relaxation oscillators, generally implemented with a resistor and a capacitor to set the oscillation frequency.

As a standalone circuit, the oscillator is a signal generator. But oscillator circuits generally are part of an overall system. As part of an overall system, the oscillator may serve as a test-tone oscillator for an audio mixer or for an RF system, the oscillator may be employed as a local oscillator for a superheterodyne radio. More important, oscillators can be varied in frequency and be part of a frequency synthesizer circuit. And in digital circuits, oscillators often are used to provide the clock signal.

It should be noted that there are no formal projects in this chapter because oscillators serve more of a support role for other projects. However, the reader is encouraged to build some of the circuits as experiments to get a feel for the various oscillators.

NOTE For this chapter and some of the succeeding ones, it is recommended that an oscilloscope be obtained to examine the waveforms from the oscillator circuits.

A Brief Overview of Oscillation Systems

Oscillation systems work on a principle of positive feedback of signals. That is, noise from the transistors or resistors acts as a "seed" to start the oscillation signal. The oscillation signal is reinforced amplitude-wise and grows to a steady-state amplitude. For this chapter, we will look into oscillators that provide both sine-wave and pulsed waveforms.

Most sine-wave oscillators use an amplifier and a phase-shift circuit. The amplifier may have a positive gain or a negative gain (e.g., inverting amplifier). See Figure 9-1.

For oscillation to occur, the system including the amplifier and phase-shift circuit has to be 0 or 360 degrees. Generally, the phase-shift circuits are implemented as high-pass, low-pass and/or band-pass filters. For an oscillation system using a noninverting amplifier, the gain of the amplifier must be large enough to overcome losses in the phase-shift circuit to provide a loop gain of at least 1. For example, if the phase-shift circuit has a loss or equivalent gain of 0.1 (1/10) at the frequency of oscillation, the amplifier's gain A must be greater than 10 such that:

Loop gain = A × phase-shift-circuit gain > 1

In this case, let the amplifier gain A = 11 and the phase-shift-circuit gain = 0.1, which says

Loop gain = 11 × 0.1 = 1.1 > 1

The reason for having the loop gain greater than 1 is to ensure that the positive-feedback signal will reinforce itself to sustain a reliable oscillating signal. Typically, the type of phase-shift circuits that "naturally" have 0 degrees of phase shift at the resonating frequency is a band-pass filter such as a parallel or series inductor-capacitor circuit or, alternatively, some combination of a low-pass and high-pass filter.

In an oscillator using an inverting amplifier, the phase-shift circuit must provide at least 180 degrees of phase shift. A low-pass or high-pass filter using two or more

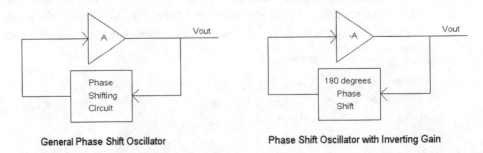

General Phase Shift Oscillator   Phase Shift Oscillator with Inverting Gain

FIGURE 9-1   Oscillation systems with noninverting and inverting amplifiers.

capacitors or an inductor-capacitor circuit with a resistor-capacitor circuit will provide the sufficient 180 degrees of phase shift for oscillation.

Most oscillators will inherently generate a distorted sine wave such as a rectangular pulse, square wave, or triangle wave. Some RF oscillators provide sine waves via a band-pass filter that reduces harmonic distortion.

When an amplifier is configured for positive feedback with a net gain of greater than 1, the oscillation starts out as a small sine wave and grows until the sine-wave signal is clipped due to limitation of the power supply. Actually, no practical power supply can ever be great enough to escape clipping from the growing signal. This growing signal increases exponentially with passing time. Therefore, one way to control the amplitude of the oscillating signal within the range of the power-supply voltage is to apply an automatic-gain-control amplifier to lessen the gain of the positive-feedback system such that the system reaches an equilibrium for the output amplitude of the signal (see Figure 9-2).

An amplitude-stabilized oscillator such as a Wein bridge oscillator uses a noninverting amplifier with a positive-feedback network that has 0 phase shift at the oscillation frequency. The oscillation signal starts out small in amplitude and then grows to a point where a resistor that depends on the output amplitude changes to reduce the overall gain of the noninverting amplifier.

In an automatic-gain-control (AGC) system, similar to the audio limiter amplifier mentioned in Figure 7-11, the output signal of the oscillator in Figure 9-2 is "measured" for amplitude, and there is a gain-control device (variable resistor) to adjust the gain electronically to provide a predetermined output amplitude for the oscillation signal. For example, the higher the signal increases, the more the gain is reduced. If the output amplitude is smaller than the predetermined setting, the oscillating signal will grow or increase until something close to the predetermined amplitude is achieved. One of the first ingenious devices used to maintain a steady and

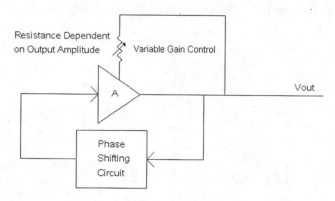

Amplitude Stabilized Oscillator via AGC

Figure 9-2 An oscillator with automatic gain control to provide a distortion-free sine wave.

undistorted sine-wave signal was a simple light bulb. The light bulb, when heated, changes to an increased resistance. We saw this effect in Table 4-2 when a #222 penlight bulb was found to have a resistance of 7 Ω at 1.2 volts and 9.6 Ω at 2.4 volts.

Because the lamp is used as one of the resistors in a negative-feedback amplifier, the increased resistance causes the amplifier to reduce in gain. For example, the gain of a noninverting op amp is given as $[1 + RF/R_{Lamp}]$, where RF is the feedback resistor from the output to the (−) input of the op amp, and R_{Lamp} is the resistance connected to the (−) input of the op amp and to ground. We can see that if the resistance of the lamp R_{Lamp} increases, the gain decreases.

Time to Sit Back and Relax for a Relaxation Oscillator (AKA Moving the Goal Posts)

Another type of oscillator known as a *relaxation oscillator* involves switching between two voltage levels. The gist of this system is that a resistor-capacitor circuit (e.g., a low-pass filter) receives a direct-current (DC) voltage. The capacitor charges via the resistor, and the capacitor's voltage increases with time. Before the capacitor can charge to the final voltage, the DC voltage is switched to 0 volt to start discharging the capacitor via the resistor. When the capacitor's voltage falls to a lower voltage, the circuit switches the DC voltage back on, and the capacitor starts charging again and its voltage increases. A cycle of charging and discharging of the capacitor is established, and thus an oscillating signal at the capacitor is provided in the form of a waveform that resembles a triangle or sawtooth waveform. Basically, as the capacitor's voltage gets up to a certain level, it changes its mind and decides to relax and "sit down" to a lower voltage before trying to get up again (see Figure 9-3).

A relaxation oscillator works on a principle that two voltages are determined for turning on and off the output DC voltage. That is, there are two trip voltages V1 and

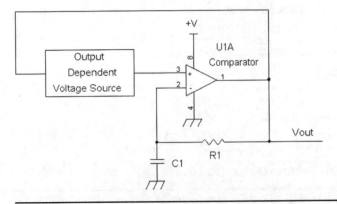

Figure 9-3 A relaxation oscillator.

V2, where V1 > V2. One trip voltage is set to turn on the output, and the other trip voltage is set to turn off or ground the output signal. The 555 timer chip works on this principle as well.

For a concrete example, let's say that V1 = 3 volts and V2 = 1 volt, which are the two different voltages available to the (+) input terminal of U1A (see Figure 9-3). The output Vout has two voltage levels, 0 volt and +V, where +V is the power-supply voltage (e.g., +5 volts). When Vout = +V, the output-dependent voltage source sends 3 volts to the (+) input of U1A, and when Vout = 0 volt, the output-dependent voltage source switches to a lower voltage and sends 1 volt to the (+) input. *Thus the voltage at the (+) input is always > 0 volt.*

Initially, capacitor C1 is not charged and is sitting at 0 volt coupled to the (–) input. Because the (+) input is greater than 0 volt regardless of the output voltage, the output has to be +V. Having the output at +V is good because, via R1, capacitor C1 now charges up, and the voltage across C1 increases with time. Eventually, the voltage across C1 charges to a voltage that exceeds 3 volts (V1). When this happens, all of the sudden, the output-dependent voltage source switches from V1 (3 volts) to the lower-voltage V2, which is 1 volt. With the capacitor voltage at about 3 volts into the (–) input terminal and the (+) input terminal switched or latched to 1 volt, the output of the comparator has to switch off and deliver 0 volt at the output terminal. At this point, the output at pin 1 of U1A is grounded.

Now R1 will discharge the 3 volts from the capacitor until C1's voltage drops to slightly below 1 volt, which then causes the (+) input terminal at 1 volt to be greater in voltage than the (–) input terminal. This then results in the output going to a high level of +V, and two things happen: the voltage-dependent voltage source switches to 3 volts, and C1 starts charging up again from 1 volt towards 3 volts via R1 connected to the output of U1A. This cycle of charging and discharging C1 between 1 volt and 3 volts repeats itself and forms an oscillating waveform similar to a sawtooth across C1. The manner in which the output-dependent voltage source provides its signal to the (+) input terminal is akin to "moving the goal posts" as to what the threshold or switching voltages are. Thus the voltage across C1 is being compared with a time-varying voltage, which is responsible for making the circuit an oscillator. We will return to relaxation oscillators in more detail later, but now let's look at some other oscillator circuits.

Phase-Shift Oscillators

One type of phase-shift oscillator includes an inverting-gain amplifier with either multiple high-pass or low-pass filters that use a resistor and capacitor for each phase-shift section. Each phase-shift section of a high-pass filter provides up to +90 degrees of leading phase shift, and each section of a low-pass filter provides up to –90 degrees of lagging phase shift. Note that when we talk about phase shift, we are applying the phase shift to a sinusoidal signal.

Although, in theory, two sections of a resistor-capacitor filter can provide up to 180 degrees of phase shift, the amplitude response at the 180-degree phase-shift frequency is generally so low that an extremely high-gain amplifier is needed. Also, the lagging phase shift inherent in the amplifier may cancel some of the leading phase shift in a phase-shift oscillator circuit using two high-pass filter sections. This will cause the total phase shift to be less than +180 degrees and prevents oscillation.

To allow for reliable oscillation, usually three or more sections of phase shift are used. The most common phase-shift oscillators involve three sections of resistor-capacitor (RC) high-pass or low-pass filters (see Figure 9-4).

By using a series of high-pass filters via C1 R1, C2 R2, and C3 R3, a phase shift of up to 3 × +90 degrees = +270 degrees is possible from the output of the inverting amplifier with gain –A connected to the C1 input, which is connected to R3. The positive or leading phase shifts will occur at frequencies below the cut-off frequency of the high-pass filter. For example, if we take a one section filter comprised of C1 and R1 and design it for a high-pass filter at 1 kHz, the filter will pass frequencies above 1 kHz (e.g., 2 kHz) but attenuate signals whose frequencies are below 1 kHz (e.g., 500 Hz). These attenuated signals below 1 kHz will exhibit a positive phase shift. With the three-section high-pass filtering that provides a leading phase shift from the output of the amplifier to the input, there will exist an *oscillation frequency below 1 kHz* such that a phase shift of +180 degrees is provided. For simplicity, some designs have C1 = C2 = C3 and R1 = R2 = R3. A variable-frequency oscillator can be made with a three-ganged variable resistor for R1 to R3, but each section should have a series resistor so that the R1 to R3 cannot go to 0 Ω. For example, a 100 kΩ three-section variable resistor with a 1 kΩ series resistor for each of the three sections provides a frequency adjustment of about 100:1.

Also shown in Figure 9-4 is a phase-shift oscillator with phase lag (negative phase shift) using three RC low-pass filters, R10 C10, R20 C20, and R30 C30, to provide a maximum phase shift of 3 × –90 degrees = –270 degrees. If we again set the low-pass filter to a cut-off frequency of 1 kHz, signals with frequencies below 1 kHz

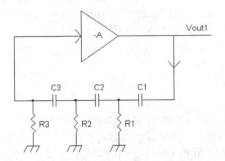

Phase-Shift Oscillator with Leading Phase Shift

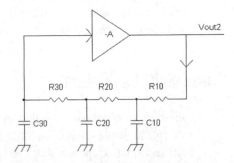

Phase-Shift Oscillator with Lagging Phase Shift

Figure 9-4 Phase oscillator system using three high-pass and three low-pass filters to form leading and lagging phase shifts.

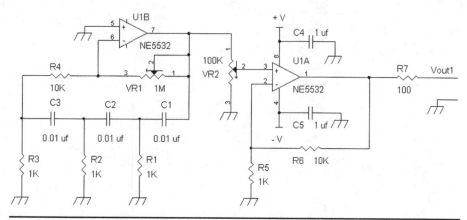

FIGURE 9-5   A phase-shift oscillator using an op amp and three-section high-pass filter.

(e.g., 500 Hz) will pass relatively unattenuated from R10 to C30, but signals that have frequencies above 1 kHz (e.g., 2 kHz) will be attenuated in amplitude and have a phase lag. The *oscillation frequency then has to be above 1 kHz* in this example to provide the necessary –180 degrees of phase shift. For a practical example of a phase-shift oscillator, see Figure 9-5.

This oscillator uses equal values for C1–C3 at 0.01 μF and 1 kΩ resistors for R1–R3, and the power supply can be anywhere from ±9 volts to ±15 volts. An inverting gain of about 43 is set by VR1 at 430 kΩ. Recall that the gain in this circuit is about –VR1/R4 = –430 kΩ/10 kΩ = 43. VR1 may be replaced with a 510 kΩ resistor with 5 percent tolerance to ensure oscillation.

The oscillation frequency was measured at about 6 kHz. Thus, at 6 kHz, the high-pass filter provides a phase shift of about +180 degrees. The approximate cut-off frequency is based on the formula for cut-off frequency of a one-section high-pass filter being $1/(2\pi R1C1)$ and is about 16 kHz, which means that the measured oscillation frequency is less than half the calculated cut-off frequency.

To change the oscillation frequency, one can replace all the capacitors C1–C3. Based on the frequency of oscillation measured at 6 kHz for 0.01 μF, the approximate frequency is then:

$$f_{osc\_HPF} = 6\ kHz \times 0.01\ \mu F/C1$$

For example, if C1 = 0.1 μF = C2 = C3, then:

$$f_{osc\_HPF} = 6\ kHz \times 0.01\ \mu F/0.1\ \mu F = 6\ kHz \times 1/10 = 600\ Hz$$

One can also vary the oscillation frequency by changing resistors R1, R2, and R3. For example, if their values are increased, the oscillation frequency should be lowered.

Note that, in practice, even after adjusting VR1 for the lowest sine-wave distortion, the output waveform is clipped or slightly distorted [approximately 3 percent total harmonic distortion (THD)]. Because there is no automatic gain control (AGC)

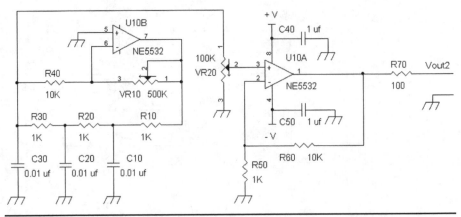

FIGURE 9-6 A phase-shift oscillator using a multiple-section low-pass filter.

mechanism in this oscillator, the sine-wave output signal has to go into some type of limiting that distorts the waveform. The output of the oscillator at U1B is connected to level control VR2, which is connected to a gain of 11 amplifier U1A.

Now let's take a look at a phase-shift oscillator using low-pass filtering (see Figure 9-6).

In Figure 9-6, we turn the high-pass filter circuit from Figure 9-5 into a three-section low-pass filter comprised of R10 C10, R20 C20, and R30 C30. Power-supply voltages may be from ±9 volts to ±15 volts. The gain was set to about –23 with VR10 set to about 220 kΩ for a gain of –(VR10/R40) = –(220 kΩ/10 kΩ) = –23. Note a 220 kΩ to 270 kΩ resistor may be substituted for VR10.

Again, the cut-off frequency based on R10 and C10 is calculated as $1/(2\pi R10C10)$ = 16 kHz. However, we will expect the oscillation frequency to be maybe twice the cut-off frequency. This guess is based on the high-pass-filter phase-shift oscillation frequency being less than half the calculated frequency. The oscillation frequency was measured at about 31 kHz, which is about twice the calculated cut-off frequency. Thus the low-pass filter network comprised of R10 C10, R20 C20, and R30 C30 provided a phase shift of about –180 degrees at 31 kHz.

We can then figure out how to scale capacitors C10, C20, and C30 to provide a new oscillation frequency. Based on what was measured, the frequency of oscillation is given by:

$$f_{osc\_LPF} = 31 \text{ kHz} \times 0.01 \text{ μF}/C10$$

For example, if C10 = 0.1 μF = C20 = C30, then:

$$f_{osc\_LPF} = 31 \text{ kHz} \times 0.01 \text{ μF}/0.1 \text{ μF} = 31 \text{ kHz} \times 1/10 = 3.1 \text{ kHz}$$

Similarly, we can change the resistance values to change the oscillation frequency. By increasing the resistor values for R10, R20, and R30, we would expect a lowering of the oscillation frequency.

We know that the output of U10B will deliver a clipped sine wave, so it would be advantageous to take the oscillator's signal after the low-pass filter section that reduces harmonic distortion signals. Thus we see that VR20 is connected to C30 instead of pin 7 of U10B. The output amplifier U10A provides a clean, nonclipped sine wave (<2 percent THD)—well, at least it's cleaner than the clipped waveform from Figure 9-5, U1A's output.

NOTE   For the phase-shift oscillators in Figures 9-5 and 9-6, the choice of the op amp is important concerning the highest-frequency oscillation. An NE5532 was chosen because it is about 10 times faster than an LM1458 or LM741. The NE5532 with a 10 MHz gain bandwidth product (GBWP) allows it to oscillate with good performance for typical audio test frequencies up to 20 kHz. Also, an LM4562 with a 55 MHz GBWP should provide about 5 times higher in oscillation frequency than the NE5532.

Now we turn to using a nonlinear resistor such as a light bulb to provide automatic gain control of an oscillator circuit so that the output sine-wave signal, without filtering, is not clipped.

Wein Bridge Oscillator with Automatic Gain Control to Provide Non-Clipped Sine Waves

A basic sine-wave oscillator includes a non-inverting-gain amplifier where one of the feedback resistors changes value such that the gain is lowered when the output voltage is increased (see Figure 9-7).

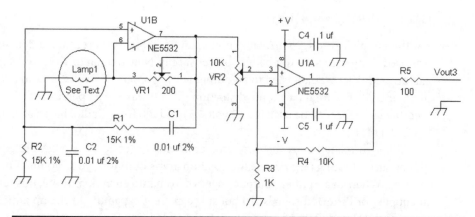

Figure 9-7   A Wein oscillator circuit using a low-current lamp to regulate the output amplitude.

The oscillator section is implemented by U1B, and its output signal is fed to VR2, a level control for output amplifier U1A. The output of U1A then provides a sine-wave signal by adjusting VR1 for an output of about 2 volts peak to peak at pin 1 of U1A. Lamp1 is rated at 6 volts, 25 mA and has RadioShack part number 272-1140. The power-supply voltages can be ± 9 volts to ±15 volts.

This oscillator includes a non-inverting-gain amplifier, U1B, whose gain = [1+ (VR1/RLamp1)]. A positive-feedback network is formed by C1, R1, C2, and R2. If we look at this network intuitively, *a very crude approximation* of it would be that C2 forms a low-pass filter with R1 and R2 forms a high-pass filter with C1.

If R1 = R2 and C1 = C2, then we will find after a bunch of messy calculations that this network will provide 0 degrees of phase shift at a resonant frequency:

$$f_r = 1/(2\pi R1 C1)$$

Given for this example that R1 = 10 kΩ and C1 = 0.01 μF, thus:

$$f_r = 1/(2\pi \times 10 \text{ k}\Omega \times 0.01 \text{ μF}) \approx 1.6 \text{ kHz}$$

In terms of providing a variable-frequency Wein oscillator, there is a way to implement this. One can replace R1 and R2 with a twin-gang variable resistor such as a stereo potentiometer. To avoid having 0 Ω for R1 and R2, a small-valued series resistor such as a 1 kΩ resistor can be connected in series with each of the two poten-tiometers.

And at the resonant frequency, there is a loss in signal of 1/3 from the output of U1B to its (+) input terminal. For example, if there are 2 volts of AC signal at the output of U1B at pin 7, the AC signal at the (+) input of U1B pin 5 has an amplitude of (2 volts)/3 = 0.667 volt.

Because we require a loop gain of greater than 1 to maintain a reliable oscillation, the non-inverting-gain amplifier, whose gain is [1+ (VR1/R$_{Lamp1}$)] must be greater than 3. That is:

$$(1/3)[1 + (VR1/RLamp1)] > 1$$

For this circuit, a 6 volt, 25 mA lamp bought at RadioShack (part number 272-1140) was used along with a ±9 volt battery power supply. Note that up to a ±15 volt supply can be used. VR1 was adjusted to an output level of about 1.94 volts peak to peak at pin 7 of U1B. The total harmonic distortion (THD) was < 0.03 percent (< –70 dB). Another lamp that can be used is a 12 volt, 25 mA bulb from RadioShack (part number 272-1141).

Although many lamps that are rated at 25 mA or less can be used, bulbs with higher current ratings may cause the 5532 op amp's output stage to go into current limiting. When an amplifier's output stage is driven into current limiting, this results in clipping or distorted signals. To boost the current capability of the op amp, two transistors, Q1 and Q2, and a resistor, R6, are added (see Figure 9-8).

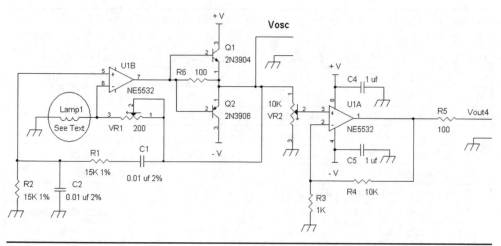

Figure 9-8 A Wein oscillator with a current-boosting output stage.

NPN transistor Q1 provides increase current to be sourced in the positive direction, while PNP transistor Q2 allows for more current to be sunk in the negative direction. Resistor R6 passes the signal from the op amp's pin 7 for small amounts of current up to about 7 mA into VR1 and Lamp1. If Lamp1 draws more than 7 mA in positive and negative directions, resistor R6 at 100 Ω develops 0.7 volt across the base-emitter junctions of Q1 and Q2 to turn them on and dump extra current into Lamp1. Note that 7 mA × 100 Ω = 0.7 volt.

Transistors Q1 and Q2 provide the extra current via the +V and –V power supplies connected to their collector terminals. The NE5532 has about 30 mA of output current before its current-limiting circuit is enabled. This means that there are at least 30 mA – 7 mA = 23 mA available to drive the bases of Q1 and Q2. The boosted current capability is then about β(23 mA), where β is the current gain of the transistors. Typically, we can assume that β =10 at "high" currents for the 2N3904 and 2N3906 transistors. So we now have about 230 mA of output current available instead of 30 mA. With this extra current capability, we can now use higher-current light bulbs that are more common.

Output voltages at V_{osc}, pin 1 of VR2, are in peak-to-peak (p-p) voltages as adjusted by VR1. The output voltages may be adjusted to higher or lower sine-wave amplitudes. The tests were made with ±9 volts for the power supply, and other supply voltages may be used, such as ±12 volts or ±15 volts. See Table 9-1 and note that p-p voltage denotes peak to peak voltage.

TABLE 9-1   Results of Various Lamps for the Wein Oscillator Circuit in Figure 9-8

Lamp1 (volts, mA)	Output V$_{osc}$ p-p (Voltage)	Distortion (THD)	Notes
47 (6.3 volts, 150 mA)	7.20 volts	<0.070%	Very stable output
49 (2.0 volts, 60 mA)	5.70 volts	<0.100%	Glows dimly
50 (7.5 volts, 220 mA)	4.72 volts	~ 0.100%	Glows red
327 (28 volts, 40 mA)	8.90 volts	<0.020%	Very stable output
428 (12.5 volts, 250 mA)	8.80 volts	<0.050%	Small modulation
756 (14 volts, 80 mA)	10.80 volts	~ 0.050%	Glows red
1487 (14 volts, 200 mA)	4.98 volts	~ 0.030%	Glows red
1828 (37.5 volts, 50 mA)	9.00 volts	~0.015%	10 seconds to settle
1847 (6.3 volts, 150 mA)	2.16 volts	~0.030%	Very stable output
1891 (14 volts, 240 mA)	9.50 volts	<0.05%	Glows

A Universal Wein Oscillator Circuit

Using 2N3904 and 2N3906 transistors is fine for driving up to about 100 mA, but beyond the 200 mA of current that they are rated for, the current gain of the transistors falls off sharply to a much lower current gain. A fix to increase the output-current capability is to simply use very high-power transistors such as the TIP41 (NPN) and the TIP32 (PNP). *These power transistors have a pin-out of BCE instead of the EBC used in the 2N3904/2N3906 transistors* (Figure 9-9).

The TIP32 transistor is a 3 amp device, whereas the TIP41 handles 6 amps of collector current. In terms of current gain β, with 23 mA into the base, the mini-

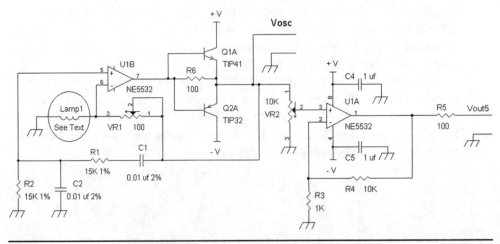

FIGURE 9-9   A universal Wein oscillator circuit with power transistors Q1A and Q2A.

TABLE 9-2 Universal Wein Oscillator Circuit in Figure 9-9 with Higher-Current PR-2 and PR-4 Flashlight and #112 and #222 Penlight Bulbs

Lamp1 (Volts, mA)	Output V$_{osc}$ p-p (Voltage)	Distortion (THD)	Notes
14, 2.47 volts, 300 mA	6.40 volts	<0.070%	Glows orange
44, 6.3 volts, 250 mA	7.90 volts	<0.030%	Glows
112, 1.2 volts, 220 mA	4.72 volts	~0.100%	Glows
222, 2.25 volts, 250 mA	3.76 volts	<0.100%	Glows
PR-2, 2.38 volts, 500 mA	2.57 volts	~0.150%	Stable output
PR-4, 2.33 volts, 270 mA	3.28 volts	<0.700%	Stable output

mum β is about 40 to deliver about 800 mA of current into the Lamp1. Other power transistors can be used, such as the MJE3055 or TIP3055 (NPN) and the MJE2955 or TIP2955 (PNP). See the results in Table 9-2 with higher-current lamps where the supply voltages are ±9 volts to the oscillator.

In terms of the notes for Tables 9-1 and 9-2, all bulbs produced stable amplitude waveforms except the #428 lamp. The #428 bulb had some small signs of "breathing," or very low-frequency amplitude modulation of the sine wave.

Relaxation Oscillators

These types of oscillators are often used for voltage-controlled oscillators at low frequencies and are also commonly the heart of analog function generators. Although the oscillation frequencies are generally below a few MHz (megahertz), very high-frequency versions can oscillate in the tens of MHz (megahertz). Let's examine them now. Figure 9-10 shows a very low-frequency oscillator.

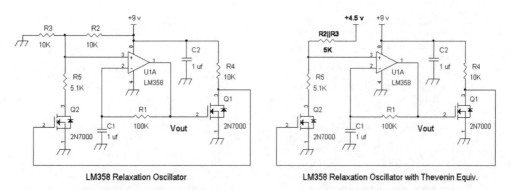

LM358 Relaxation Oscillator

LM358 Relaxation Oscillator with Thevenin Equiv.

FIGURE 9-10 A relaxation oscillator circuit using an LM358 op amp (*left*) and a circuit using a Thevenin equivalent with R2‖R3 and +4.5v (*right*).

In this example, U1A is an op amp but functions as a comparator. In Figure 9-10 (*left*) we have an RC low-pass filter R1 and C1 connected from the output terminal of U1A, pin 1. The voltage across C1 is connected to the (–) input terminal of the op amp. Also connected to pin 1 is a logic inverter circuit Q1 and R4. The output of the op amp will be a rectangular pulse waveform, and thus relaxation oscillators are not sine-wave generators like the previous circuits. The output of inverter Q1 via its drain terminal is connected to the gate of Q2. Note that Q1 and Q2 may be 2N7000, VN2222, or VN10K. As mentioned previously, a relaxation oscillator includes an output-dependent voltage source that provides two different voltages to the (+) terminal of the LM358.

In Figure 9-10 (*right*), the voltage-divider circuit R2 and R3 with a 9 volt source is simplified with a *Thevenin equivalent circuit*, and the voltage-divider circuit is expressed as a 4.5 volt source driving a Thevenin resistance of R2||R3 = 5 kΩ.

NOTE   Any two-resistor voltage-divider circuit can be expressed by a voltage source that represents the divided voltage and a series resistor whose value is the parallel resistances of the two resistors. This new circuit is commonly called a Thevenin equivalent circuit.

We now turn to Figure 9-11, which shows the two states of Q2, which are "off," an open circuit, and "on," a short to ground. Pertaining to R5, Q2 then acts as an open circuit or as a switch to ground.

Initially, capacitor C1 is sitting at 0 volts. So, regardless of whether Q2 is on or off, the voltage at the (+) terminal of U1A will be > 0 volt. We know this because when Q2 is off, the voltage at the (+) terminal is 4.5 volts, and if Q2 is on, the voltage would be roughly 2.5 volts via a voltage-divider circuit from the 4.5 volt feeding R2||R3 and R5. Because R2||R3 is about the same resistance as R5, we get roughly half the 4.5 volts into the (+) terminal or pin 3 of U1A.

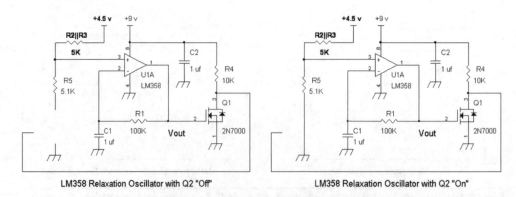

Figure 9-11   Relaxation oscillator showing the two states of Q2.

With the voltage at the (+) terminal greater than the voltage initially at the (−) terminal of the comparator, the output has to be a logic high voltage, which is typically close to 9 volts. The output voltage then charges up C1 via R1. The logic high voltage from the output Vout is also connected to Q1's gate. This produces a logic low into the gate of Q2 (see Figure 9-10), which confirms that Q2 is off, as seen on the left side of Figure 9-11, and that there is 4.5 volts connected to the (+) input terminal of U1A.

The logic high voltage from pin 1 of U1A will then continue to charge up C1 via R1 until C1's voltage into the (−) input terminal exceeds 4.5 volts. When this happens, the output of U1A goes low and sends 0 volt to the gate of inverter Q1, which then outputs a logic high voltage to the drain of Q1 via R4. Because the gate of Q2 is connected to the drain of Q1, Q2 then turns on and shorts or connects R5 to ground (see Figure 9-11, *right*).

Once R5 is connected to ground as shown, a voltage divider is set up by R2||R3 and R5, and the voltage to the (+) input terminal is changed from 4.5 volts to roughly half that, or 2.25 volts, to the (+) input terminal of U1A. Because the C1's voltage is around 4.5 volts at the (−) input terminal, which is greater than 2.25 volts at the (+) input terminal, Vout goes to 0 or ground, which starts C1 to discharge. Eventually, C1's voltage will decrease to just below 2.25 volts into the (−) input terminal with 2.25 volts still at the (+) input terminal. Once this happens, Vout goes back to a logic high state and starts charging up C1, while Q2 reverts to an open-circuit state. With Q2 at an open-circuit state, the voltage into the (+) input terminal changes from 2.25 volts back to 4.5 volts.

The cycle will just repeat itself, and a sawtooth waveform across C1 will be formed, where the positive and negative peaks of the sawtooth are at 4.5 volts and 2.25 volts. The output, Vout, will provide a periodic pulsed waveform with a frequency of about 7.9 Hz given that C1 = 1 µF (film capacitor with 5 percent or lower tolerance) and R1 = 100 kΩ.

NOTE Normally, film capacitors with 5 percent tolerance or better are used in RC oscillators. Even for large values at 1 µF or more, it is better to use film (e.g., polyester, Mylar, polycarbonate, etc.) than to use electrolytic capacitors that typically have tolerances of +80 percent and −20 percent. If the frequency or pulse width does not need to be accurate, electrolytic capacitors can be used. But, as a reminder, note the polarity. With positive pulses at the output, usually the negative side of the electrolytic capacitor is grounded.

Now that we have looked at one example of a relaxation oscillator, let's look at another. Figure 9-12 uses an LM393 comparator to allow for higher oscillation frequencies over the LM358 op amp version. Again, Q1 and Q2 may be 2N7000, VN2222, or VN10K.

We basically have the same circuit as Figure 9-10 (*left*) but with a "real" comparator to replace the slow-moving LM358. Many comparators have an open collec-

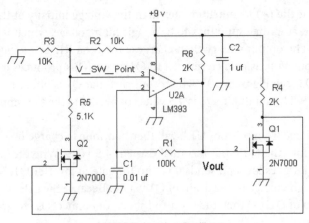

FIGURE 9-12   An LM393 relaxation oscillator.

tor output stage, which requires a pull-up resistor (R6). The pull-up resistor's value at 2 kΩ is chosen as a compromise of speed and power dissipation. Too low of a value (e.g., 50 Ω to 500 Ω) in the pull-up resistor (R6) can damage the output transistor, whereas too high a value for R6 will cause a slowness in response, even though the power dissipation into the output stage of the LM393 will be lower. C1 is about 100 times lower in this schematic to show that this circuit can oscillate at a higher frequency than the LM358 version. The switch-point voltages are still the same here at 4.5 volts and approximately 2.25 volts.

Another way to design a relaxation oscillator is to include a hysteresis resistor (R7) to generate the two switch-point voltages into the (+) input terminal of the comparator (see Figure 9-13).

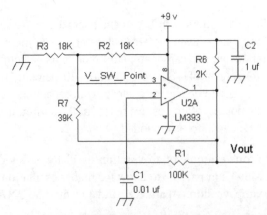

FIGURE 9-13   A simplified relaxation oscillator circuit that does not require Q1 and Q2.

In this oscillator, the current through R7 is different when Vout is a logic high and a logic low. We still have the same RC circuit R1 and C1 to charge and discharge the capacitor for providing a voltage at the (–) input terminal of the comparator. When the output is logic high at +9 volts, we have about 93 uA *flowing into* the voltage-divider circuit R2 and R3, which adds 0.843 volt to the 4.5 volts to provide 5.34 volts (4.5 volts + 0.834 volt) to the (+) input. When the output is low, we subtract 0.834 volt from the 4.5 volts = 3.65 volts because we have 93 μA *flowing out* of the voltage-divider circuit. Thus, in Figure 9-13, there is a sawtooth waveform across C1 that is bounded by 5.34 volts and 3.65 volts. To vary the oscillation frequency, R1 can be replaced with a potentiometer (e.g., 500 kΩ) or C1 can be changed to another value.

So far, we have seen a few different types of relaxation oscillators. One should note that these relaxation oscillators will change oscillation frequency if the power supply voltage varies. Thus it is recommended that these circuits (e.g., Figures 9-10 through 9-13) be connected to a regulated power supply (e.g., +9 volts or +12 volts).

Just for Fun, Build a 555 Timer with a Dual Comparator, FET, and a Flip-Flop

The following is more of an exercise to lift the block diagram of the 555 timer and implement it with other parts. We will show how to make an astable or free-running oscillator. This circuit is more for the fun of building it. Figure 9-14 shows the block diagram.

The 555 timer circuit uses two comparators instead of one, as seen in previous circuits. To operate as an oscillator, the top comparator (U1A) is fed two-thirds of

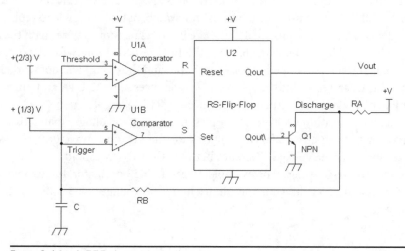

Figure 9-14   A 555 timer block diagram.

the supply voltage +V at its (–) input terminal, whereas the bottom comparator (U1B) is fed a one-third supply-voltage reference.

Initially, capacitor C is at 0 volt, and the output of U1B is logic high because there is 1/3 the supply voltage at the (+) terminal of U1B. However, the output of top comparator U1A is logic low because the +(2/3)V voltage at the (–) input terminal is greater than the initial 0 volt across the capacitor C, which is connected to the (+) terminal of U1A. So we are safe in that we do not want the RS flip-flop circuit to have a logic high at both R and S inputs. With U1B outputting a logic high voltage to the Set or S input of RS flip-flop U2, the Qout\ output is low and Q1 is off. Qout\ is the inverted logic output of Qout. The Qout\ output of the flip-flop will stay low even if the voltage drops to a logic low into the S input.

The flip-flop has a memory, and Qout\ can only change states when there is a low- to high-level signal at the R input. Therefore, via V+ connected to RA and RB in series, capacitor C is charging, and over time, the voltage increases such that eventually the capacitor voltage slightly exceeds (2/3) of the supply voltage, and the top comparator outputs a high logic level to the flip-flop circuit that causes Qout\ to go to a high logic level and turn on Q1 to start discharging C via RB. When the voltage of C drops slightly below (1/3)+V, the bottom comparator outputs a logic high voltage to the S input of U2, and Q1 shuts off and becomes an open circuit to allow C to start a charging process again via +V, RA, and RB. As a result of the flip-flop latching on and off as the capacitor voltage charges and discharges between two-thirds and one-third of the power-supply voltage, an oscillation is provided. The output voltage is taken from the Qout terminal that provides a periodic pulsed waveform. For a circuit implementation, see Figure 9-15.

For this circuit, the dual comparators are in a LM393, and the flip-flop circuit is implemented by cross-connected NOR gates U2C and U2D. U2 can be of a 74HC02, 74AHC02, or 74AC02 NOR gate. The supply is regulated via the 7805T chip to provide 5 volts to U1 and U2. Q1 may be a 2N7000, VN2222, or VN10K MOSFET.

A MOSFET (Q1) is used for the transistor switch, but it is possible to replace Q1 with a 2N3904 transistor, provided that there is a 1 kΩ resistor in series with the base so that pin 13 of U2 is driving the base via the 1 kΩ resistor. Otherwise, the output at pin 13 of U2D will be clamped to less than 1 volt, and the circuit will not work. Note the two-thirds and one-third supply-voltage references at pin 2 U1A and pin 5 U1B are implemented via three equal-valued resistors R1, R2, and R3.

In all the relaxation oscillator examples, there are two voltages that activate the comparator to form a boundary of voltages for the sawtooth waveform across the timing capacitor. The voltage difference between the two (threshold) voltages is also called the *hysteresis voltage*. For example, in the LM358 oscillator, the two voltages are 4.5 volts and 2.25 volts, which result in a hysteresis voltage of 4.5 volts – 2.25 volts = 2.25 volts.

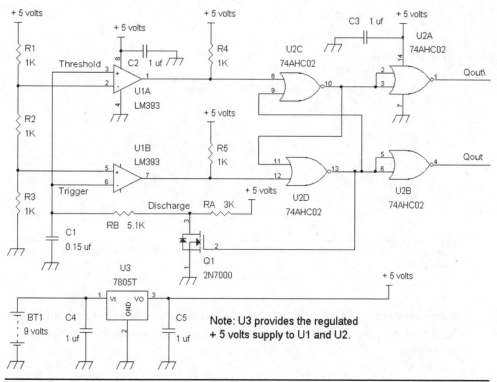

FIGURE 9-15   An equivalent 555 oscillator circuit.

To Wrap Up Relaxation Oscillators, Just Use a 74HC14 or 74AC14 Chip

After all the relaxation oscillators presented, probably the easiest circuit to use is an inverter logic gate that has a hysteresis input stage. The hysteresis input stage has two different voltages for turning on and off the gate and is equivalent to a comparator with a voltage-dependent voltage source that changes the voltage thresholds of turning on and off. For a 74HC14 gate, for example, with a 4.5 volt supply, the positive-going threshold voltage is typically 2.7 volts, and the negative-going threshold voltage is 1.8 volts, which leads to a hysteresis voltage of:

2.7 volts −1.8 volts = 0.9 volt

See Figure 9-16. The 74HC14 oscillator is capable of generating signals whose frequencies are in the MHz. If higher frequencies are desired, a 74AC14 gate can be used instead to provide an oscillation frequency well into the tens of MHz. Also, using an HC or AC 7400 series logic device such as a 74HC14 or 74AC14 produces a waveform that is nearly a 50 percent duty-cycle squarewave. Note that these chips

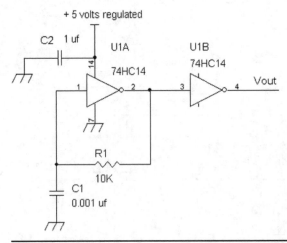

FIGURE 9-16   A minimum parts relaxation oscillator that is capable of high-frequency generation.

have a maximum voltage rating of about 6 volts and should be connected to a regulated power supply to ensure safe operation and for frequency stability of the oscillation signal.

Radio-Frequency (RF) Oscillators

For both radio transmitters and receivers, oscillators play an important role. They provide signals that are stable in frequency for transmitting and variable frequencies for tuning across the band. Although these days many of the "analog" oscillator signals are replaced with frequency synthesizer chips such as the Analog Devices AD9851, RF oscillators are still being used in low-power applications and for many do-it-yourself (DIY) hobbyist radio circuits. We will touch on some basic RF oscillator circuits here, and other types will be addressed in Chapters 10 and 11.

We will start with a Colpitts oscillator, a zero-degrees phase-shift oscillator, which is often used in fixed- and variable-frequency oscillators. Let's first take a look at a block diagram of a basic zero-phase-shift oscillator (see Figure 9-17).

Many RF oscillators use a parallel LC (inductor-capacitor) bandpass filter, where at the resonant frequency, the phase shift is 0 degrees and at below the resonant frequency there is a positive phase shift and at above the resonant frequency, there is a negative phase shift. Maximum output of the bandpass filter is at the resonant frequency. In Figure 9-17, an amplifier with noninverting gain is fed to the zero-degrees phase-shift filter. Should the amplifier have some phase lag due to fall-off in high-frequency response, an extra phase lag will be incurred in this oscillating system. So, to ensure that oscillation occurs, the oscillator will "magically" move the frequency

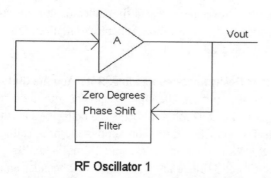

RF Oscillator 1

FIGURE **9-17** An RF oscillator having a zero-phase-shift positive-feedback network (e.g., parallel inductor-capacitor bandpass filter).

of oscillation slightly lower than the resonant frequency because the phase-shift filter has a phase lead at frequencies below the resonant frequency. This phase lead will then cancel the phase lag from the amplifier to ensure that there is always a 0-degree phase shift in the system. And this explains why an oscillator with slower transistors may oscillate at a lower frequency than one with higher-frequency transistors.

Now let's look at the example circuit shown in Figure 9-18. A Colpitts oscillator generally uses a capacitive voltage divider to provide a 0-degree phase signal into the input of the amplifier. In this case, the emitter serves as the input terminal. Normally, we would think of the base of a transistor as the input terminal or point. However, in this case, the base of Q1 is grounded AC-wise via C4, which then makes Q1 a grounded base or common-base amplifier. The use of C1 and C2 is to form an impedance-matching network that steps down the signal across L1 and provides a

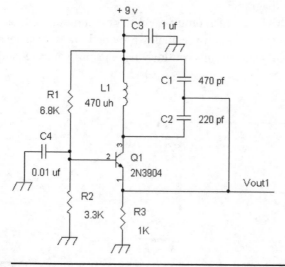

FIGURE **9-18** A Colpitts oscillator with a capacitive voltage divider formed by C1 and C2.

low-impedance drive to the emitter. R3 generally sets the collector current sufficiently high to provide gain in the system to sustain a reliable oscillation. One can experiment with changing the values of R3.

> **NOTE** **Math Stuff (Not Really Required Reading, But if You Are Curious):** In common-base amplifiers, the input resistance into the emitter is approximately $1/g_m$, where $g_m = ICQ/0.026$ volt and ICQ is the collector current. In this example, via voltage divider R1 and R2 and a 9 volt supply, the base voltage is about 3 volts, and with a 0.7 volt VBE, there is about 2.3 volts DC at the emitter. The emitter current is then 2.3 volts/R3 = 2.3 volts/1 kΩ = 2.3 mA $\approx$ ICQ because emitter and collector currents are essentially the same in this case. The input resistance into the emitter is then about 11 Ω = 0.026 volt/2.3 mA.

The oscillation frequency is given by $f_{osc} = 1/(2\pi\sqrt{L1C})$ where L1 is the 220 µH coil in Figure 9-18, and C = C1C2/(C1 + C2), the series capacitance of C1 and C2. For this circuit:

C = (470 pF × 220 pF)/(470 pF + 220 pF) = 150 pF

Thus:

$f_{osc} = 1/(2\pi\sqrt{220\ \mu H 150\ pF}) = 876$ kHz

The output amplitude can be controlled by changing R3 and/or by adding a resistor in parallel with L1 (e.g., 10 kΩ to 100 kΩ). Because of the bandpass filtering characteristics of L1, C1, and C2, the output waveform is sinusoidal, even though the collector current resembles a pulsed waveform.

Let's now move onto another oscillator that is used in AM and shortwave radios. Figure 9-19 shows the block diagram.

Typically, oscillators of this kind use a common-emitter or common-source configuration. The 180-degree phase shift is actually provided by a transformer. Oscillator coils made for solid-state circuits routinely have primary and secondary

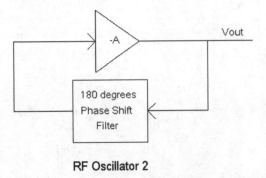

RF Oscillator 2

FIGURE 9-19   An oscillator with an inverting amplifier and a 180-degree phase shifter.

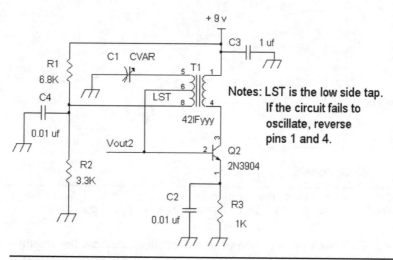

FIGURE 9-20 A typical AM or shortwave radio oscillator circuit using a transformer and common-emitter amplifier.

windings. The transformer itself includes inductance to resonate with a capacitor (Figure 9-20).

Transformer T1 allows a phase inversion to occur by the connections of the secondary winding, denoted as pins 1 and 4. Q2 serves as a common-emitter amplifier because C2 is low impedance at the oscillating frequency. R3 determines the amplitude of the sine-wave output, and its value can be in the range of 10 kΩ to 1 kΩ.

In this circuit, the variable capacitor has a range of about 10 pF to 60 pF. This variable capacitor with an oscillator coil (~440 µH), T1 a 42IF300 or 42IF110 (Xicon) or equivalent, will provide a frequency range of about 990 kHz to about 2.1 MHz, the frequency range of an AM radio's local oscillator. On the primary side of T1, pins 6 and 8 are the low-side tap of the primary winding. By *low-side tap*, we mean that the primary winding is not tapped symmetrically and that there is a lower-resistance winding between pins 6 and 8 than at the higher-side tap at pins 5 and 6. The low-side tap is required to load into the base of Q1, which has a medium input resistance (e.g., 300 to 3 kΩ).

Sometimes building one of these circuits with the transformer is tricky, and the circuit may not oscillate. If this happens, confirm the low-side tap or reverse pins 5 and 8 to see if the circuit then oscillates. Also try reversing pins 1 and 4. Pins 5 and 8 of T1 (~440 µH) form the inductance for variable-capacitor C1.

The Colpitts oscillator in Figure 9-18 and the one in Figure 9-20 are probably the most common types of oscillator circuits. But there are other types. Figure 9-21 shows a block diagram of a different type of oscillator.

The type of oscillator seen in Figure 9-21 uses an emitter-follower or source-follower amplifier. Oscillation occurs when there is a phase-shift lag in one circuit that is coupled to another circuit that provides a leading-phase shift. The net phase

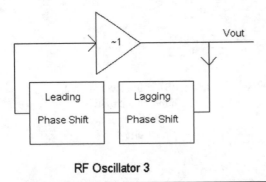

RF Oscillator 3

FIGURE 9-21 An oscillator with noninverting gain of approximately +1, where the phase-shift network is divided into two stages, a phase-lag and a phase-lead circuit.

shift cancels out from output to input of the amplifier. Because the amplifier itself has a gain approximately +1, we need to provide some type of voltage gain. The leading phase-shift circuit, a high-pass filter, has a peaky response that provides voltage gain at the resonant or oscillation frequency.

In some instances, this type of oscillator sometimes represents an amplifier going into a very high-frequency parasitic oscillation. The output of the amplifier is connected to a capacitive load, and the internal capacitance of the transistor between the base and emitter (or gate and source) is connected to wire or trace to form a small-valued inductor that is eventually AC grounded. The result is a *parasitic oscillation*.

In general, though, this type of configuration does provide reliable oscillation when used with inductors (see Figure 9-22).

Here is a circuit that oscillates at about 700 kHz. Q3 is an emitter follower that is biased at about 100 µA such that its output resistance at the emitter approximately $1/g_m \approx 260\ \Omega$, where $g_m = ICQ/0.026$ volt and ICQ is the collector current. The out-

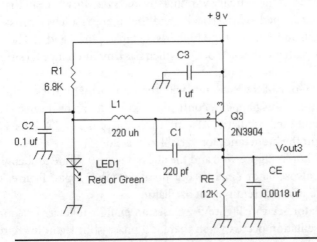

FIGURE 9-22 An emitter-follower oscillator.

put resistance at 260 Ω and CE = 1,800 pF forms a low-pass filter at $1/(2\pi\ 260\ \Omega$ 1,800 pF) = 340 kHz. A low-pass filter at 340 kHz provides a lagging phase shift at the frequency of oscillation, which is approximately ~ $1/(2\pi\sqrt{L1C1})$ = 723 kHz.

To determine a more exact oscillation frequency, CE must be part of the equation because it is in series with C1. CE and C1 in series is (220 pF × 1,800 pF)/(220 pF + 1800 pF) = 196 pF, which leads to an oscillation frequency of:

$$1/(2\pi\sqrt{220\ \mu H\ 196\ pF}) = 766\ kHz.$$

Again, the lagging circuit is formed by the output resistance of Q3 and CE. Capacitor C1 is then coupled to L1 to form a high-pass filter with a positive or leading phase. The net phase shift from the emitter of Q3 to its base is then 0 degree because the lagging and leading phase shifts cancel out.

But the main question is, if an emitter follower has a gain of less than 1, where does the extra gain come from? If we look carefully, L1 is AC grounded via C2, and thus Q3's emitter drives a series-resonant circuit C1 and L1. At resonance, the impedance of a series LC circuit is low. Thus there are huge amounts of current flowing through C1 and L1 at the resonant frequency, as provided via the emitter of Q3. The large AC current flowing through L1 steps up the voltage from the emitter of Q3. By stepping up the voltage from the emitter of Q3 to the base, we get a net gain of more than 1 in the system. So it turns out that C1 and L1 behave like a step-up transformer for signals at the resonant frequency.

Voltage reference diode LED1 provides a "stable" 1.8 volts DC into the base of Q3 via L1. And the emitter current is then 1.8 volts − VBE ≈ 1.2 volts across RE, given that the VBE turn-on voltage of Q3 is approximately 0.6 volt. The collector current ICQ ≈ 1.2 volts/RE = 1.2 volts/12 kΩ = 100 μA.

Finally, we turn our attention back to a common-emitter amplifier oscillator. This type of oscillator is commonly used also for the Pierce crystal oscillator, where the inductor is replaced by a crystal and the capacitors at the collector and base are adjusted to cause the crystal to allow for oscillation. Let's take a look at the block diagram first (see Figure 9-23).

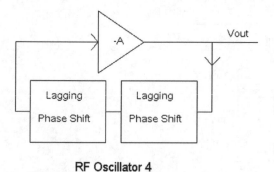

RF Oscillator 4

FIGURE 9-23 A two-stage phase-lagging network to provide oscillation.

Generally, an oscillator such as this one requires the equivalent of three stages of phase shifting in a lagging manner. It is very similar to one of the original phase oscillators mentioned earlier and shown in Figure 9-4, where there are three phase-shift low-pass-filter sections implemented by C10, R10, C20, R20, R30, and C30.

For this oscillator, the first section is usually an RC or one-section low-pass filter connected to the output of the inverting-gain amplifier. This provides close to –90 degrees of phase shift. The next section of low-pass filtering is set up by an LC (inductor-capacitor) low-pass filter. The LC low-pass filter is phase-wise equivalent to two sections RC filtering. At the resonant frequency, this LC low-pass filter provides –90 degrees of phase shift but also can provide up to –180 degrees at beyond the resonant frequency. In practice, this filter provides typically between –150 degrees and –90 degrees of phase shift. Now let's take a look at an example circuit (see Figure 9-24).

Transistor Q4 forms a common-emitter amplifier with C4, a 1 µF capacitor, to effectively AC ground the emitter of Q4. Note that choosing C4 too small, such as a 0.01 µF capacitor, will lead to "squegging," where C4 at 0.01 µF causes a low-frequency oscillation that partially mutes out the oscillating signal. It was found the 1 µF for C4 produced reliable oscillations without "squegging." However, if squegging (low-frequency chirping) occurs, C4 may be reduced to 0.1 µF.

The first stage of the approximately –90 degrees of phase shift is provided by C1 and R4, which produce a –3 dB cut-off frequency of about 48 kHz. The target oscillation frequency is about 750 kHz, so a roll-off from an RC filter at 48 kHz will

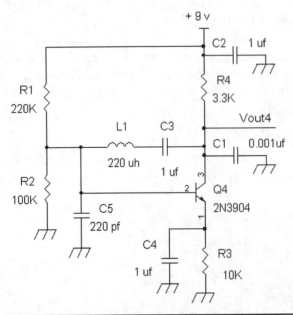

Figure 9-24 A common-emitter oscillator with multiple sections of lagging phase shift.

ensure approximately –90 degrees of phase shift. The next stage of phase shifting is done via L1 and C5. DC blocking capacitor C3 ensures that L1 does not short the collector to the base of Q4 DC-wise, which will upset the DC biasing. The resonant frequency is determined approximately by:

$$f_{osc} = 1/(2\pi\sqrt{L1C}) = 1/[2\pi\sqrt{(220 \text{ uh})(180 \text{ pF})}] \approx 799 \text{ kHz}$$

where C is the series capacitance of C1 and C5, which is:

$$(1,000 \text{ pF} \times 220 \text{ pF})/(1,000 \text{ pF} + 220 \text{ pF}) = 180 \text{ pF}$$

The output waveform is a distorted sine wave, as expected, because there is no AGC circuit here. About 2 volts peak to peak was measured at a frequency of approximately 790 kHz, which is pretty close to the calculation.

References

1. Ronald Quan, *Build Your Own Transistor Radios*. New York: McGraw-Hill, 2013.
2. *Linear Circuits Data Book*. Dallas, TX: Texas Instruments, 1984.
3. *CMOS Logic Databook*. Santa Clara, CA: National Semiconductor, 1998.
4. Paul R. Gray and Robert G. Meyer, *Analysis and Design of Analog Integrated Circuits, 3rd ed*. New York: Wiley, 1993.

PART THREE

Advanced Electronics

CHAPTER 10

Amplitude Modulation Signals and Circuits

In this chapter, I will try to keep the math to a minimum, but it is inevitable that it will creep in to describe exactly what's going on with amplitude-modulated (AM) signals. This chapter will cover some basics of amplitude modulation (AM). Actually, there are many types of amplitude modulated signals. In the standard broadcast or medium wave band from about 530 kHz to about 1.7 MHz, a carrier signal within that band is varied in strength. This variation in carrier strength makes the resulting signal an amplitude-modulated signal.

Other forms of amplitude modulation can be found in cell phones and digital television transmission. In cell phones, a pair of amplitude modulated signals is used in presenting the voice or data information in a signal that possesses both amplitude modulation and phase modulation. In digital television signals (DTV), there are two types. One is similar to standard broadcast AM signals (vestigial sideband AM), and the other is very much the same as cell phone amplitude-modulated (QAM or quadrature amplitude modulation) signals.

This chapter will cover various AM signals and some of the circuits used in AM radios, such as mixers, amplifiers, and AM demodulators. Instead of separately describing single circuits such as oscillators, mixers, or demodulators, they will be explained in the context of an actual radio system. Again, it is recommended that an oscilloscope and an RF (radio-frequency) signal generator (or a function generator with AM capability) be at hand for the projects and experiments.

AM Signals—What Are They? "Hey, Can You Please Turn Down the Volume ... Now Can You Turn It Back Up?"

In the case of standard AM radio signals, the best analogy is just volume variation. When we turn up the volume or loudness of music on our radio or stereo, we are in

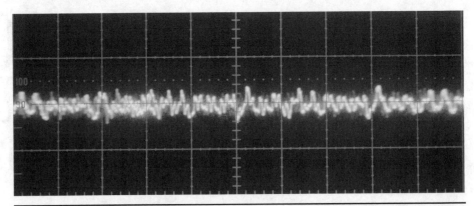

FIGURE 10-1   A music signal turned to low volume.

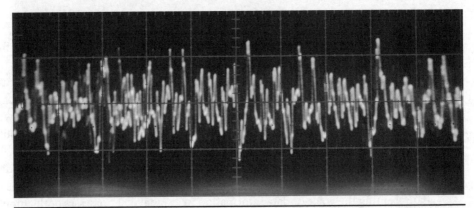

FIGURE 10-2   A music signal turned up in volume.

fact modulating the amplitude of the music. And when we turn it down, we are again modulating the music in terms of reducing the amplitude of the music signal (see Figures 10-1 and 10-2).

For standard AM signals, we "replace" the music with a single tone. Although we can use a volume control to vary the intensity of the single tone, that is limiting in the sense that it's hard to turn a knob that varies the resistance of the potentiometer fast enough to mimic voice or music. The limits of the volume control allow us to turn the music completely off or to a very loud volume, which depends on the amplifier's output voltage rating.

In the earlier days of telephone and radio, carbon microphones served the purpose of a variable resistor that could vary the intensity of an electric signal for voice and music. Later on, it was found that a voltage or current controlled amplifier yielded the same results in terms of varying the signal strength (see Figure 10-3).

In the top diagram in Figure 10-3, the carbon microphone acts as a variable resistor in series with the antenna. Depending on the sound pressure (e.g., via voice

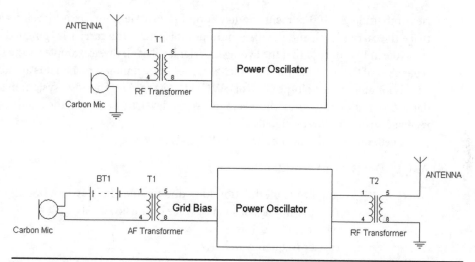

Figure 10-3   A block diagram of methods to modulate a signal via a carbon microphone and voltage-controlled amplifier.

or music) to the carbon microphone, the RF signal is varied. In the bottom diagram, the carbon microphone forms a varying direct-current (DC) signal with battery BT1 into the primary winding of T1. The secondary winding of T1 provides an alternating-current (AC) signal to vary the grid bias voltage to a power oscillator to vary the oscillator's output signal. Figure 10-4 illustrates what an AM signal looks like for a single high-frequency carrier tone and a low-frequency modulating tone.

Zero percent modulation of an AM signal represents the carrier amplitude when silence is being transmitted. Normally, when we tune an AM station between music or voice passages, we can still sense that the carrier is there via a drop in the hiss level. When there is no station on the air or the radio is tuned to a frequency where no RF is being transmitted, usually hiss, or random noise, is heard.

At 50 percent modulation, the peak carrier level is one and a half times the unmodulated carrier level, and its minimum is 50 percent of the unmodulated car-

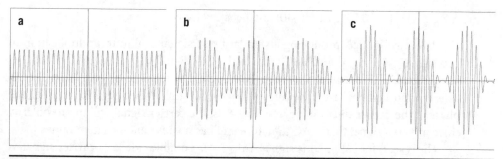

Figure 10-4   (*a*) An unmodulated AM signal or an AM signal at 0% modulation. (*b*) An AM signal at 50% modulation. (*c*) An AM signal at 100% modulation.

rier amplitude. At 100 percent modulation of the carrier, the carrier level peaks at twice the carrier level, and the minimum point is where the carrier is "pinched" off.

Note in Figure 10-4 that the average carrier level of all three examples is the same regardless of the modulation. This fact becomes important later when it is applied as a basis for automatic volume control (AVC) for AM radios. The AVC system in standard AM radios measures the average carrier level to even out the loudness of received strong and weak signals.

A general formula for broadcast AM signals is

$$AM_{DSBC}(t) = [1 + m(t)]cos(2\pi f_c t) \tag{10-1}$$

where AM_{DSBC} = amplitude modulation for a double sideband carrier (DSBC)
    $m(t)$ = modulating signal such as voice or music signals,
        where $|m(t)| \leq 1$ and
    f_c = carrier frequency in hertz

For example, radio station KCBS in the San Francisco Bay Area has a frequency of 740,000 Hz = f_c. And the modulating signal $m(t)$ is generally voice signals such that the modulation of the AM transmitter does not exceed 100 percent in the negative peaks.

There is a little quirk in United States AM radio broadcasting, though. Because music or voice is not a symmetrical waveform, the Federal Communications Commission (FCC) in the United States allows the positive peaks to go as high as 125 percent modulation.

In a specific case of a sinusoidal modulation signal, the AM signal can be described as:

$$AM_{DSBC}(t) = [1 + m(cos(2\pi f_{mod}t))] \; AM_{DSBC}(t) = [1 + m(cos2\pi f_{mod}t)] \; cos(2\pi f_c t) \tag{10-2}$$

where the modulation index is m, such as the preceding examples, where m = 100%, 50%, and 0%. The modulation frequency is = f_{mod}. For example, if a radio station at f_c = 1 MHz is modulating at m = 75% at a modulating frequency of 440 Hz= f_{mod}, the AM signal would be expressed as:

$$AM_{DSBC}(t) = [1 + 0.75(cos(2\pi(440Hz)t)] \; cos[2\pi(1MHz)t]$$

If you will notice, producing a standard AM signal is akin to multiplication of two signals, $[1 + m(t)]$ and $sin(2\pi f_c t)$. In the first signal, $[1 + m(t)]$, the added DC offset signal "1" is essential to ensure that the carrier signal never goes lower than 0. This is important because if the carrier signal is multiplied by a negative number, the phase of the carrier flips 180 degrees. [That is, the carrier signal will flip in polarity since m(t) is a signal that is AC in nature and has positive and negative values.]

The reason for calling this signal $AM_{DSBC}(t)$, a double sideband carrier signal, is because this signal contains three signals: a lower sideband (LSB), a carrier, and an

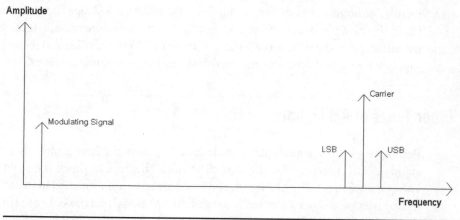

Figure 10-5   A spectrum of an AM signal modulated at 100%, (*from left to right*) LSB, carrier, and USB.

upper sideband signal (USB). When a spectrum analyzer is connected to an AM signal, the three signals appear (see Figure 10-5).

Figure 10-5 shows for convenience the spectrum of the modulating signal, which does not appear at the output of the AM RF signal. The only signals at the output are the LSB, carrier, and USB.

The frequency spacing is equal from the LSB and carrier and the carrier to the USB. For example, taking the preceding example for:

$$AM_{DSBC}(t) = [1 + 0.75(\cos(2\pi(440Hz)t)] \, [\cos(2\pi(1MHz)t]$$

the frequencies for the LSB, carrier, and USB are, respectively:

1 MHz – 440 Hz
1 MHz
1 MHz + 440 Hz

NOTE   To transmit a 440-Hz signal for standard AM, the spectrum occupied is (1MHz + 440 Hz) – (1MHz – 440 Hz) = 880 Hz, which is twice the bandwidth of the modulating signal at 440 Hz.

Thus, the sideband signals are comprised of difference and sum frequencies of the carrier frequency. The more common term is that the frequencies of the upper and lower sideband signals are the sum and difference frequencies from the carrier frequency. Thus, multiplication of two signals having two different frequencies f_1 and f_2 results in two signals that have sum and difference frequencies of $(f_1 + f_2)$ and $(f_1 - f_2)$.

A major advantage of transmitting a standard AM signal to a radio is that the demodulator or detector to recover the modulated signal m(t) is typically a simple rectifier or nonlinear amplifier circuit. Initially, standard AM signals were transmitted in this manner to simplify and economize the manufacture of radio receivers. As we will see later, there are more "efficient" types of AM signals that also require more sophisticated signal demodulators to recover m(t), the modulating signal.

Other Types of AM Signals

Before we leave AM signals that include the carrier signals, there is one other class of standard AM signals with a slight modification. This type of signal is known as the *vestigial sideband* (VSB) AM signal, and it was and commonly is used in transmitting analog television signals. It is really a standard AM signal that has either the lower or the upper sideband filtered off, but not completely, thus leaving a vestige of that sideband (see Figure 10-6).

In a standard AM signal, the upper and lower sidebands occupy a spectrum of twice the modulation signal. To save on spectrum space, a portion of the upper or lower sidebands is chopped off. This portion relates to the higher-frequency components of the modulating signal. The lower-frequency components of the modulating

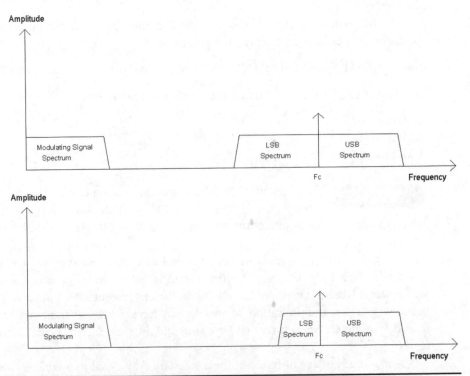

Figure 10-6 Spectra of a standard AM signal and a vestigial sideband AM signal.

signal are not affected sideband-wise, and thus there is a standard double sideband carrier AM signal at the lower modulation frequencies in a VSB signal. The lower-frequency signals in a sense bias the AM carrier to a minimum level such that the negative modulation never goes below 100 percent.

One may ask how can we always guarantee that there is a low-frequency signal to bias the AM envelope? The answer is that a video signal always has synchronizing signals that are low frequency in nature. Even when a TV program or show fades to black, the synchronizing signals are still being transmitted.

The question is, then, what other feature do we have in VSB? The answer is that the VSB signal is also compatible with a simple detector circuit such as a half-wave rectifier that is commonly used in demodulating standard AM signals. This makes the receiver circuits simpler and more cost-effective. To illustrate how a TV signal is modulated with VSB for a fade-to-black signal, see Figure 10-7, which is an oscilloscope trace of an analog cable TV signal.

The top trace in Figure 10-7 shows an inverted video signal, which is normal for modulating an RF carrier for the 525-line (480i) TV analog system. The VSB RF signal on the bottom trace is then detected or demodulated with a simple diode

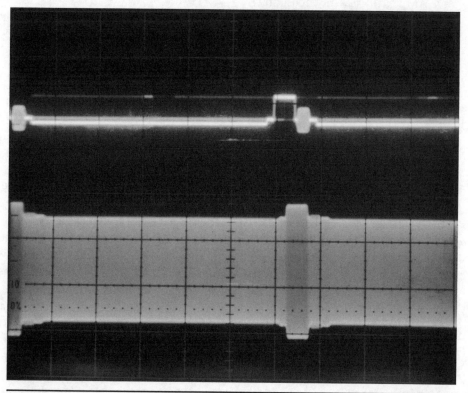

Figure 10-7 A "black level" television signal (*top*). The resulting VSB RF modulated signal (*bottom*).

detector. When analog television transmission became commercialized in the 1940s in the western hemisphere, vacuum tubes were primarily used. There were no *inexpensive* phase lock loop or synchronous detector circuits. So the VSB signal was invented such that a small signal rectifier (e.g., a 6AL5 vacuum tube or a 1N60 germanium diode) could be used for demodulating the VSB signal back into a baseband video signal. This baseband video signal, when amplified, was generally sent to the TV's picture tube's cathode for displaying the television image.

Suppressed Carrier AM Signals

There are quite a few AM signals where the AM signal's carrier is suppressed. These types of suppressed carrier signals are found everywhere today. For example, they are used in cell phone signals, digital TV signals, single sideband signals, and analog color TV signals, just to name a few.

First, what does a suppressed carrier AM signal mean? Let's look again at Equation (10-1), which states

$$AM_{DSBC}(t) = [1 + m(t)]\cos(2\pi f_c t) \tag{10-1}$$

We know that the carrier signal is *always on* regardless of m(t), the modulating signal, which can go to zero to have

$$AM_{DSBC}(t) = [1 + 0]\cos(2\pi f_c t) = \cos(2\pi f_c t)$$

where $\sin(2\pi f_c t)$ is the carrier signal at frequency f_c.

An AM signal with suppressed carrier is a signal where there is no carrier at all by eliminating the DC offset "1," and it is expressed as:

$$AM_{DSBSC}(t) = [m(t)]\cos(2\pi f_c t) \tag{10-3}$$

Note that the new subscript in Equation (10-3) is DSBSC (double sideband *suppressed* carrier) instead of DSBC (double sideband carrier) in Equation (10-1). A typical double sideband suppressed carrier signal is illustrated in Figure 10-8.

In a DSBSC signal, if one looks carefully, during the positive cycle of the modulating signal, the DSBSC signal is in phase with the reference carrier signal. However, during the negative cycle of the modulating signal, the DSBSC signal is 180 degrees out of phase with the reference carrier signal. With a double sideband suppressed carrier signal, we cannot use a simple diode detector to demodulate the signal. A comparison between a standard AM signal and a double sideband suppressed signal when a simple half-wave rectifier is used for demodulation is shown in Figures 10-9 and 10-10.

From Figure 10-9, we can reproduce the modulating signal fine with a standard AM signal. But with the double sideband suppressed carrier signal as shown in Figure 10-10, we get a demodulated signal that will look like a full-wave rectified version of the modulating signal itself (also see Figure 10-11).

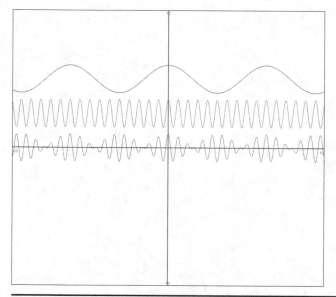

FIGURE 10-8   A double sideband suppressed carrier signal (*bottom trace*) with sinusoidal modulation (*top trace*) of the reference carrier signal (*middle trace*).

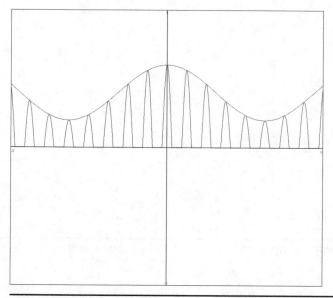

FIGURE 10-9   A standard AM signal with half-wave rectification.

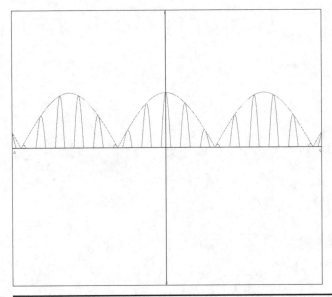

FIGURE 10-10   A double sideband suppressed carrier signal that has been half-wave rectified.

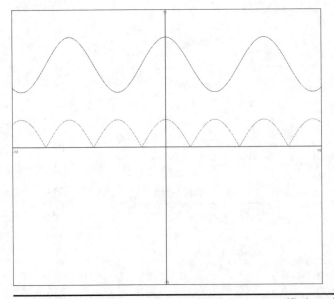

FIGURE 10-11   A modulating signal and a full-wave rectified version of it.

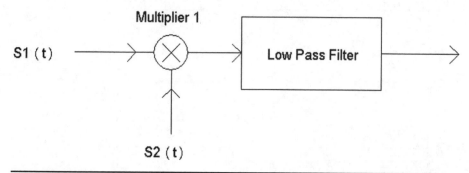

FIGURE 10-12   A product detector to demodulate a double sideband suppressed carrier signal.

To demodulate a double sideband suppressed carrier AM signal, we have to use a *product detector* (see Figure 10-12). The product detector provides a multiplying function. We multiply the incoming double sideband suppressed carrier signal by another signal of the same frequency and the correct phase φ. In general, the detection of this signal is characterized as:

$$AM_{DET\_DSBSC} = AM_{DSBSC}(t) \times \cos(2\pi f_c t + \varphi) = [m(t)]\cos(2\pi f_c t) \times \cos(2\pi f_c t + \varphi)$$

where φ is the phase angle or a phase shift that is different from the phase of the original transmitted carrier signal.

After filtering out the high-frequency signals at 2fc, we are left with the following:

$$AM_{DET\_DSBSC} = AM_{DSBSC}(t) \times \cos(2\pi f_c t + \varphi) = (½)[m(t)] \times \cos(\varphi) \qquad (10\text{-}4)$$

Figures 10-13 through 10-15 illustrate demodulation with a product detector for phase angles φ of 0 degrees, 45 degrees, and 90 degrees.

As one can see if the demodulating phase angle is incorrect (at 90 degrees), the detected DSBSC AM signal can drop to zero.

On another note, a "perfect" product detector does not leak through any of the input signals to the output. So when two signals such as an RF signal and an oscillator signal are multiplied via a perfect product detector, the output contains neither the RF signal nor the oscillator signal.

In terms of bandwidth requirements, the double sideband suppressed carrier signal takes up the same amount of spectrum or bandwidth as a standard AM signal—only that the carrier signal is missing. So, for example, if the carrier frequency is 1,000 kHz and the modulating frequency is 5 kHz, the upper and lower sidebands will be 1,005 kHz and 995 kHz. The bandwidth (BW) is just (1,005 kHz – 995 kHz) = 10 kHz = BW. Again, the bandwidth of the DSBSC AM signal is twice the bandwidth of the modulating signal.

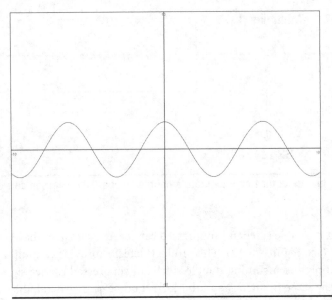

Figure 10-13 Demodulated output of a DSBSC signal with a 0-degree phase shift leads to the correct and full output from the detector.

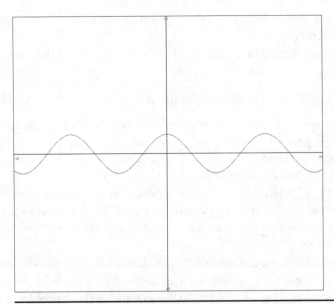

Figure 10-14 Demodulated output of a DSBSC signal with a 45-degree phase shift leads to reduced output at the demodulator.

Figure 10-15   Demodulated output of a DSBSC signal with a 90-degree phase shift leads to zero output from the detector.

It should be noted that in practice, all product detectors or multiplying circuits (even ones that are implemented via digital signal processing) produce some other leak-through and/or distortion signals along with the intended sum and difference frequency signals. Also, the receiver's demodulation signal $\cos(2\pi f_c t + \varphi)$ usually has to be regenerated or recovered somehow from the transmitter. Because the carrier is not sent in the usual manner like standard AM signals, generally the receivers that demodulate suppressed carrier AM signals require more circuitry.

Moving on, what else can we do with a double sideband suppressed carrier signal?

Knowing Your I's and Q's

A combination of double sideband suppressed carrier signals can lend itself to transmitting two channels within the same bandwidth. For example, if the modulating signals of each channel have a spectrum of 5 kHz, two independent channels of 5 kHz bandwidth will be provided. The bandwidth of the RF signal is still 10 kHz, and two 5 kHz channels amount to 10 kHz of information.

By taking advantage of the fact that Equation (10-4) shows if the demodulating signal is 90 degrees out of phase from the original transmitting signal, the output is zero, that is:

$$AM_{DET\_DSBSC} = AM_{DSBSC}(t) \times \cos(2\pi f_c t + \varphi) = (\tfrac{1}{2})[m(t)] \times \cos(\varphi) \qquad (10\text{-}4)$$

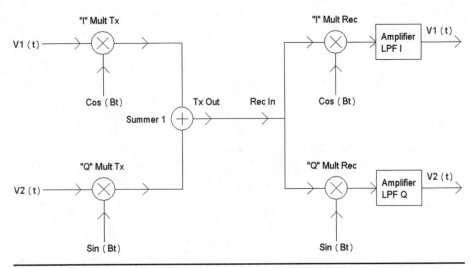

Figure 10-16 A two-channel information transmission and receiving system.

We can then send two signals that occupy the same bandwidth but are 90 degrees out of phase from each other. At first, the combining of the two signals at the transmission end makes them seem inseparable. But at the receiving end, we can use two product detectors with demodulating signals that are 90 degrees out of phase from each other to demodulate two separate channels of information (see Figure 10-16).

In Figure 10-16, the first channel is arbitrarily defined as the I channel, with the cosine waveform defined as 0 degrees or in phase (I). The second channel, a sine waveform, is then phase shifted 90 degrees from the cosine waveform and is called the *quadrature* (Q) channel. *Quadrature* in this case means a "quarter cycle" or 90 degrees shifted from the reference signal, a cosine waveform. V1(t) and V2(t) are usually two separate channels of information to be transmitted. For example, V1(t) and V2(t) can be the red color difference (R − Y) = Pr and the blue color difference (B − Y) = Pb video signals such as the video signals you may notice at the output of a DVD or Blu-ray player, Pr and Pb. Figure 10-17 shows for cosine and sine waves.

The two separate channels are demodulated with the correct phases at 0 and 90 degrees referenced to the original transmitting signal. We also get "demodulation" of the "opposing" channel in each case, but the demodulation signal referenced to the transmitting channel is 90 degrees out of phase, which results in zero output. That is, the I and Q channels transmitted are decoded without cross talk as long as the demodulating signals are perfectly aligned and 90 degrees from each other. If channels are amplitude matched and there is an imperfect 90-degree separation, such as when the I channel is 0 degrees and the Q channel is 89 degrees, then there is an error of 1 degree. And the resulting cross talk or leakage of "contamination" into

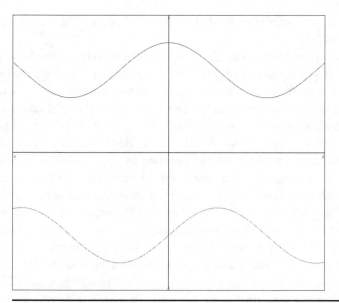

FIGURE 10-17   Cosine and sine waveforms on the top and bottom, respectively.

each channel will be on the order of 1 percent. The amount of leakage for small angles (<10 degrees) is about 1 percent per 1 degree of error.

And now for something related to single sideband signals (SSBs)—we can use a special form of I and Q signal modulation to generate SSB signals. If instead of independent signal sources V1(t) and V2(t) at the input of the transmitter V2(t) is phase shifted 90 degrees from V1(t) and combined with the two mixers, then we get a single sideband signal. In essence, a single sideband signal is a frequency-translated signal. Figure 10-18 shows an input signal Vin(t) that is split out to a 0 degree and a

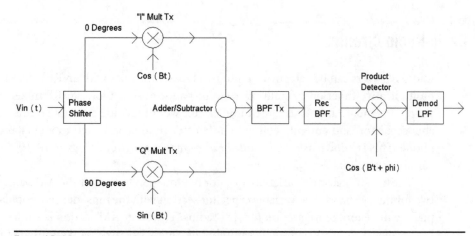

FIGURE 10-18   A single sideband transmitter and receiver.

90 degree signal into two multipliers, I Mult Tx and Q Mult Tx. The carrier signals are Cos (Bt) and Sin (Bt)

When the adder or subtractor is set for addition of the two signals from the multipliers, the result is a lower sideband signal, which frequency translates the input signal's spectrum but causes an inversion in frequency spectrum as well. For example, if the input signal is from music and has a spectrum of 20 Hz to 20 kHz and we set the adder/subtractor for addition with a 1 MHz RF signal, the resulting spectrum will be a lower sideband signal from (1 MHz – 20 Hz) to (1 MHz – 20 kHz). Alternatively, while using an adder, we can generate an upper sideband signal by flipping the phase by 180 degrees of just one of the carrier signals to either mixer.

If the adder/subtractor is set for subtraction, we instead get an upper sideband signal that is then a straight frequency translation of the input signal's spectrum. There is thus no spectrum inversion. For example, if we have a voice signal whose spectrum spans 100 Hz to 4 kHz, and if we want to frequency translate that voice spectrum upward by 100 kHz, the resulting spectrum will span from 100.100 kHz to 104 kHz. In this example, we have produced an upper sideband signal at 100 kHz. Alternatively, while using a subtractor, we can generate a lower sideband signal by flipping the phase by 180 degrees of just one of the carrier signals to either mixer.

Demodulating the SSB signal is simpler in that almost any type of mixer circuit can be used. Fortunately, we do not require two mixers in the receiver. That is, a multiplier circuit is not the only circuit that can be used to recover the modulation signal. A simple diode and an oscillator signal with the same frequency as the transmitting frequency are combined and rectified. Another advantage to demodulating an SSB signal is that it is phase independent. The reason is that we are just "beating" or frequency translating the (RF) SSB signal downward back to the baseband signal, such as a voice signal. If the demodulating frequency at the receiver end is incorrect, the pitch of the recovered signal will be shifted high (e.g., "Donald Duck–sounding") or shifted low (e.g., a deep voice).

Basic Radio Circuits

Radio receivers can be categorized into two types—the ones with an RF mixer (e.g., multiplicative function or multiplier circuit) and those without an RF mixer. The simplest standard broadcast AM radio circuit would include a tuned circuit, variable capacitor, and antenna coil with a detector diode connected to a crystal earphone. This is known as a TRF (*tuned radio-frequency*) circuit. Figure 10-19 shows variable capacitors.

The top air dielectric variable capacitor has four gangs, two for the AM band and two for the FM band. The variable capacitor section with the most densely populated plates will be for the RF section for AM radios. The other AM section is similar but has smaller plate areas or a smaller number of plates. The two FM sections are iden-

Figure 10-19 Air dielectric variable capacitors.

tical and have three rotor plates (rotor plates are connected to the shaft and usually are grounded or AC grounded) and two stator plates. The middle two-gang variable capacitor is designed just for AM radios. Again, the section with higher capacitance is for the RF section, and the other section is for the oscillator circuit.

The variable capacitor shown at the bottom of Figure 10-19 is a single-gang 365 pF version. This variable capacitor is commonly found on eBay or at electronics supply stores (e.g., www.tubesandmore.com).

For TRF radios, one or more sections (e.g., two sections connected together in parallel) of the four- or two-gang variable capacitors may be used to match the antenna coil or RF inductor. Note that the frame of an air dielectric variable capacitor is normally grounded or AC grounded.

In Figure 10-20, the variable capacitor on the left has two dissimilar sections and is generally for superheterodyne radios. The middle lead is the ground or common lead. The top lead is for the RF section and is usually connected to an antenna coil, and the bottom lead is a smaller-capacity variable capacitor that is used for the oscillator section of the radio and is connected to an oscillator coil.

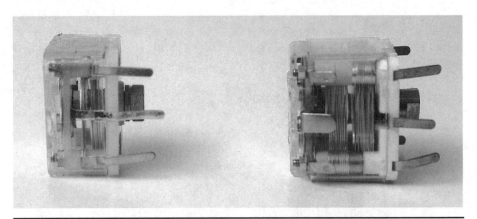

Figure 10-20   Polyvaricon variable capacitors.

The polyvaricon variable capacitor on the right is normally used for AM and SW (shortwave) superheterodyne radios. One section is used with the antenna coil for the RF filter, and the other is used with the oscillator coil. For AM bands, there is usually a series capacitor with one of the twin sections of the polyvaricon variable capacitor. The series capacitor is usually about 110 percent of the maximum capacitance of a twin variable capacitor. For example, if you are using a twin 270 pF variable capacitor, the series capacitor for the oscillator coil is about 300 pF.

Generally, antenna coils are available with ferrite rods or bars, as seen in Figure 10-21. The top and middle antenna coils have separate primary and secondary windings for a total of four leads. The bottom antenna coil has three leads, which is a tapped inductor or autotransformer. The low-side tap has about 5 to 10 turns of wire. A TRF crystal radio circuit is shown in Figure 10-22.

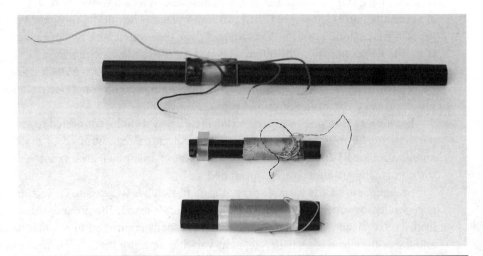

Figure 10-21   Antenna coils.

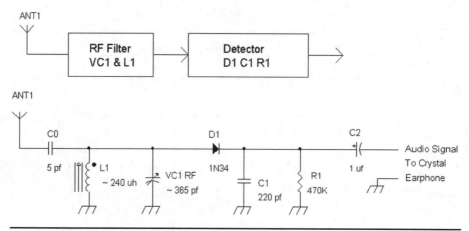

A crystal radio circuit.

A crystal radio typically requires a long wire antenna, typically at least 20 feet, strung up above the ground. The antenna should be coupled into the tuned circuit via a small capacitance value, typically about 5 pF, to isolate the low source resistance of the antenna. A low source resistance will load the tuned circuit (L1 and VC1) down and reduce the radio's ability to separate the various radio stations. For example, if the antenna is connected directly to L1/VC1 by shorting out the 5 pF capacitor C0, then the radio will receive many stations at once, and tuning VC1 will have little effect. With C0 in series with the antenna, the low source resistance of the antenna is buffered or isolated via C0, and VC1 and L1 are essentially allowed to form a narrow band-pass filter that allows tuning to be more selective.

One question arises: why do we need an antenna coil (in the first place) when a small inductor can be used instead? The answer is that most small inductors of the same inductance value have resistive and other types of losses that antenna coils generally do not have. One can replace the antenna coil with an inductor, but even with a 5 pF capacitor connected to the antenna, the user will notice that the tuning is less sharp with the inductor versus the antenna coil.

Also, the antenna coil on its own without the long wire antenna is capable of receiving radio stations better than an inductor. For example, having a 6- or 8-inch ferrite-rod antenna with Litz wire provides very high sensitivity, and without a long wire antenna, usually one or two local stations of strong signal strength can be heard.

To improve on the crystal radio's sensitivity, one can add an audio amplifier. However, due to the turn-on voltage of the diode detector, the improvement in sensitivity is limited. The audio amplifier must also have a high input resistance (e.g., >100 kΩ) so as to not load down the tank circuit (VC1 and L1).

Instead of a diode, one can use the "fuzz" circuit seen in Chapter 8 to act as an AM power detector. See Figure 10-23. Although the transistor amplifier input resistance is moderately high at 45 μA collector current with about 57 kΩ for a current gain β =

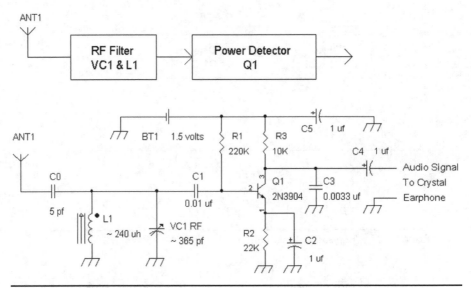

Figure 10-23   Crystal radio with power detector Q1.

100, use of the one-transistor power detector circuit provides adequate selectivity and sensitivity.

The power detector crystal radio works at a minimum of 1.5 volts and better at higher voltages such as 3 volts. Q1 is really a common emitter or AC-grounded emitter bipolar transistor amplifier. We know that for small signals < 5 mV, Q1 amplifies the signal with low distortion. But when the input signal into the base is >5 mV, Q1 starts to distort. We take advantage of the distortion feature because with a large enough amplitude AM signal (see Figure 10-23) into the base of Q1, Q1 distorts and becomes a demodulator by clipping off almost half of the AM signal. The load resistance to the tank circuit L1/VC1 is about 50 kΩ for a current gain of 100 = β. A higher-current-gain transistor such as a 2N5089 (β = 800) may be used for better selectivity and sensitivity by raising the load resistance to about 200 kΩ. To achieve still better selectivity and sensitivity (see Figure 10-24). An emitter-follower Q0 is added to raise the load resistance to >500 kΩ using a transistor such as 2N3906.

At some point, amplifying the detected audio signal leads to diminishing returns. The reason is that in order for the power detector to work, we have to drive the base-emitter junction of Q1 into distortion. If the RF signal from the tank circuit VC1 and L1 is too small, Q1 merely acts as an amplifier, and there is no detection of the AM signal. But maybe we should amplify the RF signal and then demodulate the AM signal afterwards (see Figure 10-25).

Figure 10-25 keeps Q0 as an emitter follower but makes Q1 into an RF amplifier by removing the 0.0033 μF capacitor from the collector to ground, as shown in Figure 10-20. The amplified RF signal via Q1 and R3 is then coupled to power detector circuit Q2, R6, and C7.

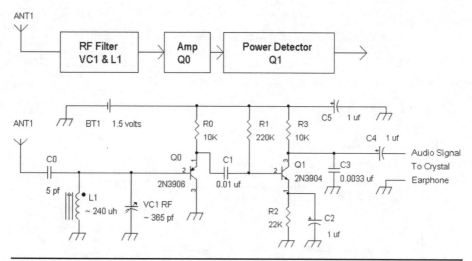

Figure 10-24 Higher-selectivity version with Q1 and emitter-follower Q0.

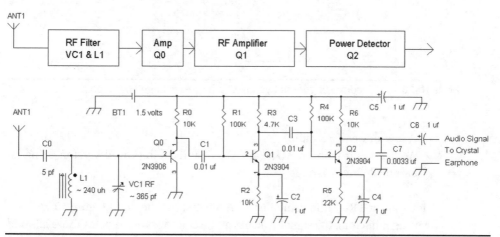

Figure 10-25 Converting the power detector into an RF amplifier and adding a power detector afterwards.

An advantage of power detectors is that the demodulated output signal is usually higher than a diode detector for the same input signal. Thus the power detector demodulates and amplifies the demodulated (e.g., audio) signal.

Tuned Radio-Frequency Radio Projects

So far we have described TRF (tuned radio-frequency) receivers and for projects, the following circuits can be built:

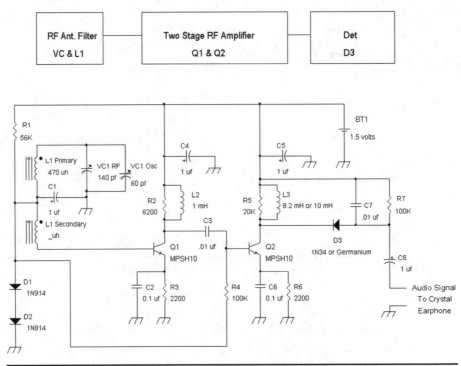

Figure 10-26 TRF radio using inductive loading for low-voltage operation.

- A low-power TRF radio
- A TRF radio with an MK484/TA7642 and its improved version using an emitter-follower circuit

A low-power TRF radio using just an AA battery is shown in Figure 10-26.

By using inductors, the signal voltages at the collectors at Q1 and Q2 are allowed to swing *above* the 1.5 volt supply. The collectors also have a quiescent voltage at 1.5 volts, which allows for reliable biasing of the bases of Q1 and Q2, low-capacitance transistors MPSH10. Note the pin-out for an MPSH10 is base, emitter, collector instead of the emitter, base, collector configuration for a more common transistor such as the 2N3904. Diodes D1 and D2 form a voltage reference for the base biasing. The voltage at the D1 anode maintains a voltage of about 1 volt and thus keeps a reliable voltage at the bases even when the battery voltage drops from 1.5 volts to 1.2 volts. Q1 and Q2 are RF amplifiers, and detection of the AM signal is done via D3. A germanium diode is chosen for AM detection because it has very low turn-on voltage (~0.1 volt) compared with a silicon diode (~0.5 volt).

When AM radios of this type are built, try to keep the wires short (within a couple of inches). See Figure 10-27. The prototype here is built to mimic the flow of the schematic diagram. The reader can build this circuit or others in this book,

Figure 10-27   TRF radio using inductive loading.

although they are not designated as official projects. An intermediate to advance skill level of electronics may be required.

First Project: A TRF Radio

The MK484/TA7642 integrated circuit (IC) that has a multiple-stage amplifier and a peak detector is generally available on the web, including eBay. We present multiple versions of this radio, as shown in Figures 10-28 through 10-31.

Parts List

- C1, C3, and C4, 1 μF electrolytic capacitor (10 to 50 volt rating); C4 is optional
- C2, 0.1 μF film or ceramic capacitor
- R1, 100 kΩ resistor, any type available
- R2 and R3, 1 kΩ resistor, any type available
- L1, 240 μH or 250 μH antenna coil
- VC1, 330 pF to 365 pF variable capacitor
- U1, MK484 or TA7642 three-terminal IC (integrated circuit)

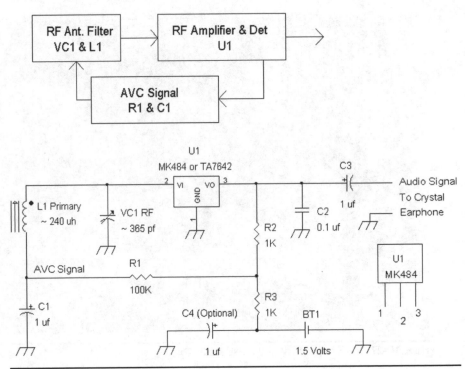

FIGURE 10-28   MK484/TA7642 TRF radio, version 1.

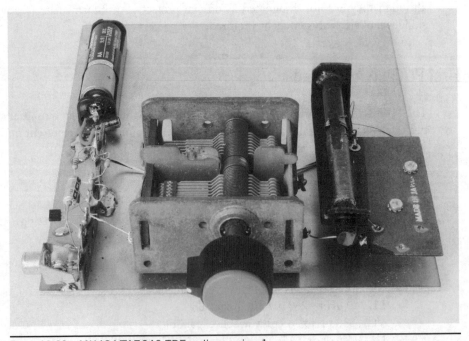

FIGURE 10-29   MK484/TA7642 TRF radio, version 1.

- 1.5 volt supply via a battery or equivalent (do not exceed 2 volts on the supply)

The 365 pF variable capacitor was bought on eBay. However, the single-gang 365 pF variable capacitor shown in Figure 10-19 can be used instead. The antenna coil shown is a 250 μH ferrite-rod type, which is available at the time of this writing at www.tubesandmore.com. The TA7642 IC was available on eBay for less than $1.

This radio pulled in about seven stations. The radio can be improved slightly by coupling a long wire antenna to the antenna coil L1 via a 5 pF to 10 pF capacitor, as shown in Figure 10-30. By coupling a long wire antenna, sensitivity is increased.

However, it was found that with some TA7642 chips, the input resistance was lower than expected, which loaded down the LC tank circuit. An emitter-follower amplifier was added, and sensitivity and selectivity were improved, with the radio picking up more than 10 stations (see Figure 10-31). Note that a different antenna coil and variable capacitor combination is presented in this version. But the reader can just as well use 365 pF for VC1 and 250 μH for L1.

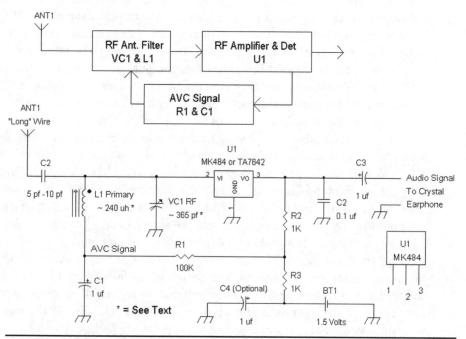

Figure 10-30 MK484/TA7642 TRF radio, version 2.

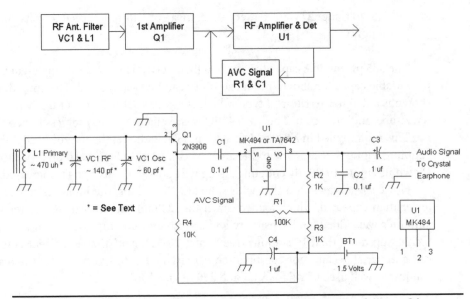

FIGURE 10-31   MK484/TA7642 TRF radio, version 3, with the emitter follower Q1.

Solid-State Regenerative Radio

The TRF radios we have shown so far can be improved further by adding *positive feedback* to the tank circuit. By doing this, two things are accomplished. One is that if there is a sufficient RF signal at the tank circuit, a portion of an amplified RF signal is fed back to the inductor-capacitor tuned circuit in a manner to reinforce the RF signal, which then provides very high gain. The second thing is that if the amount of positive feedback is adjusted just below the threshold of making the circuit self-oscillate, the selectivity of the tuned circuit is also increased. The positive-feedback signal "cancels" out some of the resistive and other losses in the antenna coil to allow better separation of tuned radio stations (see Figures 10-32 and 10-33).

VR1 adjusts the amount of positive feedback into the antenna coil via its bottom lead. The tap of the antenna coil is AC grounded via C1 and has a DC bias voltage via D1. Diodes D2 and D3 provide a "regulated" voltage to the bases of Q2, Q3, and Q4.

Q1 is an emitter-follower amplifier that has a gain of about 1 and allows for a high-resistance load to the tank circuit L1 and VC1. It drives VR1 to Q2, which supplies positive-feedback signal current into the antenna coil. Q1's emitter also supplies the regenerated RF signal to RF amplifier Q3. The output of Q3 supplies sufficient amplitudes of the AM RF signal to power detector circuit Q4. Most regenerative radios require an outdoor antenna. This design allows for a built-in ferrite antenna coil to receive sufficient RF signal levels such that the regeneration control is useful. For example, if there is insufficient signal level into the antenna coil, the regeneration control would have to be turned up to the point where the circuit turns into an oscillator. Recall that a regenerative circuit is essentially the same as an oscillator

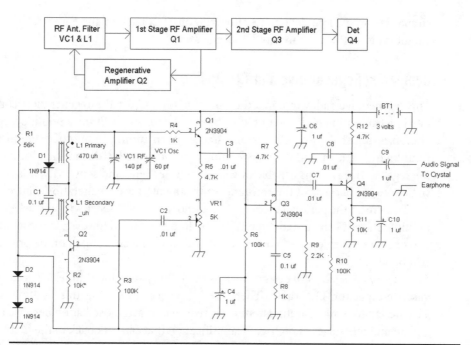

FIGURE 10-32   Schematic of a low-power regenerative radio.

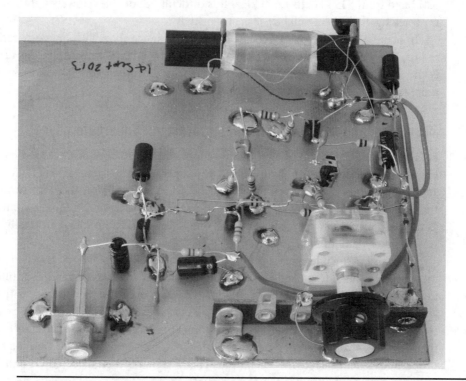

FIGURE 10-33   The regenerative AM radio.

circuit, but we will be deliberately lowering the feedback or the gain such that the circuit is just below the threshold of oscillation.

Comparing Regenerative and TRF Radios

The regenerative radio requires the operator to readjust the regeneration control (VR1) when tuning to different radio stations. For example, if the amount of regeneration is adjusted just right for one station, then tuning to another station may result in oscillations, and the regeneration control has to be "dialed back." Although the regenerative radio is capable of very good selectivity and sensitivity, it is inconvenient to use because of the constant readjustment of the regeneration control.

To increase the performance of a TRF radio in terms of sensitivity and selectivity, multiple stages of the tuned circuits can be implemented. However, tracking of the two or more variable tuned circuits is difficult over the entire span of the broadcast AM band.

There were successful designs such as the 1930's J.W. Miller Model 570 four-gang variable-capacitor TRF radio. There are "obstacles" in getting this multiple-stage TRF design to work. Careful attention is required to avoid oscillation because there will be multiple stages of high-gain amplification that can leak back to the first stage. Another problem is just aligning the two or more RF stages to track throughout the AM band from 535 kHz to 1,600 kHz. It is difficult but doable. However, if there is mistracking, then there will be a loss of sensitivity.

In about 1918, to provide much better performance than TRF or regenerative radios, a new type of radio architecture was design by Edwin Armstrong. This radio is known as the *superheterodyne radio*, which includes a mixer, oscillator, and one or more fixed-frequency filters.

Mixing and Beating Signals (aka Heterodyning or Superheterodyne)

Let's take a look at the superheterodyne architecture first for a typical AM radio (see Figure 10-34). As shown in Figure 10-34, the superheterodyne (superhet) radio has an RF filter circuit for the incoming RF signals, a local oscillator, an RF mixer that frequency translates the incoming RF signal frequencies to a lower frequency known as the IF (*intermediate frequency*), at least one IF filter and amplifier, and finally, an AM detector/demodulator.

The basic principle of a superheterodyne radio is to frequency translate the incoming RF signal to a fixed IF signal and apply fixed-frequency band-pass filtering. Amplifying the IF signal with multiple IF filters provides increased sensitivity and selectivity. Whereas a multiple-stage TRF radio has many variable tuned circuits to provide selectivity, the superhet uses multiple fixed-frequency tuned circuits to do the same job. Fixed tuned circuits are easier to adjust, and once they are aligned properly, excellent selectivity is achieved. The IF signal is then detected via a diode or power detector to provide the audio signal.

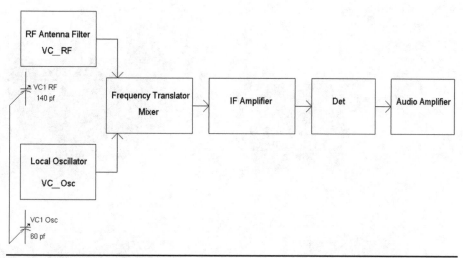

Figure 10-34 Block diagram for a superheterodyne radio.

A typical standard AM radio has a variable tuned RF filter and local oscillator. There are usually two variable capacitors that track each other such that the resonant frequency of the RF signal minus the resonant frequency of the oscillator signal is a fixed frequency such as 455 kHz. For example, if the RF section is tuned to 600 kHz, the oscillator has a frequency of 1,055 kHz, which is 455 kHz above the 600 kHz RF signal. When the RF and oscillator signals are fed to an RF mixer, there will be signals at (1,055 kHz – 600 kHz) = 455 kHz and (1,055 kHz + 600 kHz) = 1,655 kHz. The output of the mixer will be fed to a 455 kHz IF band-pass filter to pass only the 455 kHz signal and not the 1,655 kHz signal.

Alternatively, if one tunes to a station at 1,000 kHz for the RF signal, the oscillator will track accordingly and provide an oscillator signal at 1,455 kHz. Hence, again, the difference-frequency signal, 1,455 kHz – 1,000 kHz, will be 455 kHz. Once the 455 kHz signal is extracted from the mixer and amplified, demodulation of the amplitude-modulated 455 kHz signal is done typically by rectification or power detection.

Let's now take a look at the various components that are key to making an AM superhet radio.

Two-Gang Variable Capacitors

For the RF and oscillator circuits, a two-gang variable capacitor is normally used. See Figure 10-35. Generally, the two-section (two-gang) variable capacitors are not identical for standard AM radios. The larger-capacity section (e.g., 10 pF to 140 pF polyvaricon) is used for the RF section, whereas the smaller-capacity section (e.g., 10 pF to 60 pF) is used for the local oscillator circuit.

FIGURE 10-35 (*a*) Air dielectric variable capacitors. (*b*) Polyvaricon variable capacitors.

For AM band and shortwave (SW) radios from 530 kHz to about 30 MHz, twin-gang variable capacitors are generally used. When a twin variable capacitor is used for the AM band, the oscillator circuit has a series capacitor between the oscillator coil and one section of the variable capacitor so that the maxiumum capacitance is cut approximately in half. For example, twin 270 pF polyvaricon types are common among the amateur radio community. To use this for an AM radio, connect a series capacitor whose value is between 300 pF and 330 pF to VC1B of the variable capacitor as shown in Figure 10-36 (*right side*).

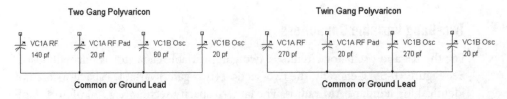

FIGURE 10-36 Schematic of a dissimilar variable capacitor and a twin-variable capacitor for AM radios.

Oscillator Coils

For most transistor radios, the traditional oscillator coil is a tapped transformer. The tapping in the winding is not a center tap. Thus there is a low-side tap that is identified by using an ohmmeter. From the center pin to one of the outside terminals will show higher or lower resistance than the other outside terminal. The two terminals that include the center pin with the lower resistance are the low-side tap.

Typical oscillator coil part numbers for AM band transistor radios are 42IF100, 42IF110, and 42IF300. All these coils have five pins, three on one side (including a tapped winding terminal in the middle) and another two on the other side. The transistor radio oscillator coils and IF transformers have a slug tune adjustment that requires a special screwdriver so that the brittle ferrite slug is not damaged while it is turned (see Figures 10-37 and 10-38). Typically, the color code for AM radio oscillator coils is a painted red tuning slug.

IF Transformers

The transistor radio IF transformers are similar to the oscillator coils. However, typically, they have internal capacitors across the outer pins of the tapped winding.

NOTE The schematic drawings in this book will not include the internal capacitor in the IF transformers. When a particular IF transformer part number is specified, unless otherwise noted, assume that there is already an

FIGURE 10-37 Transistor radio oscillator coils.

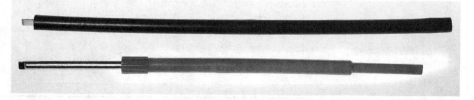

FIGURE 10-38 Alignment/adjustment tools.

internal capacitor across the outer pins on the side of the coil that has three pins. Also note that oscillator coils do not have internal capacitors because they are generally connected to variable capacitors.

The slugs are color coded for the Xicon IF transformer as:

- Yellow for the first IF filter, 42IF101 *or it may be substituted with 42IF301*
- White for the second IF filter, 42IF102 *or it may be substituted with 42IF302*
- Black for the third or last IF filter, 42IF103 *or it may be substituted with 42IF303*

The first and second IF transformers are interchangeable and can be substituted for each other. However, the third IF filter, 42IF103, should not be replaced with a 42IF101 or 42IF102 part. The primary to secondary turns ratio of the 42IF103 is the correct one to deliver the proper voltage to the AVC (automatic volume control) circuit. If a 42IF101 or 42IF102 IF transformer is used as the last IF filter for the detector and AVC circuit, there will be insufficient AVC voltage, and thus the AVC circuit will not be effective in turning down the gain when a strong signal is received. Like the oscillator coils, the IF transformers also have low-side taps.

For superhet radios similar to the ones in my book, *Build Your Own Transistor Radios* (McGraw-Hill, 2013), see Figures 10-39 to 10-42 for examples of radios using the oscillator and IF coils.

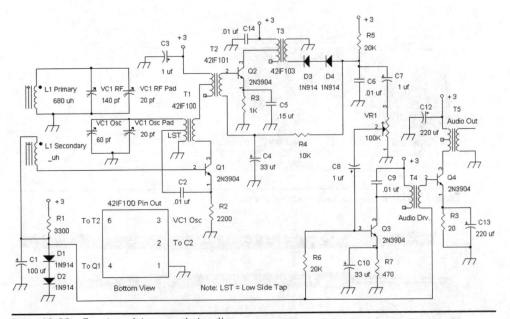

Figure 10-39 Four-transistor superhet radio.

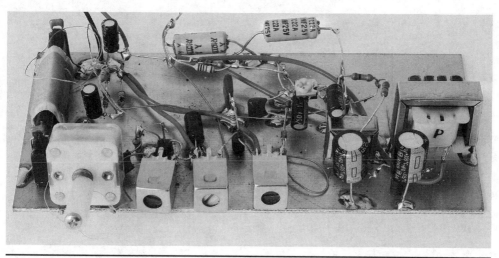

Figure 10-40 Four-transistor radio prototype circuit.

The radio in Figure 10-39 uses a mixer-oscillator circuit via Q1. The oscillator is a common-base amplifier with the signal being fed back from the collector to the emitter. Using the low-side tap coupled to the emitter of Q1 via C2 allows the oscillator tank circuit from VC1 Osc and T1 to be essentially unloaded. If, for some reason, the circuit does not oscillate, try switching leads 6 and 4 of T1. The base of Q1 is fed to a stepped-down voltage of RF signal via L1's secondary winding. The RF signal is thus added to the base-emitter voltage of oscillator Q1 to form a combination mixer-oscillator circuit. The oscillator's frequency varies between about 990 kHz and about 2,050 kHz and typically has at least 200 mV peak to peak at the emitter of Q1. The RF signal at the base is much smaller than the oscillator signal, and thus the oscillator signal drives Q1 into distortion to provide an amplitude-modulation effect on the RF signal, which is also a form of multiplying the RF and oscillator signals. Thus, at the output of Q1, there are signals whose frequencies are related to the RF signal, the oscillator signal, and sum and difference frequencies of the RF and oscillator signals. The IF signal is extracted via T2 and is amplified by IF amplifier Q2. A second IF transformer is connected to the output of Q2 via the collector of Q2. Demodulation of the AM signal is accomplished by half-wave rectifiers D3 and D4. The two diodes are used not only for rectification but also to establish an initial positive AVC voltage at the anode of D4 that supplies a bias voltage (R4 and C4) to IF amplifier Q2. When the signal strength is high, a negative voltage is added to the initial positive AVC voltage, which results in a smaller positive voltage that in turn biases a lower collector current to Q2. This then lowers the gain of Q2 to establish AVC (automatic volume control). The audio signal from D4 is coupled to volume control and audio amplifier Q3. The output of Q3 is fed to interstage audio transformer T4 on the center-tap primary winding (e.g., 10 kΩ center-tapped primary, 600 Ω secondary). From the secondary winding of T4, it is coupled to audio

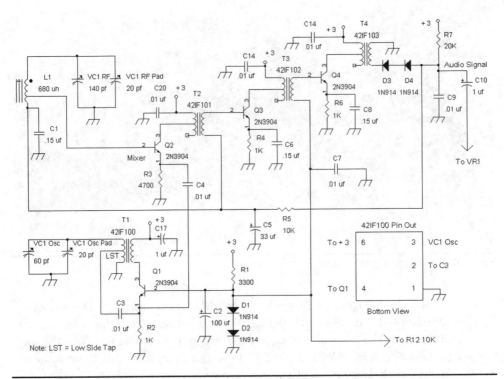

Figure 10-41 Tuner section of an eight-transistor radio.

power amplifier Q4. The output collector signal from Q4 is fed to T5 (120 Ω primary and 8 Ω secondary windings). A speaker or low-impedance headphone can be connected to the secondary winding of T5.

We can achieve more sensitivity and selectivity by adding an extra IF transformer and amplifier. Also, we can provide more AVC range by applying the AVC voltage over two stages, a mixer and the first IF amplifier (see Figure 10-41).

In this radio, there are separate oscillator and mixer circuits. The oscillator circuit via Q1 is again a common-base design, as previously described in the four-transistor radio. A separate RF mixer is provided via Q2, which is a common-emitter amplifier that is overdriven with an oscillator signal at >200 mV peak to peak into its emitter. The base of Q2 receives a tapped-down RF signal. The tapping down or stepping down of the RF voltage is essential to keep the RF tank circuit from being loaded down. The output of the mixer is fed to the first IF transformer T2 and first IF amplifier Q3. The output of Q3 is connected to a second IF transformer and IF amplifer T3 and Q4. Demodulation of the AM IF signal (converted RF signals to IF signals) is done via diodes D3 and D4. The audio signal is filtered via R5 and C5 to provide AVC signals to the bases of mixer Q2 and first IF amplifier Q3.

The audio signal level is adjustable via volume control VR1 in Figure 10-42 and is fed to the first audio preamplifier transistor Q5. A second preamplifier transistor

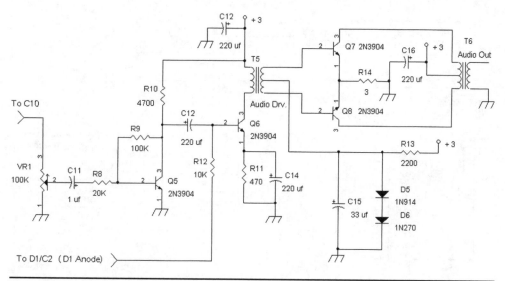

FIGURE 10-42   Audio amplifier section of the eight-transistor radio.

Q6 is connected to interstage audio transformer T5 that provides a push-pull or balanced signal to the bases of Q7 and Q8. The outputs of Q7 and Q8 are connected to output transformer T6 to provide an audio signal to a loudspeaker or headphone.

To reiterate, in the four-transistor radio, the oscillator circuit also serves as a mixer. Sometimes this circuit is called a *converter circuit* because it converts the incoming RF signals to IF signals all in one step. When the RF signal is much smaller than the oscillator signal, it can be added or piggy-backed on top of the oscillator signal. The transistor "sees" the two signals, RF and oscillator, across the base-emitter junction, and because of the nonlinearity of the transistor, RF mixing action is achieved. The IF signal is extracted from the collector of the converter transistor, even though the collector signal has a combination of a strong oscillator signal, the small RF signal, and sum and difference frequency signals from the RF and oscillator signals.

In the eight-transistor radio, Q2 serves as the RF mixer. In both cases, where the oscillator's signal is >200 mV peak to peak, the conversion transconductance at the intermediate frequency is just the same as the small-signal transconductance of the converter (Q1 in Figure 10-39) or mixer transistor (Q2 in Figure 10-41) based on the collector quiescent current ICQ. The small-signal transconductance of a common-emitter or common-base amplifier is $g_m = ICQ/0.026$ volt.

Second Project: A Superhet Using the MK484/TA7642 Chip

By using a one-transistor converter circuit with an oscillator coil and IF transformers, we can build a superhet radio with the MK484/TA7642 chip. Previously, the chip

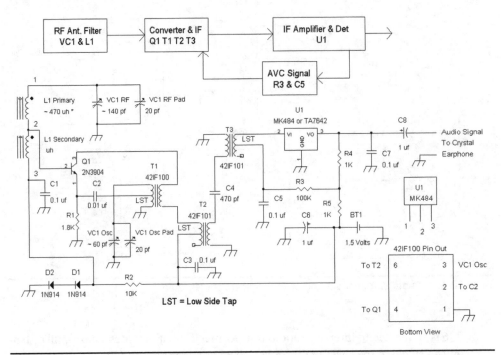

FIGURE 10-43   A 1.5 volt superhet radio using an MK484/TA7642 chip.

amplified RF signals in the TRF radios, but this time the chip will only be amplifying the IF signal (see Figures 10-43 and 10-44).

A mixer-oscillator circuit via Q1 provides about 200 mV peak to peak of oscillation signal into the emitter of Q1. The oscillator voltage amplitude can be increased by lowering the value of R1 (e.g., 1.8 kΩ → 1.5 kΩ or 1.2 kΩ). However, this also raises the conversion transconductance and may overload the MK484/TA7642 IC. The collector output of the mixer-oscillator or converter circuit is coupled to two IF transformers T2 and T3. The two IF transformers are coupled via C4, a 470 pF capacitor, to allow for some "independence" between the two coils so that improved selectivity is achieved and filtering out of the oscillator signal prior to inputting to the MK484/TA8642 chip. If there is too much oscillator signal along with the IF signal, the AM envelope detector in the MK484/TA7642 chip will output distorted demodulated audio signals. The output of T3 is coupled to the input of U1, which then amplifies the IF signal and provides an AVC signal via R3 and C5. Operation of this radio is at 1.5 volts. Diodes D1 and D2 provide a voltage reference so that Q1 provides a stable oscillation signal even if the battery drops in voltage (e.g., from 1.5 volts to 1.2 volts).

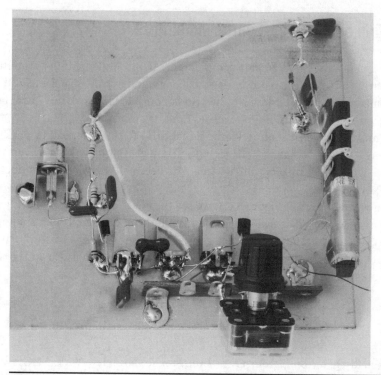

Figure 10-44   Picture of the MK484/TA7642 superhet radio.

Parts List

- C1, C3, C5, and C7, 0.1 μF film or ceramic capacitors
- C2, 0.01 μF film or ceramic capacitor
- C4, 470 pF silver mica or ceramic capacitor
- C6 and C8, 1 μF, 50 volt electrolytic capacitor
- D1 and D2, 1N914 or 1N4148 silicon diodes
- R1, 1.8 kΩ, ¼ watt 5% resistor
- R2, 10 kΩ, ¼ watt 5% resistor
- R3, 100 kΩ, ¼ watt 5% resistor
- R4 and R5, 1 kΩ, 5% ¼ watt resistor
- Q1, 2N3904 or 2N4124 transistor
- T1, 42IF100 oscillator coil
- T2 and T3, 42IF101 455 kHz IF (intermediate frequency) transformer
- U1, MK484 or TA7642 IC (integrated circuit)
- L1, 680 μH antenna coil with a 470 μH high-side tap
- VC1, two-gang variable capacitor, 140 pF and 60 pF

An Observation on Common-Emitter Amplifiers

In this chapter, we see that when a grounded or AC-grounded emitter amplifier is overdriven, it can be used as a power detector for a standard AM signal and also used as an RF mixer in superheterodyne radios. So not only does it work as a "fuzz" guitar audio circuit, but it also has other purposes in radio circuits.

As a mixer with sufficient oscillator signal drive (>200 mV peak to peak), the conversion transconductance = IF signal current/RF input voltage is very high compared with other mixers and is equal to the small-signal transconductance ICQ/0.026 volt.

When we look back at Figure 8-23, we see the effect of a small signal riding on top of a larger signal. The smaller signal is amplitude modulated, which amounts to a multiplying effect. A simple common-emitter amplifier serves well as an RF mixer, but it is not a balanced mixer, and the oscillator and the RF signal also appear at the output along with the IF signal.

References

1.   Ronald Quan, *Build Your Own Transistor Radios*. New York: McGraw-Hill, 2013.

2.   K. Blair Benson and Jerry Whitaker, *Television Engineering Handbook*. New York: McGraw-Hill, 1992.

CHAPTER 11

Frequency Modulation Signals and Circuits

In Chapter 10, we explored some of the different types of amplitude-modulated (AM) signals such as amplitude modulation with carrier, suppressed carrier, and single sideband. The modulating signal such as a voice or music signal varied the amplitude of the carrier in standard AM, or it varied the amplitude of the sideband(s) of the suppressed carrier AM signal. Frequency-modulated (FM) signals do not vary the amplitude of the carrier signal. Instead, the actual frequency of the carrier changes according to the modulation signal. This chapter will look at broadcast FM signals and circuits for transmitting and receiving FM signals. Two broadcast FM radio projects will be presented, which should be workable for the reader, even though we are venturing into the very high-frequency (VHF) range of about 100 MHz.

What Are FM Signals?

See Figure 11-1 for an illustration of an FM signal. Figure 11-1 shows an FM signal where the positive peak of the low-frequency modulating signal (top waveform) causes the carrier signal (lower waveform) to increase in frequency, whereas the negative peak of the modulation waveform is at a minimum and causes the high-frequency carrier signal to decrease in frequency, as shown by the expansion in period of the high-frequency signal during that time.

An FM signal can be characterized by the following equation:

$$FM(t) = A \cos(\omega_c[1 + k \, m(t)]t + \varphi) \qquad (11\text{-}1)$$

where $\omega_c = 2\pi f_c$
$\qquad f_c$ = the carrier frequency measured in hertz
$\qquad\quad k \ll 1$, typically $k < 1\%$ $(k < 0.01)$

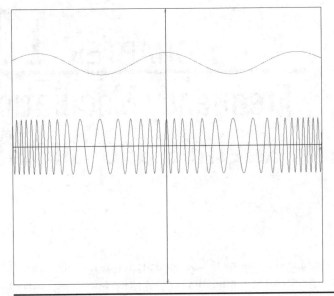

FIGURE 11-1   An FM signal.

m(t) = modulating signal, where |m(t)| < 1

φ = constant or static phase shift (for practical purposes, φ → 0 degrees)

Another way of expressing the FM signal is by the following equation:

$$FM(t) = A\cos[2\pi(f_c[1 + k\,m(t)])t + \varphi] \qquad (11\text{-}2)$$

We see that the carrier signal f_c is varied over *time* by the factor of $[1 + k\,m(t)]$, where m(t) is an alternating-current (AC) signal so that $f_c[1 + k\,m(t)] = f_c + f_c[k\,m(t)]$. Since k m(t) is AC in nature, it has positive and negative numbers, $\pm k\Delta f_c$, which means that the instantaneous carrier frequency is $f_c \pm k\Delta f_c$.

An example of a frequency-modulated (FM) signal is when one uses a signal generator and varies the frequency while having the amplitude set for a fixed level. Another example of frequency modulation is the sound of sirens or horns from a vehicle as it approaches you and goes away from you. The change in pitch is caused by the Doppler effect. In addition, when one listens to a violinist who applies vibrato by modulating the finger position on the neck of the violin, an FM effect occurs. Thus, when the frequency of a particular signal is changed with time, this leads to an FM signal.

For an FM broadcast signal, the carrier frequency is in the 100 MHz range, and the maximum deviation or shift in frequency at the carrier frequency is ±75 kHz. The modulation frequency ranges from 50 Hz to 15 kHz for the monaural channel known as the *left plus right channel* (L + R) and from 23 kHz to 53 kHz to form a double sideband suppressed carrier signal carrying the difference channel (L − R).

In AM signals, there are two aspects of the modulation signal. One is the amplitude of the modulation signal, which determines the depth of the amplitude-modulating effect on the carrier. The second is the frequency of the modulating signal, which determines the rate at which the AM signal's carrier is varying with time.

For FM signals, there are also two aspects related to the modulating signal. One is the deviation in frequency from the nominal carrier frequency that is determined by the amplitude of the modulating signal. The higher the amplitude of the modulating signal, the higher is the frequency deviation. In a wideband FM channel, usually the maximum frequency deviation is approximately 75 kHz. For example, this means for a United States FM channel at 100.1 MHz, when the carrier is frequency modulated to its maximum, the carrier will vary from 100.100 MHz – 75 kHz to 100.1 MHz + 75 kHz. Stated another way, the radio station's frequency varies from 100.025 MHz to 100.175 MHz. The second aspect of the modulating signal for FM is the frequency of the modulating signal that determines how "fast" the carrier swings back and forth.

From these two aspects, frequency deviation and modulation frequency, we can define two terms related to FM, which are the deviation ratio and the modulation index. The *deviation ratio* is the maximum peak-frequency deviation divided by the highest modulation frequency. A more general term is the *modulation index*, which is equal to the maximum peak-frequency deviation divided by a *modulation frequency*, which is less than or equal to the maximum modulation frequency.

In terms of sideband generation, the spectrum of an FM signal is different from an AM signal. In AM signals, the amplitude of the sidebands around the AM carrier is proportional to the amplitude of the modulating signals. Also, the sideband frequencies are just the sum and/or difference frequency such as $f_c \pm f_{mod}$, where f_c is the AM carrier frequency and f_{mod} is the modulation frequency into the AM system.

However, in a (wideband) broadcast FM signal, the spectral content is much more complicated. The FM sidebands are generally a cluster of multiple sidebands whose frequencies are

$$f_{fm\_carrier} \pm f_{mod}$$

$$f_{fm\_carrier} \pm 2f_{mod}$$

$$f_{fm\_carrier} \pm 3f_{mod}$$

$$f_{fm\_carrier} \pm 4f_{mod}$$

That is, the sidebands include the first pair of sidebands like an AM signal but also include sidebands that are related to the harmonics of the modulating signal. The higher the frequency deviation, the more the higher-order sidebands are generated. Thus FM is not a linear process in terms of sideband generation, whereas AM has a linear relationship with its AM sidebands.

The actual analysis of FM sidebands takes a combination of higher-order mathematics including *Taylor series expansion* and *Bessel functions*. Fortunately, we will not cover this type of analysis in this chapter.

However, there is an approximation one can use to determine the bandwidth of an FM signal, which is *Carson's rule*. The FM bandwidth BW_{FM}, based on knowing the peak frequency deviation $f_{peak\_deviation}$ and the modulation frequency f_{mod} is

$$BW_{FM} = 2(f_{peak\_deviation} + f_{mod}) \tag{11-3}$$

For example, for an FM station with a peak frequency deviation of 75 kHz and a maximum modulation frequency of 15 kHz, the bandwidth is

$$BW_{FM} = 2(f_{peak\_deviation} + f_{mod}) = 2(75 \text{ kHz} + 15 \text{ kHz}) = 2(90 \text{ kHz}) = 180 \text{ kHz} = BW_{FM}$$

For analog TV signals, the audio FM carrier frequency after the video signal has been recovered in an analog TV tuner is 4.5 MHz in the United States. The maximum frequency shift or deviation is ±25 kHz, with typically a maximum modulating frequency of 15 kHz. Analog TV signals have been replaced with the digital TV transmission standard [e.g., Advanced Television Systems Committee (ATSC)], and therefore, FM audio signals are not used with TV signals except for some cable and amateur TV applications.

In the United States, the FM signal was invented by Edwin Armstrong as an alternative to amplitude modulation. FM signals have the advantage of providing higher-quality demodulation of the transmitted signal. This is true when the signal is fading in and out. In the case of an FM signal, once a sufficient level of the FM RF signal is received, for the most part, no fading is heard. In AM signals, fading or slow-moving changes in the strength of the AM signal cause an equivalent effect on the demodulated signal, which changes in volume.

For generating an FM signal, basically an oscillator is electronically varied in frequency, usually by a varactor or varicap tuning diode. There are other ways to achieve frequency modulation of an oscillator, such as electronic variable inductance (reactance modulator or electronically controlled gyrator), resistance-capacitance oscillators (e.g., an NE555 timer chip), or phase modulation. For this chapter, we will look at just using electronic tuning diodes as time-varying capacitors resonating with inductors to form tank circuits in LC oscillators.

Experiment No. 1: FM Oscillators Using Voltage-Controlled Capacitance Diodes

For a first experiment on frequency modulation, let's build a low-power FM transmitter. This circuit will provide an FM signal within the FM band of 88 MHz to 108

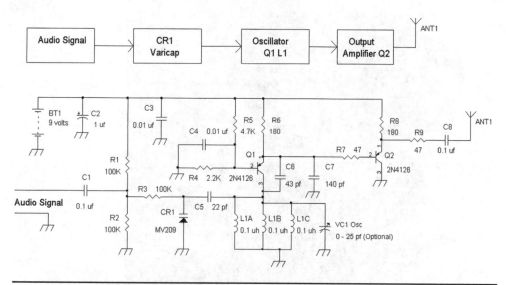

FIGURE 11-2 An FM oscillator with an MV209 tuning diode.

MHz. An optional variable capacitor allows the transmit frequency to work below 88 MHz if desired (see Figure 11-2).

The FM oscillator in Figure 11-2 is a Colpitts oscillator using positive-feedback capacitors C6 and C7. With the values shown, the oscillation frequency is about 97 MHz. If the optional variable capacitor VC1 is installed, the carrier frequency can be lowered. For a ±75 kHz deviation or shift of the carrier frequency, an audio signal of about 350 mV peak to peak is connected to C1 and coupled over to varactor diode CR1 via R3. Varactor diode CR1, an MV209 tuning diode, provides the necessary voltage-dependent capacitance to electronically vary the oscillator's frequency and generate the FM signal. Note that the cathode of MV209 is the lead on the right side when observing the flat portion of the TO-92 case with the leads pointing down.

Q1 forms a common-base amplifier via capacitor C4 providing an AC ground to its base. The collector current is set by voltage divider circuit R4 and R5 that places about +2.9 volts at the base of Q1. The emitter voltage is then 0.7 volts *above* the base voltage or +3.6 volts. Because the collector current is determined by the emitter current, which is the voltage across R6 divided by the resistance of R6, the collector current is then (9 volts − 3.6 volts)/180 Ω = 30 mA. The power across R6 is then I^2R = (30 mA)^2 × 180 Ω = 162 mW ≈ 1/6 watt. Although the suggested wattage rating for all resistors is ¼ watt, R6 and R8 would be better served with ½ watt resistors.

CR1 provides the variable capacitance for the FM effect and is reverse biased at about +4.5 volts. If more deviation is required, C5's value can be increased. However, the carrier frequency will drop. Adding another inductor (e.g., 0.1 μH to 1 μH) in parallel with L1C will increase the carrier frequency. The output of the oscillator via

Figure 11-3 Prototype circuit of a Colpitts FM oscillator.

Q1's emitter is fed to an emitter-follower Q2 to provide a buffered FM RF signal. Buffering the RF signal provides stability to the oscillation frequency when the output terminal is connected to various antennas with different equivalent capacitances to ground. For example, hooking up an antenna straight to the emitter or collector of Q1 will cause the oscillation frequency to shift dramatically because of the parasitic capacitance of the antenna. Figure 11-3 shows a prototype breadboard of the circuit in Figure 11-2.

As mentioned in Chapter 5, power rectifiers also can be used as varactor diodes. Small amounts of voltage-controlled capacitance can be provided with just about any diode or rectifier. Chapter 5 suggested that power diodes can be used. For example, a 1N4002 diode does exhibit varactor diode characteristics. However, this 1-amp diode only provides small amounts of capacitance changes as a function of reverse-bias voltage. A larger-current diode such as the 1N5401 or 1N5402 provides variable capacitances comparable with some standard varactor diodes. Note that CR1 may be a 1N5400–1N5405 diode (see Figure 11-4).

The 1N5401 or 1N5402 power rectifier will have somewhat less capacitance than the MV209, but nevertheless, it can provide an FM effect with less frequency devia-

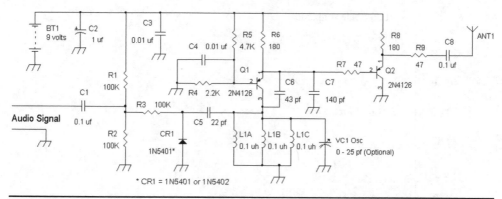

FIGURE 11-4   FM oscillator using a 1N5401 or 1N5402 power rectifier as the electronic tuning diode.

tion or shift than the MV209 varactor diode. To increase the frequency deviation while using the 1N5401 or 1N5402 power rectifier as a varactor, C5 can be increased from 22 pF to 100 pF or as high as 0.01 μF.

In both FM oscillators from Figures 11-2 and 11-4, using signals containing music may result in the sound being a little rolled off in the higher frequencies, but the circuits are adequate for voice transmission. With music, one will notice that both FM oscillators have a treble loss or treble cut when receiving the oscillator's signal with a standard FM stereo receiver. The reason for this treble cut is that *all* FM stereo receivers have a built-in *de-emphasis* filter, which is a simple RC low-pass filter at 2.12 kHz (75-μs time constant, where μs = μsecond = microsecond).

Pre-Emphasis and De-Emphasis Transmitting and Receiving Equalization Curves

One of the characteristics of oscillators is *phase noise*. All oscillators have jitter or some time instability between each cycle of oscillation. For oscillators using crystals instead of inductors and capacitors, the phase noise is very low. But to generate an FM signal, inductor-capacitor oscillators are used. If a phase detector is used to convert the phase noise into a voltage, we will find, for example, that the noise from the output of the phase detector is flat in frequency spectrum. However, if we use a frequency-modulation detector instead, we will find that the noise voltage out of the FM detector is a boosted at high frequencies. This is equivalent to having the same noise from the phase detector with an added audio middle- to high-frequency boost.

That is, when any FM demodulator is used prior to a de-emphasis network, such as the FM detector in your stereo receiver or tuner, that FM detector adds a treble boost to white noise (flat frequency spectrum noise) (see Figure 11-5). Notice that the noise spectrum shape from an FM detector has a rising or upward ramp emphasis, which is commonly called a *triangular-noise spectrum*.

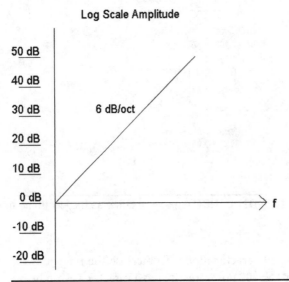

FIGURE 11-5   Noise spectrum from the FM detector on a zero-modulation FM carrier with random phase noise.

If we apply a simple low-pass filter after the FM detector, we get to attenuate the noise by reshaping the triangular noise spectrum back to mostly a flat noise spectrum (see Figure 11-6). However, if we now decide to modulate the FM transmitter with signals beyond the cut-off frequency of the simple RC filter after the FM demodulator, the frequency response of the transmission will suffer accordingly and roll off (see Figure 11-7).

In one sense, we "fixed" the noise problem on the receiving end of the FM oscillator but caused another problem by having the demodulated FM signal suffer with a treble cut frequency response. So how do we get out of this mess? We can add a

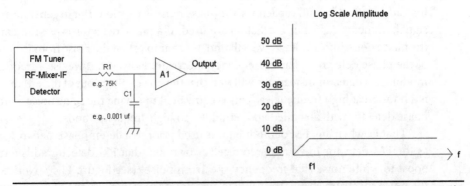

FIGURE 11-6   Simple RC filter to transform the triangular noise spectrum back to a mostly flat noise spectrum.

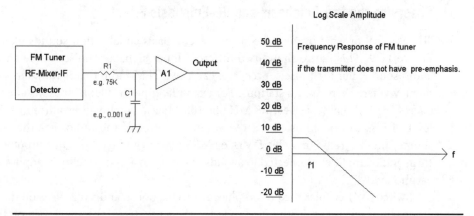

FIGURE 11-7   Frequency response with an RC filter on the receive side to reduce triangular noise.

treble boost to the audio signal prior to its being frequency modulated. For example, if the RC filter at the receiver end has a 75-μsecond time constant for the roll-off at 2.12 kHz, we use an RC treble boost filter circuit to boost the treble starting at 2.12 kHz for the modulating signal at the transmitter end to cancel out the roll-off on the receiving end.

The treble boost at the FM modulator or transmitter side is called *pre-emphasis*, and the treble cut or low-pass filtering at the receiver is called *de-emphasis*. Figure 11-8 presents pre-emphasis and de-emphasis equalization curves. Note that the "corner frequency" for the boost (f1) at the transmitter side must be matched at the receiver side in order to maintain a flat response for the received modulating signal.

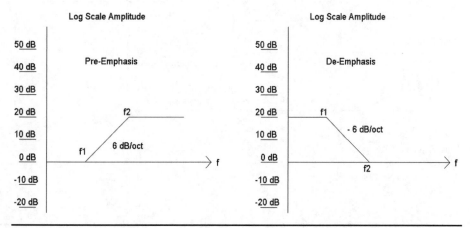

FIGURE 11-8   The transmitter's pre-emphasis and receiver's de-emphasis frequency responses.

Observing the Pre-Emphasis and De-Emphasis Effect

To get a feel of what pre-emphasis and de-emphasis equalization sounds like, see Figure 11-9. This figure shows two different input signals, a noise source and an audio source. For the noise source, one can use an FM radio that is tuned to a frequency where there are no stations. For example, tune to "blank" stations at either end of the FM dial (e.g., near 88 MHz or 108 MHz), which will produce hiss. The level of hiss can be adjusted via the volume control of the radio. For the audio source, another radio or a CD, DVD, or MP3 player can be used. The output is fed to an audio amplifier connected to a loudspeaker or to a high-fidelity headphone or earphone.

Switch SW1 enables the pre-emphasis or treble boost, and SW2 turns on the de-emphasis or treble cut. With noise added to the audio (music) source, turn off both SW1 and SW2, and listen to the amount of noise in the background of the music. Now turn on both SW1 and SW2, and you should not notice a change in fidelity, but the noise level should be lower. R8 can vary from about 22 Ω to 220 Ω to change the overall audio level to the headphones. If you want to listen to just pre-emphasis, turn SW1 on and SW2 off. And turning SW1 off and SW2 on will let you listen to what de-emphasis sounds like to a flat frequency-response signal.

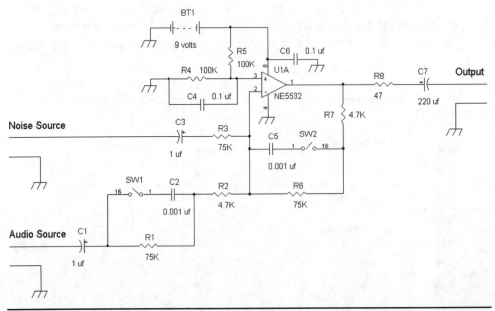

Figure 11-9 Experiment with pre-emphasis (via SW1 turned on) and de-emphasis (via SW2 turned on) networks.

FM and a Pre-Emphasis Network

Pre-emphasis of the audio signal is achieved by adding an RC network. See Figures 11-10 and 11-11. With the addition of R10 (91 kΩ), R11 (12 kΩ), and C9 (820 pF) in Figure 11-10, the modulator now has a pre-emphasis network. R11 (12 kΩ) via AC

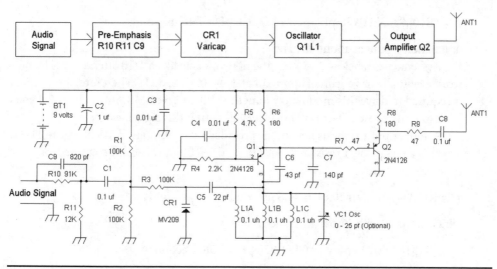

FIGURE 11-10   Two-transistor FM modulator with pre-emphasis network.

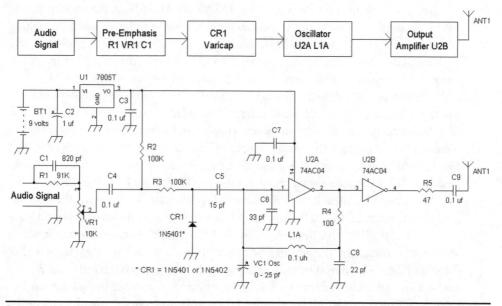

FIGURE 11-11   FM modulator with pre-emphasis network, inverter gates, and 1N5401/1N5402 varactor diode.

"short" C1 is in parallel to R1 (100 kΩ) and R2 (100 kΩ), which results in a total resistance of about 9.7 kΩ. The first time constant τ_1 for the pre-emphasis is

$$\tau_1 = R10C9 = (91 \text{ k}\Omega)820 \text{ pF} = 74.6 \text{ μseconds} \approx 75 \text{ μseconds},$$

which starts a treble boost at 2.12 kHz. The second time constant τ_2 is where the treble boost flattens out, and this is

$$\tau_2 = (R10\|9.67 \text{ k}\Omega)820 \text{ pF} = (8.74 \text{ k}\Omega)820 \text{ pF} = 7.16 \text{ μseconds}$$

or a flattening out at about 22 kHz.

At frequencies below 2.12 kHz, the network acts like a 1/10 attenuator. The network eventually has a gain of approximately 0.707 at 22 kHz. Therefore, more signal is required to drive this modulator, something like 1 volt RMS AC or about 2.8 volts peak to peak. Most CD or DVD players will output this signal level.

A similar pre-emphasis network is provided by R1, C1, and VR1 in Figure 11-11. There is some parallel resistance with R2 and VR1, but this can be ignored, and the time constants are

$$\tau_1 = R1C1 = (91 \text{ k}\Omega)820 \text{ pF} = 74.6 \text{ μseconds} \approx 75 \text{ μseconds}$$

which starts a treble boost at 2.12 kHz. Then:

$$\tau_2 = (R1\|10 \text{ k}\Omega)820 \text{ pF} = (9.0 \text{ k}\Omega)820 \text{ pF} = 7.38 \text{ μseconds}$$

which flattens out at about 21.5 kHz and is essentially the same as the pre-emphasis network of Figure 11-10. Now see Figure 11-11, CR1.

This oscillator uses a power rectifier 1N5401 or 1N5402 as the varicap diode. Again, the diode CR1 is reverse biased, so there is close to infinite resistance looking into the anode of CR1 at audio frequencies. The oscillator uses a logic inverter gate U1A and is direct-current (DC) biased via R4 and L1A to use the logic inverter gate as an inverting amplifier. Oscillation occurs because the phase-lag network via R4 and C8 adds negative phase shift to the phase-lag network of L1A and C6 in parallel with VC1 and the series capacitance of C5 and voltage-controlled capacitance of CR1. This type of oscillator takes at least a few minutes to stabilize in frequency, and the 7805 voltage regulator U1 ensures a stable power-supply voltage to minimize frequency drift. Full ±75-kHz frequency deviation is achieved by applying about 2.8 volts RMS (8 volts peak to peak) at the audio signal input terminal and turning VR1 to maximum.

The output of U1A is then buffered with logic inverter U1B, which then is coupled to the antenna. This oscillator circuit will also emit harmonics of approximately 200 MHz, 300 MHz, and beyond. With VC1, a trimmer variable capacitor set to minimum capacitance, the approximate oscillation frequency is 94 MHz. A higher frequency can be achieved by paralleling an inductor between 0.33 μH and 3.3 μH across L1A. An oscillator using a 74AC04 gate has a "top speed" or oscillation frequency in excess of 150 MHz. Figure 11-12 shows a prototype breadboard of the 74AC04 FM oscillator without the pre-emphasis network.

Figure 11-12   Prototype breadboard of the 74AC04 FM oscillator without the pre-emphasis network.

FM Radios

Unlike AM radios that can be implemented in a non-superheterodyne manner such as a crystal, TRF, superegenerative, reflex, or regenerative radio, FM radios from the beginning of commercial manufacture were superheterodyne in nature. Yes, there are some rare designs that use a superregenerative architecture, but these designs are generally sold only as hobbyist kit radios.

Virtually all commercially available FM radios come from a superheterodyne circuit (see Figure 11-13). The basic FM radio has many common elements with the AM superhet. For example, there are the local oscillator, mixer, intermediate-frequency (IF) amplifier, and audio circuits. In many FM receivers, there are no AVC (automatic volume control) systems such as the ones that are common in AM superhet radios. Generally, the last IF amplifier causes the IF signal to limit its amplitude. This can be done with a differential-pair amplifier to ensure that the waveform is a symmetrical squarewave signal that will be inputted to the FM detector. By presenting a clipped sine wave that is transformed into a squarewave, any small amplitude variation that causes fading of the RF signal between the transmitter and receiver is processed to be ignored by the limiter or clipping amplifier.

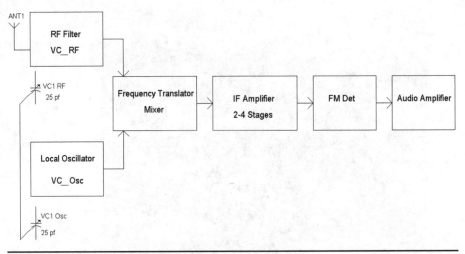

Figure 11-13   Block diagram of a superheterodyne FM radio.

In terms of a receiving antenna, there are no air-core or ferrite-core antenna coils as used in AM radios. Instead, most FM radios use a wire antenna in the form of a telescopic whip antenna or a piece of wire at least 6 inches long. Another difference in the FM radio is that there are generally more IF stages in terms of amplifiers and IF filters than those found in AM radios. The reason is that the signal from the wire antenna is very low in strength compared with the signal from an AM antenna coil. For example, a ferrite-bar antenna may output millivolts of RF signals, while the FM antenna will generally receive only tens of microvolts. In an AM radio, one IF amplifier stage with one or two IF filters will suffice. But in an FM radio, at least two stages of amplification are needed with two IF filters.

Most AM radios did not have an RF amplifier stage. For an FM radio, an RF amplifier, generally a common-base stage, provided isolation from the local oscillator such that the wire or telescopic antenna did not radiate signals from the oscillator circuit.

Finally, the most important distinguishing feature of an FM radio versus its AM counterpart is the detector or demodulator. AM radios use simple diode rectification or power detectors to recover the envelope information of the amplitude-modulated signal. In FM radios, however, the IF signal should have no form of amplitude modulation. Therefore, standard AM detectors will generally not work for demodulating the FM signal straight from the IF filter.

What most FM detectors (e.g., stagger tuned, Foster Seeley FM discriminator, differential peak, and ratio detector) do is to first convert the FM signal into an AM signal and then use the diodes as half-wave rectifiers to recover the audio signal. We will look into some more details of the different types of FM detectors later in this chapter.

In summary, differences between FM and AM radios include

- There are tuned RF stages in FM radios, whereas AM radios typically run the RF signal to the converter or mixer directly.
- There are generally more IF amplifiers and IF filters in FM radios than in AM radios.
- FM detectors are more complicated and use simple AM detection only after converting the FM signal into an AM signal.
- Telescopic or wire antennas are used to capture the FM signal versus using antenna coils for AM radios.

Historically, in the 1950s and 1960s, commercially manufactured transistor FM radios were hardly ever sold as FM-only radios. One example in 1963 is that Hitachi made a nine-transistor radio, the KH-915, that was FM only. This radio had an RF amplifier, three stages of IF amplification, three 10.7 MHz IF transformers, and a ratio (FM) detector. However, almost all radios that had FM also had AM, and some added shortwave (SW) as well.

Let's take a look at a multiband portable AM-SW-FM radio from the late 1960s, the Ross Model RE-1915-N. Figure 11-14 is a partial schematic of the radio showing the FM section. As shown in the RE-1915-N AM-FM-SW radio schematic transistor,

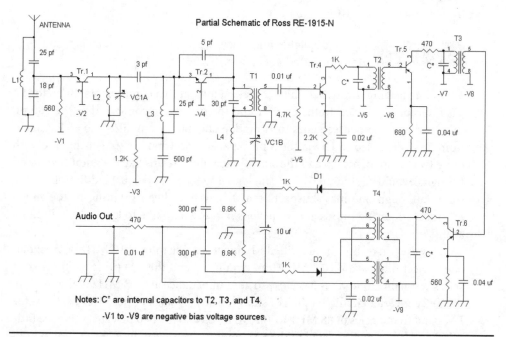

Figure 11-14 A partial schematic diagram of a Ross RE-1915-N radio.

Tr.1 is a grounded base amplifier that receives the RF signal via a telescopic rod antenna into the emitter. Because the antenna is a low-impedance voltage source and is connected directly to the parallel inductor-capacitor circuit with L1 and capacitors, the bandwidth is actually large enough to cover the entire frequency range of 88 MHz to 108 MHz.

The collector (current source) output of Tr.1 is then connected to a tunable RF filter formed by the FM RF variable capacitor section that is in parallel with inductor L2. In order to preserve the Q, or quality, factor of this parallel LC circuit, a small-valued 3 pF capacitor is used to couple the amplified tuned RF signal from the collector of Tr.1 to the emitter of Tr. 2, an oscillator-mixer circuit otherwise known as a *converter circuit*. By using the 3 pF coupling capacitor, loading on the tuned RF tank circuit with L2 is minimized, and as the RF section variable capacitor is tuned across the FM band, good variable-frequency band-pass filtering is achieved. This band-pass filtering is essential for rejecting out-of-FM band signals and for providing image rejection.

Transistor Tr.2 is a variable-frequency Colpitts oscillator with the FM oscillator section of the variable capacitor in parallel with inductor L4. The collector of Tr.2 is connected to positive-feedback capacitors whose values are 5 pF across L4 and emitter and a 25 pF capacitor from the emitter to a large-valued 500 pF capacitor. The other end of the 500 pF capacitor is tied to ground. The collector of Tr.2 is also connected to 10.7 MHz IF transformer T1 that extracts the difference-frequency signal $(f_{osc} - f_{RF}) = f_{IF} = IF = 10.7$ MHz.

NOTE f_{osc} = oscillator frequency, f_{RF} = tuned RF frequency, and f_{IF} = intermediate frequency or IF.

Before leaving the Tr.2 circuit, let's take a look at two components. Inductor L3 has one side connected to the emitter and the other side connected to the 500 pF capacitor whose other lead goes to ground. Note that the amplified tuned RF signal from Tr.1 is coupled to L3 via the 3 pF capacitor. So what are we getting into with these two components, L3 and the 500 pF capacitor? Seems like the amplified tuned RF signal would be singing, *"We're caught in a trap,…"* but I'm just kidding about the song—although one may "suspect" that I may have been listening to too much music … enquiring minds and *suspicious minds* would like to know—just kidding again!

However, in reality, the amplified RF signal is going into a 10.7 MHz or IF signal-band reject filter, aka a 10.7 MHz trap circuit. L3 and the 500 pF capacitor form a series resonant circuit that has close to zero impedance at 10.7 MHz at the input of the Tr.2 converter circuit to notch or short out any signal at or near 10.7 MHz. At the FM band frequencies of 88 MHz to 108 MHz, L3 and the 500 pF capacitor have little effect and pass signals related to the FM band.

There are two good reasons for having a notch filter prior to the input of the converter. One is so that 10.7 MHz signals from other FM radios will not leak through the RF front-end circuit and, second, because there are multiple stages of IF amplifiers that deliver very high voltage gains at 10.7 MHz. The 10.7 MHz trap reduces the chance that IF signals can oscillate by leaking back to mixer Tr.2, which is still an amplifier.

Moving on, we now need to amplify the IF signals. In this radio, Tr.4, Tr.5, and Tr.6 form a three-stage IF amplifier system. IF transformers T1, T2, and T3 provide band-pass filtering at 10.7 MHz. The alternate channel selectivity (amount of attenuation 400 kHz away from 10.7 MHz) is about 12 dB per stage. With three stages, the alternate channel selectivity is about 36 dB. For most instances, 36 dB is acceptable but will have trouble separating close-by stations.

What I would like to share with you is that just by adding one more IF amplifier and IF filter, this radio would be improved immensely. The selectivity will go from about the mid-thirties to the middle to high forties decibel (dB)wise, which then allows very good separation of closely packed FM stations. The sensitivity also will go up. Better yet, instead of using an extra IF transformer, use a 10.7 MHz ceramic filter to take the selectivity >50 dB, which is then comparable with high-quality stereo tuners selectivity-wise.

We now turn to the FM detector of the RE-1915-N radio, which is a *ratio detector*, via T4, a special ratio detector transformer with four windings. Basically, this type of detector works on two principles:

1. One of the windings has a 90-degree phase-shifted 10.7 MHz IF signal when compared to a reference winding that is defined to have 0 degrees when the 10.7 MHz IF signal has no FM modulation.
2. As the 10.7 MHz IF FM signal shifts below and above the nominal 10.7 MHz frequency, there are added and subtracted phase shifts from the 90 degrees. For example, at 10.65 MHz, the phase shift is 105 degrees; at 10.70 MHz, it is 90 degrees; and at 10.75 MHz, the phase shift is 75 degrees. In this case, the subtracted and added phase shifts are +15 degrees and –15 degrees or +φ and –φ, where φ is the added or subtracted phase shift to the 90-degree signal.

Eventually, the two signals, the reference 0-degree signal and the second signal that has 90 degrees minus or plus the phase shift, φ, are added together.

By way of a graphing calculator, let's see what happens to the amplitude of the added signals for ±15 degrees (see Figures 11-15 and 11-16). That is, let's plot the resulting amplitudes for the following three signals:

- $\cos(x) + \cos(x + 90 \text{ degrees} - 15 \text{ degrees}) = \cos(x) + \cos(x + 75 \text{ degrees})$
- $\cos(x) + \cos(x + 90 \text{ degrees})$
- $\cos(x) + \cos(x + 90 \text{ degrees} + 15 \text{ degrees}) = \cos(x) + \cos(x + 105 \text{ degrees})$

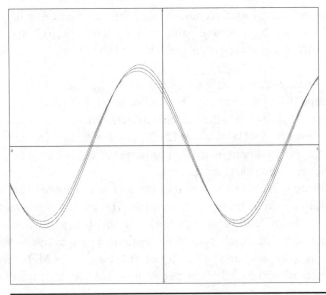

Figure 11-15 Resulting amplitudes of cosine waveforms with 0 degrees summed with 75, 90, and 105 degrees from largest to smallest amplitudes. Waveform in the middle is 0 degrees summed with exactly 90 degrees.

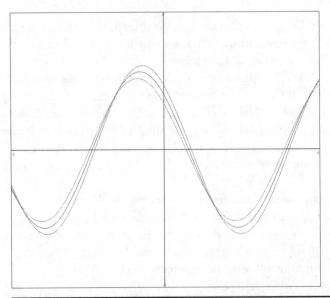

Figure 11-16 Resulting amplitudes of cosine waveforms with 0 degrees summed with 60, 90, and 120 degrees from largest to smallest amplitudes. Waveform in the middle is 0 degrees summed with 90 degrees.

As seen in Figure 11-15, the waveforms are of three different amplitudes with combinations of

- Largest amplitude: cos(x) + cos(x + 75 degrees)
- Middle amplitude: cos(x) + cos(x + 90 degrees)
- Smallest amplitude: cos(x) + cos(x + 105 degrees)

Intuitively, this makes sense. The phase-shifted cosine waveform closest to 0 degrees, which is 75 degrees, will be the largest one because two waveforms of 0 degrees will just add normally like two positive numbers.

The waveform furthest away from 0 degrees would be the one closest to 180 degrees and should be the smallest because equal amplitude waveforms of 0 degrees and 180 degrees added together would cancel out to a zero amplitude waveform. Figure 11-15 shows the amplitudes of an FM detector prior to half-wave rectification for a 15-degree phase-shift example. Suppose that the 10.7 MHz IF signal shifts are 30 degrees of phase shift for a deviation of 100 kHz instead of the 50-kHz example? And suppose that we get twice the phase shift from 90 degrees to have 120, 90, and 60 degrees for 10.6 MHz, 10.7MHz, and 10.8MHz? See Figure 11-16.

As we can see, increased frequency deviation of the IF signal will result in the ratio detector's 90-degree phase-shifting circuit to also increase the amount of phase shift added to the 90 degrees. In Figure 11-15, we see that the ratio detector's phase-shifting network "maps" frequency shift or frequency modulation to amplitude variations or amplitude modulation. For example, in Figure 11-16 the largest cosine waveform represents the IF signal shifted to 10.80 MHz, the middle-amplitude waveform represents the IF at 10.70 MHz, and the smallest waveform represents the IF signal shifted to 10.60 MHz.

A simple representation of converting the addition of the phase-shifted and non-phase-shifted signals is given in Figure 11-17. As shown in the figure, there is a 90-degree phase-shifting network that includes a series resonant circuit R1, C1, L1, and L2. At the IF (intermediate frequency) of 10.7 MHz, the nominal phase shift at

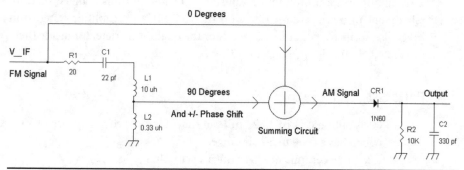

FIGURE 11-17 Conceptual diagram of converting FM to AM via a phase-shifting network.

L1 and L2 is 90 degrees. L1 and L2 form a voltage divider of approximately 1/30 because "gain" of the circuit at C1/L1 is about 30 at the resonant frequency. The phase-shifted signal and the reference FM signal V_IF are added together via the summing circuit. The output of the summing circuit produces an amplitude-modulated signal such as the amplitude variations seen in Figure 11-15. The amplitude-modulated signal is then detected or demodulated with an envelope detector circuit CR1, R2, and C2 to provide the recovered modulating signal such as audio (e.g., voice, music, etc.). It should be noted that the 90-degree phase-shifting circuit must provide additive and subtractive phase shifts outside the resonant frequency.

NOTE   If the phase shift were to be the same for all frequencies at the resonant frequency and above and below the resonant frequency, the output of the summing circuit would be a constant-amplitude waveform, free of amplitude modulation and hence free of any demodulated signal.

A balanced AM detector circuit is shown in Figure 11-14 as diodes D1 and D2. There are two diodes to provide a push-pull or balanced FM detection curve, which allows for providing 0 volts when the IF frequency is at 10.7 MHz and positive or negative voltages to output a demodulated audio signal when FM is present in the IF signal.

Historically, the ratio detector was the last of the double-tuned circuits required to demodulate FM signals. By the time FM was used for the sound portion of analog television receivers, the TV sets did have double-tuned FM detectors, but as a cost savings, eventually a single-tuned circuit was used for FM demodulation, called the *quadrature FM detector*.

For the next project, we will present an FM radio project in which a quadrature detector is used.

An FM Tuner Project with Ceramic Filter and Resonator

This circuit uses a minimum of adjustable coils or IF transformers. Instead, it will rely on one or more ceramic filters for the 10.7 MHz IF amplifier system and a 10.7 MHz discriminator ceramic resonator for the quadrature detector inside the TA2003 integrated circuit (IC) (Figure 11-18).

Parts List

- C1 is optional (1 pF to 22 pF) but may not be needed because of VC1A Pad, which has a 0 pF to 20 pF range.
- C2, 15 pF ceramic or silver mica 5% or 10%
- C3, C4, C6, C7, 0.001 μF, 5% or 10% ceramic or film capacitor

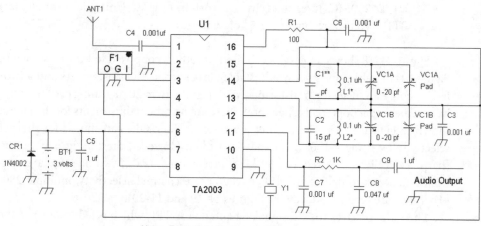

Figure 11-18 An FM tuner circuit with minimal coils and no IF transformers.

- C5, C9, 1 μF, 5% to 20% film or ceramic capacitor
- C8, 0.047 μF, 5% film capacitor
- CR1, 1N4002 or any 1N4000 Series rectifier
- L1, L2, 0.1-μH, 5% or 10% fixed or variable inductors
- Unless specified, all resistors are ¼ watt, 5%
- R1, 100 Ω
- R2, 1 kΩ
- VC1A, VC1B, a twin-gang variable capacitor 0 pF to 20 pF with two 0 pF to 20 pF pad capacitors
- Y1, 10.7 MHz ceramic resonator-discriminator (Murata Part Number CDALF10M7GA048-B0 or, alternatively, Murata Part Number CDALF10M7GA040-B0)
- Mouser part numbers, respectively, are 81-CDALF10M7GA048-B0 and 81-CDALF10M7GA040-B0 (www.mouser.com); an alternative part from www.digikey.com is the CDALF10M7GA096-B0
- F1, 10.7 MHz ceramic filter by Murata or Toko with a –3 dB bandwidth of 150 kHz to 280 kHz at 330 Ω (available at www.digikey.com or www.mouser .com)
- Typical part numbers are SFELF10M7LFTA-B0 (280 kHz) and SFELF10M7JAA0-B0 (150 kHz)

NOTE   Markings on the ceramic filter may vary and one should read the datasheet for the pin out identification.

- U1, TA2003 IC (integrated circuit) (Toshiba or UTC, Unisonic Technologies)
- BT1, 3 volt source via two 1.5 volt batteries

Figure 11-18 shows an easier-to-build FM radio as opposed to the one shown in Figure 11-14, which uses many adjustable parts such as IF transformers and a two-ferrite-core ratio detector transformer. With just the twin 20 pF tuning capacitor and its trimmer capacitors (pad capacitors), there are no other adjustments for the circuit shown in Figure 11-18. The IF filter is a ceramic filter. One can choose for sharper selectivity by using a 150-kHz bandwidth filter. However, the tuning may be too sharp. Typically, a 230-kHz or 280-kHz filter is used, and if two or more are connected in series to deliver much better alternate channel selectivity, multiple 280-kHz filters are recommended (see Figures 11-19 and 11-20).

The radio receives the RF signal via antenna ANT1, which is fed to pin 1 of the TA2003 chip via C4. The RF amplifier has a collector output that drives a tunable parallel inductor-capacitor tank circuit VC1A and L1. The amplified RF signal is coupled to a separate RF mixer. The local oscillator tank circuit is formed by L2 and VC1B. Tuning the local oscillator frequency is accomplished by VC1B, and the output of the oscillator is connected to the RF mixer. The "balanced" RF mixer in the TA2003 is more sophisticated than a simple one-transistor mixer used in AM radios.

A single-transistor RF mixer not only outputs the IF signal and the RF signal, but also the oscillator signal. However, the oscillator signal is generally the largest signal from the output of the single-transistor RF mixer or converter circuit. Generally, a ceramic filter cannot be connected directly to the output of the single transistor because the large-amplitude oscillator signal will leak through the ceramic filter.

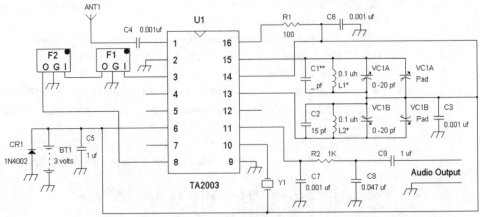

Note: F1 & F2 = 10.7 MHz Ceramic Filter

Y1 = 10.7 MHz Ceramic Resonator & Quadrature Detector

FIGURE 11-19 Improving selectivity by cascading two ceramic filters F1 and F2 (same type 10.7 MHz filters as mentioned in the Parts List).

FIGURE 11-20   A TA2003 FM tuner with two ceramic filters.

Instead, prefiltering is required, and an LC filter via an IF transformer must precede the ceramic filter.

The Gilbert mixer used in the TA2003 has at least four transistors that form a multiplier circuit to provide the IF signal while balancing or nulling out the oscillator signal. From the mixer's output connected to ceramic filter F1, the IF signal is amplified and sent to an FM detector, which is a quadrature detector. Ceramic resonator Y1 is essentially a 10.7 MHz band-pass filter resembling a parallel LC circuit. Y1 provides two characteristics to the 10.7 MHz IF signal, one is a 90 degree phase-shift signal at 10.7 MHz, and the other is a phase-shift signal in degrees below and above the 10.7 MHz IF, which varies and can be characterized as $(90 + \varphi)$ and $(90 - \varphi)$. For quadrature detection to occur, the non-phase-shifted signal and the phase-shifted signal are multiplied, which results in a signal that demodulates the FM signal. The math behind this involves sines and cosines and other trigonometric functions. Figure 11-21 provides a conceptual diagram of the quadrature FM detector.

A quadrature detector is illustrated in Figure 11-21. A multiplying circuit receives 0- and 90-degree signals from the IF amplifier. By multiplying the two signals, the additive and subtractive phase shifts caused by the IF signal shifting outside the resonant frequency (e.g., above and below or below and above 10.7 MHz) form a demodulated FM signal at the output of the multiplying circuit (e.g., NE602, SA612, or MC1496). The high-frequency "junk" that comprises signals at multiples of the

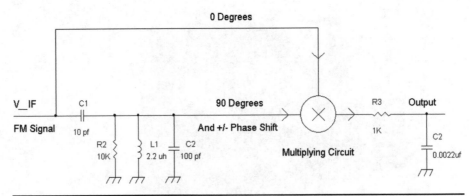

Figure 11-21   Example of an FM quadrature detector.

intermediate frequency (IF) is filtered out by R3 and C2, leaving only signals from about 20 Hz to about 80 kHz. The 80-kHz bandwidth is required to receive the main audio channel signal from 50 Hz to 15 kHz, the stereo difference channel signal from 23 to 53 kHz, and any other signal up to about 76 kHz.

Now let's look more closely at the LC circuit (see Figure 11-22). Typically, in quadrature detectors, a 90-degree phase-shifter circuit is implemented by converting a parallel resonant circuit that normally has 0 degrees of phase shift at the resonant frequency into a series resonant circuits that has a 90-degree phase shift. At first glance, it would appear that we are only dealing with parallel resonant circuit L1 and C2. But, on closer inspection, we see that it is driven by a small-value (*high-impedance*) capacitor. If we just look at the LC circuit, we find out that we can make a *Thevenin* equivalent circuit of C1, C2, and V_IF.

From the dashed lines shown on the left side circuit in Figure 11-22, we can "separate" the LC circuit and rearrange it as a capacitive voltage divider of C1 and C2. This capacitive voltage divider circuit then drives R2 and L1. But the capacitive divider can be seen as a Thevenin equivalent circuit as a voltage source divided down by (the voltage dividing effect of) C1 and C2 and, more important, as an equivalent *series* capacitor C1 ∥ C2 that is driving R2 and L1. Thus, by capacitively coupling the LC circuit of L1 and C2 with C1, we actually get the equivalent of a series resonant circuit, which then provides the required 90-degree phase shift plus additive and subtractive phase shifts from the 90 degrees when the IF is outside the resonant frequency (e.g., 10.7 MHz). As we observe the series resonant circuit on

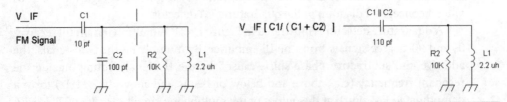

Figure 11-22   A Thevenin equivalent LC circuit from parallel to series resonant.

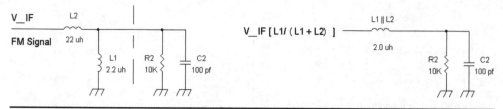

Figure 11-23 An extra inductor L2 is used to transform a parallel resonant circuit to a series resonant circuit.

the right side of Figure 11-22, we see that it is a high-pass filter that has a high Q factor (e.g., narrow bandwidth around the resonant frequency). The phase shift is leading to +90 degrees at the resonant frequency. Also note that C1 is generally 10 percent or less of C2. Note that R2 determines the Q of the circuit, and for wider-band FM detection, R2 should be lowered in value (e.g., 1 kΩ < R2 < 10 kΩ).

Similarly, we can use a large-value inductor (*high impedance*) in series with a parallel LC circuit to also make a 90-degree phase shifter (see Figure 11-23). As seen in Figure 11-23, we can again form the series-driven LC circuit as an inductive voltage divider to form a Thevenin equivalent circuit. This Thevenin equivalent circuit is shown on the right side of Figure 11-17*b*, which shows a series resonant circuit forming a high Q low-pass filter. Again, there is a 90-degree phase shift with this circuit, but the phase shift is lagging. Thus, at the resonant frequency, the phase shift is –90 degrees. For choosing the value of L2, normally it has at least 10 times the inductance of L1. Note that R2 determines the Q of the circuit, and for wider-band FM detection, R2 should be lowered in value (e.g., 1 kΩ < R2 < 10 kΩ).

A Second FM Radio Project Via Silicon Labs

Making FM radios for the do-it-yourself (DIY) hobbyist can be very challenging because of the very high frequencies involved. I deliberately avoided showing as projects the more "traditional" FM radios using discrete transistor designs with IF transformers and ratio detector coils. The alignment procedure for these types of radios requires sweep generators, calibrated FM modulators, oscilloscopes, and possibly spectrum analyzers. Also, it is very hard to obtain ratio detector coils easily unless one buys an FM radio kit such as those available from www.elenco.com via this specific link from Elenco: www.elenco.com/product/productdetails/radio_kits= MTc=/am--fm_radio_kit(combo_ic_&_transistor)=NTk1.

Therefore, a really easy FM radio project that also includes the capability to add an AM section for standard broadcast and shortwave bands can be implemented via a Silcon Labs 4836 chip, which is available at www.mouser.com. This IC (integrated circuit) is only available in a standard small outline surface-mount form factor, but it can be soldered to a 16-pad adapter board for easy access to the 16 pins. The

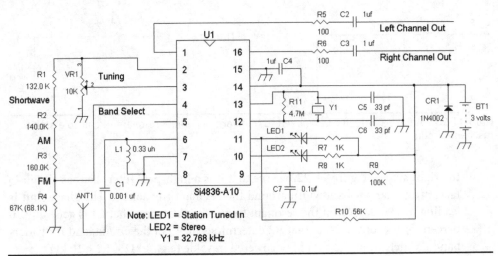

FIGURE 11-24 Silicon Labs FM stereo tuner schematic.

original prototype was built on copper-clad with the chip "floating" above the copper ground plane. Figure 11-24 shows the schematic.

Parts List

- C1, 0.001 μF, 5%, 10%, or 20% ceramic or film capacitor
- C2, C3, C4, 1 μF, 5% or 10% film or ceramic capacitor
- C5, C6, 33 pF, 5% or 10% ceramic or silver mica capacitor
- C7, 0.1 μF, 5%, 10%, or 20% ceramic or film capacitor
- L1, 0.33 μH, 5% or 10% inductor (alternatively, L1 can be any fixed inductor between 0.27 μH and 2.7 μH)
- R1, 132 kΩ, 1% or 133 kΩ 1% resistor
- R2, 140 kΩ, 1% resistor
- R3, 160 kΩ 1%, 162 kΩ 1%, or 158 kΩ 1% resistor
- R4, 67 kΩ 1%, 66.5 kΩ 1%, or 68.1 kΩ 1% resistor
- R5, R6, 100 Ω, 5% resistors
- R7, R8, 1 kΩ, 5% resistor
- R9, 100 kΩ, 5% resistor
- R10, 56 kΩ, 5% resistor
- R11, 4.7 MΩ, 5% resistor
- Y1, watch crystal, 32.768 kHz
- CR1, 1N4002 rectifier or any 1N4000 series diode
- BT1, 3 volt source from two batteries such as two AA cells
- LED1, LED2, any green, yellow, or red (preferably high-brightness) light-emitting diodes (LEDs) *Note:* Do not use a white or blue LED because the

power supply is already 3 volts, and these LEDs require at least 3 volts to turn on brightly.

- U1, Silicon Labs IC (integrated circuit) Si4836-A10
- VR1, 10-kΩ pot, single turn

NOTE   The schematic shown in Figure 11-24 is different from the data sheet circuit of the Si4836 chip. The data sheet shows that the 32.768-kHz crystal oscillator circuit is optional. However, it is *not*. If the radio is built without the crystal oscillator circuit Y1, C5, C6, and R11, the radio will not work. Hence the crystal oscillator circuit is essential.

From the schematic of the demo board, the extra parts Y1, C5, C6, and R11, along with the LEDs and R10, are shown and are to be loaded. Once the parts are connected as shown in Figure 11-24, you should be able receive FM stations in stereo. This chip uses the latest software-defined radio techniques as of the year 2013. The RF signal is mixed down to form 0 degree and 90 degree signals, which are converted via two analog-to-digital converters into a digital signal processor to set the bandwidth of the IF and demodulate the FM signal all in the digital domain. Two digital-to-analog converters provide left and right channel analog audio outputs.

Tuning and band selection are accomplished by a variable resistor and a voltage divider circuit. As shown in the figure, a four-resistor voltage divider circuit via R1, R2, R3, and R4 provides various voltages to select shortwave, AM, or FM bands. Tuning is accomplished via a single 10 kΩ potentiometer or variable resistor.

This schematic is just for receiving FM stations. To receive shortwave and AM signals, an antenna coil is required. The spec sheet can be downloaded at www.silabs .com/Support%20Documents/TechnicalDocs/Si4836-A10.pdf, and the demo board information can be found at www.silabs.com/Support percent20Documents/ TechnicalDocs/Si4836DEMO.pdf. As with the Si4836 and TA2003 tuners, the audio output should be connected to an audio power amplifier such as an LM386 circuit or to a stereo amplifier.

When one sees the Silicon Labs tuner chip and compares it with the tuner section (mid-1960s circuit design) in Figure 11-14, a radio that uses discrete transistors, coils, variable capacitors, IF transformers, ratio detector coils, one sees that progress certainly has been made over the span of 50 years in radio technology.

Figure 11-25 shows a prototype breadboard of the Silicon Labs FM tuner, and Figure 11-26 shows some surface-mount IC adapters that can be used to more easily use the Silicon Labs Si4836 surface-mount SO-16 IC (integrated circuit).

The adapter boards in Figure 11-26 are from ElectroBoards and SURFBOARDS. There are other companies that also make these types of boards. Note that one can solder the corner leads (e.g., pins 1, 8, 9, and 16) of the SO-16 (small outline) surface-mount chip first and then solder the rest of the leads. Use a very small soldering tip, and the wattage of the soldering pencil should not exceed 40 watts. Typically, a 15-watt to 25-watt soldering iron should work.

FIGURE 11-25   Prototype radio using the Si4836 chip that is "hanging" in midair.

FIGURE 11-26   Various surface-mount IC adapter boards.

Understanding "Other" FM Detectors from a Top View

We have seen that a 0-degree and a 90-degree IF signal can be used for detecting FM signals via a ratio detector or quadrature detector circuit. Actually, another circuit, the Foster Seeley discriminator circuit, which preceded the ratio detector, works off the same principle of combining 0-degree and 90-degree signals, but with the diodes connected in the same direction and the detected FM signal taken across the two diodes in differential mode. In a ratio detector, as shown in Figure 10-1, the two diodes, demodulated FM signals from D1 and D2 are summed together via two 6.8 kΩ resistors. The other types of FM detectors we will be looking at are the phase-lock loop detector, differential peak (detector), and staggered tuned FM demodulators.

Let's start with the phase-lock loop detector (see Figure 11-27). A phase-lock loop circuit is a negative-feedback system very much like an op amp. In an op amp circuit, as shown in Figure 11-27, the input voltage Vin "forces" the output voltage Vout to be the same as Vin. With phase-lock loop circuits, we are working with the frequency and phase of the input signal Vin(f) and the output signal Vout(f). Ideally, if the input signal has a frequency f, the output from the voltage-controlled oscillator (VCO) also will have a signal whose frequency is f. Depending on the phase detector, if the frequencies are the same, the output Vout(f) typically either has a 0 degree or 90 degree phase difference from the input signal.

The phase-lock loop operates in the following manner: the input signal Vin(f) is considered a reference signal, and it is connected to one of the inputs of the phase detector. The output of the voltage-controlled oscillator is connected to the remaining input of the phase detector. A voltage-controlled oscillator is simply an oscillator whose frequency can be changed by applying a control voltage. For example, the FM modulator circuits using varactor or varicap diodes are voltage-controlled oscillators (see Figure 11-2). The only difference between the FM modulators from the earlier part of this chapter and a voltage-control oscillator for a phase-lock loop is that there is no AC coupling capacitor (see DC blocking capacitor C1 in Figure 11-2). Instead, the control voltage, which includes a DC voltage, would be connected directly to R3 in Figure 11-2.

The phase detector is typically a multiplier circuit or an Exclusive OR logic gate to provide phase detection and a 90-degree-offset phase shift at the output of the voltage-controlled oscillator (VCO) when compared with the input signal's phase. If

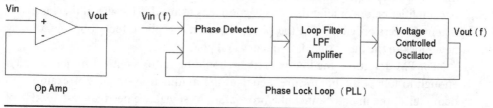

Figure 11-27 Two negative-feedback systems, an op amp and phase-lock loop circuit.

the phase-lock loop uses flip-flops or ramp and sample circuits for the phase detector, the phase offset typically will be 0 degrees, and the VCO's output signal will match the phase of the input signal.

When the phase-lock loop circuit receives input signal Vin(f), VCO may not be locked on frequency or phase with the input signal. The phase detector will generate a phase error signal that is low-pass filtered because the output of the phase detector contains frequencies that are the sum and difference of the input and VCO frequencies. By filtering out the signal that has a sum frequency of the VCO and input signal, we are left with a signal related to the difference between the frequency and phase of the input and VCO signals:

$$V_{pd}\cos[2\pi(f_{in} - f_{VCO})t + (\varphi_{in} - \varphi_{VCO})]$$

where V_{pd} is a voltage determined by the gain of the phase detector circuit, f_{in} and φ_{in} are the frequency and phase of the input signal, and f_{VCO} and φ_{VCO} are the frequency and phase of the VCO.

If the VCO is not locked to the input signal frequency-wise, we see that the phase detector's output is a sinusoidal waveform of frequency $(f_{in} - f_{VCO})$. Those who have played with phase-lock loop circuits will know this effect very well when the input signal and voltage-controlled oscillator's signals are not locked or in sync. One may ask, but shouldn't the output of the phase detector show DC voltage? Or how can a phase detector give out a sinusoidal waveform anyway?

The reason that a phase detector can give sinusoidal waveforms is that the term $2\pi(f_{in} - f_{VCO})t$ is really an angle. This seems hard to believe, but it's true. For any frequency f, $(2\pi f)t = \omega t$, and ω is measured in radians per second and t is measured in seconds. The dimensional unit, then, for ωt is [radians/second][second] = radians … and a radian is an angle (e.g., approximately 57.3 degrees). Thus the ωt term has to be an angle.

If the VCO is locked in frequency with the input signal, then $f_{in} = f_{VCO}$, and we see a familiar phase detector output signal as:

$$V_{pd}\cos[2\pi(f_{in} - f_{VCO})t + (\varphi_{in} - \varphi_{VCO})] = V_{pd}\cos[0 + (\varphi_{in} - \varphi_{VCO})] = V_{pd}\cos(\varphi_{in} - \varphi_{VCO})$$

because $[(2\pi(f_{in} - f_{VCO})t] \to 0$ when $f_{in} = f_{VCO}$.

Okay, enough digression into math for now, and back to the phase-lock loop. As mentioned earlier, the output voltage from the phase detector is filtered to remove high-frequency signals that would frequency modulate the VCO at high frequencies. An amplifier before or after the filter section may be needed because if there is insufficient conversion gain from the phase detector, the phase-lock loop's VCO may never lock to the input signal, frequency and phase–wise.

For FM detection, the low-pass filter should be set to a cut-off frequency high enough to follow the highest frequency of the modulation signal. For example, to demodulate the main channel of a broadcast FM signal, we need at least 15 kHz of bandwidth. This means that the low-pass filter has to cut off at 15 kHz or higher, and

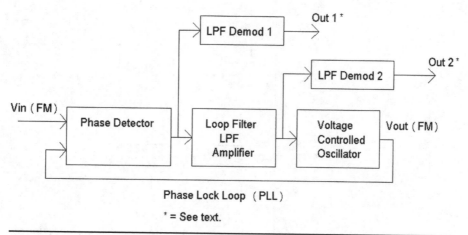

Figure 11-28 A phase-lock loop FM detector with separate filters to permit optimal recovered bandwidth of the modulating signal.

this filter is generally a simple one-section RC filter. Generally, the capacitor has an extra resistor in series with it of about 1/10 (or less) the value of R to form a lag-lead network so that the phase-lock loop does not parasitically oscillate at a low frequency.

To detect FM signals, one can optimize the low-pass filter for the VCO to track the FM signal. The voltage that drives the VCO is then a replica of the modulation signal (see Figure 11-28).

In this block diagram, the loop filter LPF is optimized to allow the VCO to track the frequency deviations from the input signal Vin (FM). The loop filter may not be optimal as far as recovering the full bandwidth of the FM channel, which can be as high as 15 kHz or up to 76 kHz for subcarrier stereo and other information signals transmitted in the FM signal. Thus one can take the signal from the output of the phase detector and filter it accordingly for a wide bandwidth via LPF Demod 1 or at the output of the loop filter-amplifier block via LPF Demod 2.

Phase-lock loop circuits are still used for demodulation of FM signals, but their signal-to-noise ratio is generally poorer than a ratio detector or Foster Seeley discriminator circuit. The reason is that the VCO itself introduces phase noise to the demodulation process.

We now enter the world of FM slope detection, which was improved to staggered tuned and differential slope detection. Let's see what is meant by slope detection (see Figure 11-29) and note that the vertical axis is at 10.7 MHz.

Single-slope detection is perhaps the simplest way to demodulate FM because most IF systems include a band-pass filter. This type of detection deliberately uses a filter to vary the amplitude of the FM signal depending on the instantaneous frequency of the FM carrier. In Figure 11-29, if the FM signal has no modulation, the frequency is at the center frequency, 10.7 MHz, which means that the amplitude of

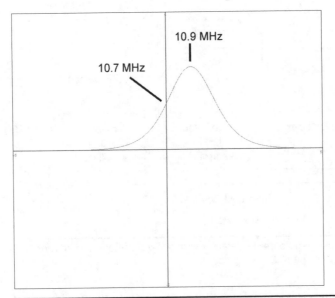

Figure 11-29 Band-pass filter mistuned to form a slope through the center frequency, with the vertical axis denoting amplitude and the horizontal axis showing frequency.

the signal is at a nominal value. If the FM signal shifts upward from 10.7 MHz, the amplitude can increase to twice that nominal value, and if the FM shifts downward from 10.7 MHz, the amplitude from the band-pass filter is near zero. Thus the "mistuned" band-pass filter transforms the incoming FM signal into an AM signal.

There are at least two ways to achieve single-slope detection. One is to make use of the existing IF band-pass filters in the tuner and mistune the local oscillator to move the IF off-center from the IF of the band-pass filters. For example, an FM tuner has its IF filters set to 10.7 MHz. If the local oscillator is mistuned slightly to provide the signal from the mixer or converter to be 10.9 MHz or 10.5 MHz, the FM signal will slope detect. The problem with this approach is that all the preceding IF filters will cause a form of slope detection also, and the demodulated FM signal may be distorted.

However, if the local oscillator is tuned to provide 10.7 MHz, and *only* the last IF filter is mistuned to resonate above (or alternatively below) 10.7 MHz, and a half-wave rectifier circuit is added as shown in Figure 11-30, FM detection is achieved.

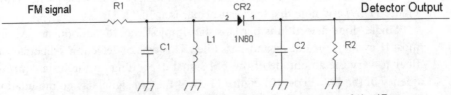

Figure 11-30 A band-pass filter that is "off-tuned" from the center of the IF.

But is there a way to improve the performance of the simple slope detector? The answer is yes.

If two band-pass filters are used instead and tune above and below the IF, a more linear or lower distortion FM slope detector can be made (see Figure 11-31). By implementing two half-wave rectifiers as shown in Figure 11-32, a differential slope detector is provided.

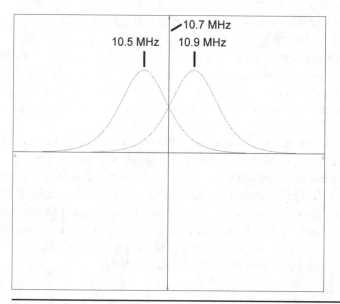

FIGURE 11-31   Frequency responses of two separate band-pass filters tuned to above and below the center frequency, with the vertical axis denoting amplitude and the horizontal axis showing frequency.

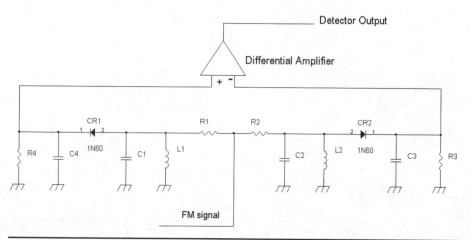

FIGURE 11-32   One implementation of a differential slope FM detector.

As shown in Figure 11-32, resonant circuits L1/C1 and L2/C2 are tuned to below and above the center IF of 10.7 MHz (e.g., 10.5 MHz and 10.9 MHz). The spread of the bandwidth of each filter is set by resistors R1 and R2, which determine the Q factor of the two parallel LC tank circuits.

Note that the slope detection with the L1/C1 that is tuned below 10.7 MHz converts the changes in frequency due to the FM signal to create an AM signal that is rectified for the positive cycle of the created AM signal. Likewise, the slope detection for L2/C2 is at frequencies above 10.7 MHz with rectification on the positive cycle. A differential amplifier takes the two demodulated AM signals and provides a subtractive process. The resulting FM detection curve is shown in Figure 11-33.

The resulting detection transfer function curve shows a wider frequency-detection range with a straighter line for demodulation of the FM signal as the frequency of the IF deviates below and above the center frequency. Also, at the resting frequency, 10.7 MHz, the detector's output is zero, which may be useful in driving a center tune meter.

Differential peak detector chips were implemented as the CA3075 and CA3065 ICs (integrated circuits) for FM radio and FM TV sound. In those circuits, a combination band-reject and low-pass filter was used in place of the two LC tank circuits (see Figure 11-34). The circuit shown in Figure 11-34 works on the principle that a single inductor L1 and two capacitors C1 and C2 form two resonant frequencies to mimic two separate band-pass filters. Again, this circuit is a form of differential slope detection. It relies on two slopes, a positive- and negative-going slope when the frequency response is swept from below the IF to above the IF.

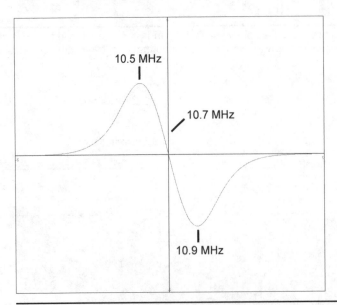

Figure 11-33   By subtracting the two peaks as shown in part *a*, an inverted or upside-down S detection curve is achieved.

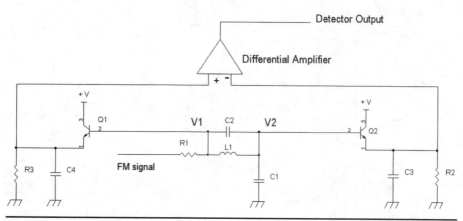

FIGURE 11-34   Using a low-pass filter to form a differential peak FM detector.

From the RCA CA3075 data sheet, typical values for the R1, L1, C1, and C2 are 2.7 kΩ, 5.5 μH (preferably an adjustable inductor), 6.8 pF, and 33 pF, respectively. The detection range can be modified by adding a 33 kΩ resistor in parallel with L1.

R1, L1, C2, and C1 form two types of filters. At node V1, which goes to the base of Q1, there is a notch or band-reject filter effect. At node V2, this is the output of a two-pole low-pass filter with a "transmission zero" or notch frequency formed by L1 and C2. At V1 and V2, there are two notch frequencies, and the notch frequency at V1 is lower than the one at V2. In principle, the frequency response curves of V1 and V2 will intersect or cross each other. That is, from 10.6 MHz to 10.8 MHz, the frequency response at V1 shows an upward slope due to the band-reject response coming back up after the notch frequency, whereas the frequency response at V2 shows a downward slope due to the low-pass filtering effect.

Transistor peak detectors Q1 and Q2 provide the same half-wave rectification as diodes except that they provide high-impedance inputs at their bases so as not to load the filter circuit (R1, L1, C1, and C2) down. The emitters of Q1 and Q2 are then connected to peak-hold capacitors C3 and C4 with their current discharge rate controlled by R2 and R3. A differential amplifier then takes the difference voltages from C4 and C3 to provide a demodulated FM signal at its output.

In practice, this circuit is rather tricky to implement because capacitor C1 is only 6.8 pF, and any parasitic capacitances from the board layout or input of the transistors must be taken into account. Also, the separation between the two notch frequencies may be excessive, and their frequency-response curves may not "exactly" intersect at the center frequency of 10.7 MHz. If the notch frequencies are too far apart, (only) one slope (either at V1 or V2) will be taken advantage of rather than both slopes (at V1 and V2) for FM detection.

And now we turn to the original differential slope detection, which is known as the *stagger-tuned detector*. This circuit preceded the ratio detector, the Foster Seeley FM discriminator, and the differential peak detector that used a differential ampli-

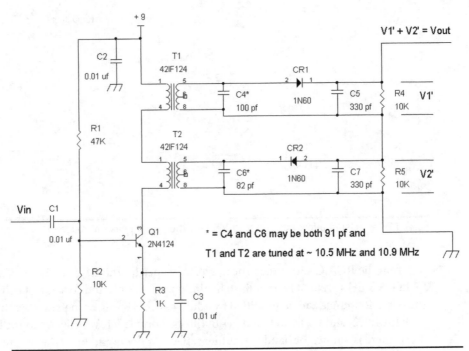

FIGURE 11-35   Stagger-tuned FM detector.

fier. Instead of having to provide a differential voltage V1–V2, it merely inverted one of the voltages and stacked or added the two voltages in series. Let's see how that works (see Figure 11-35).

The stagger-tuned circuit shown in Figure 11-35 uses two 10.7 MHz IF transformers T1 and T2. The 42IF124 transformers do not have the internal capacitors, and the external resonating capacitors C4 and C6 are shown for clarity so as to indicate that two resonating frequencies are involved. T1 is tuned to slightly below the center IF of 10.7 MHz, such as 10.5 MHz, to provide slope detection of the FM signal, which results in a "transformed" AM signal across C4. This AM signal is rectified via CR1 for the positive peaks of the transformed AM signal, and the demodulated signal is provided across C5 and R4 for a detected voltage V1'.

Resistor R4 also plays a role in determining the Q factor of the tank circuit via the secondary inductance of T1, which is nominally 2.4 µH. Higher or lower Q is possible by increasing or decreasing the value of R4 to determine the steepness of the slope detection. However, both R4 and R5 should be the same value for best detection linearity or lowest distortion for the demodulated FM signal.

Likewise, T2 is set to resonate above the center IF at about 10.9 MHz. The voltage across C6 is another "transformed" AM signal, which is rectified by CR2 for the negative cycles instead of the positive cycles. By rectifying negative cycles, the detection curve is mirror-image flipped upside down to provide a detected voltage V2'. Also, R5 plays the same role of affecting Q, but for T2.

The combined series voltages V1' + V2' form an FM detection curve similar to the one shown in Figure 11-33 because V2' is a negative voltage. When the IF FM signal is sitting at rest or at 10.7 MHz, Vout = 0 volts. As the IF signal swings below 10.7 MHz, there is a positive voltage, and when the IF signal swings above 10.7 MHz, there is a negative voltage.

To tune for T2 at approximately 10.9 MHz, connect an RF generator at 10.9 MHz to the Vin terminal. Set the generator's amplitude level to about 100 mV peak to peak. Monitor the DC voltage at the anode of CR2, and adjust the slug in T2 for maximum negative voltage. To tune for T1, first temporarily short the anode of CR2 to ground. Set the generator for 10.5 MHz, and measure the DC voltage at the cathode of CR1. Adjust T1's tuning slug for maximum DC voltage. Remove the short circuit from the anode of CR2 to ground.

Vout preferably should be buffered with an amplifier to ensure that the Q values of the two LC tank circuits are not affected by loading downstream. Also notice that the primary windings are connected in series. This is done so that there is truly independent tuning for T1 and T2. Recall that a current source that drives two circuits in series "forces" the signal current equally into the two circuits. If the primary windings of T1 and T2 were connected in parallel, then tuning T1 would affect the tuning of T2, and vice versa.

References

1. Ronald Quan, *Build Your Own Transistor Radios*. New York: McGraw-Hill, 2013.

2. K. Blair Benson and Jerry Whitaker, *Television Engineering Handbook*. New York: McGraw-Hill, 1992.

3. "RCA Linear Integrated Circuits and MOS/FET's," RCA Solid State, Somerville, NJ, 1982.

4. Robert L. Shrader, *Electronic Communication, 2nd ed*. New York: McGraw-Hill, 1967.

5. Robert G. Meyer, "EE140 Class Notes on Nonlinear Electronic Device Models and Circuits," University of California, Berkeley, CA, 1975.

6. Robert G. Meyer, "EE240 Class Notes on Nonlinear Analog Integrated Circuits," University of California, Berkeley, CA, 1976.

7. Kenneth K. Clarke and Donald T. Hess, *Communication Circuits: Analysis and Design*. Reading, MA: Addison-Wesley, 1971.

8. A. Bruce Carlson, *Communication Systems: An Introduction to Signals and Noise in Electrical Communication, 3rd ed*. New York: McGraw-Hill, 1986.

CHAPTER 12

Video Basics, Including Video Signals

What can be said about transmitting motion pictures electronically? This is a process that started back in early 1900s that we now know as television (TV). At first, there were mechanical television systems such as those pioneered by Charles Jenkins. One of the first images captured, sent, and later received and displayed with less than 100 lines was a picture of Felix the cat, a cartoon character.

However, an all electronic television system proved to be much more practical than a mechanical system. For example, the number of scanning lines could be changed easily by modifying an oscillator's frequency. The number of pixels across the screen also could be increased easily by increasing the bandwidth of the video signal. If all these terms (i.e., pixels, bandwidth, etc.) seem unfamiliar now, in this chapter we will explain how lines, pixels, and other signals work in making a television signal. The basic concepts of resolution, number lines and pixels, and frame and field rates will be discussed. Also, we will explore how the common controls we adjust for a computer monitor or TV set work, such as contrast, brightness, and sharpness. We will find out that the frequency response of a video channel affects the resolution and sharpness of the displayed picture. Just like a treble boost control in an audio amplifier that emphasizes the middle to high audio frequencies (e.g., 1 kHz to 20 kHz), a sharpness control for standard 480i or 576i line standard definition TV system has frequency boost at about 1 MHz to 5 MHz.

Once we know about the characteristics of a picture—how is it sent in an orderly manner—we will find that embedded in every television signal (analog or digital) are synchronizing signals or clock regeneration signals.

Finally, we will look at analog color TV signals. There were at least seven types of analog color TV standards prior to the switchover to digital television. Most of these analog color TV signals used a form of I and Q modulation, which are amplitude- and phase-modulated signals.

Examining Still Pictures First for Contrast, Brightness, Resolution, and Sharpness

Television is the transmission or sending of multiple images that are formatted in a particular manner in two dimensions, the X and Y axes, otherwise known as the vertical and horizontal axes, respectively. For still images, such as those taken with a digital camera, the concepts of contrast, brightness, resolution, and sharpness are the same as for television pictures.

For now, we will work with black and white (aka *monochrome*) pictures. In a still picture represented by 8 bits per pixel, the brightness typically has up to 256 levels of gray starting from black at 0 and peak white at 255 (peak white = maximum brightness in a scene). Figure 12-1 shows pictures with high contrast, normal contrast, and low contrast.

An increased-contrast picture looks "punchier" by increasing the gray levels above middle gray (e.g., 128 of 255) to a brighter or higher level of gray approaching the white levels. Conversely, increased contrast pushes the gray levels below middle gray toward the black levels. An extreme case of very high contrast occurs when one makes a photocopy of a picture. We see primarily only tones that are all black or all white, with very little in terms of different shades of gray. In a low-contrast picture such as the one depicted in Figure 12-1c, the brighter levels have been pulled down to the middle-gray level.

Now let's take another look at high-, normal-, and low-contrast pictures. When one "brackets" the photographic exposure of a subject to +1 and –1 F-stops of exposure compensation, the equivalent video gain is at 2× the normal signal range for a +1 F-stop of overexposure that results in increased contrast and (½) × the normal signal range for a –1 F-stop of underexposure that causes a low-contrast picture. Figure 12-2 shows the effects of overexposure, normal exposure, and underexposure of photos taken at night. Note in Figure 12-2a that the contrast is higher among the last three houses on the right side compared with Figures 12-2b and 12-2c.

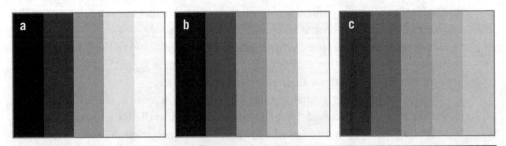

Figure 12-1 (*a*) A high-contrast picture illustrating increasing contrast via Adobe Photoshop. (*b*) Original picture with normal contrast. (*c*) A low-contrast picture produced by rendering the normal contrast picture with Adobe Photoshop.

Figure 12-2 (*a*) An overexposed picture that results in higher contrast. (*b*) A normally exposed picture with normal contrast. (*c*) An underexposed picture that results in lower contrast.

It should be mentioned that more commonly, pictures that are taken in the daytime show lower contrast when under- or overexposed compared with a normally exposed photo that shows a fuller range of contrast. The overexposed pictures tend to look too bright with a white-washed out look, whereas the underexposed pictures tend to look too dark.

From the pictures in these figures, we see that contrast is a function of peak-to-peak signal level from the dark or black levels to the peak white levels. Pictures whose brightness levels show a great deal of deep blacks to peak whites are higher contrast than the same pictures with less frequency or occurrence of deep blacks and peak whites. High-contrast pictures thus have high occurrences of the 0 to 255 brightness range.

Low-contrast pictures may have brightness levels congregrating near the middle grays or lower grays or even near the peak white levels. In all cases for low-contrast pictures, the peak-to-peak brightness level is a fraction of the full brightness range. For example, if the full brightness or luminance range is 0 through 255, a low-contrast picture will have a peak-to-peak range of less than 128 (Figure 12-3).

We now turn to the brightness control, which is a direct-current (DC) offset added to the brightness levels. Actually, the peak-to-peak brightness or luminance levels do not change, but the brightness control can place the luminance levels in a different range to give the appearance of higher or lower contrast. Let's see what we mean by this (see Figure 12-4).

In Figure 12-4a, a normal-contrast picture is brightly washed out by adjusting the brightness control in Adobe Photoshop. This is equivalent to adding a positive DC offset voltage to bring the black levels to gray and the gray levels to near white. In Figure 12-4c, a normal-contrast picture with the brightness control set to adding a negative DC offset results in pulling the middle gray parts of the picture to near black and the peak white portions to an upper middle gray. Note that lowering the brightness causes loss of details in the shaded or dark areas. This is also known as a *loss of shadow detail*, which is quite apparent in Figure 12-4c.

We now can show how varying the video signal amplitude (via VR1) and changing the DC offset (via VR2) of video signal can be used to adjust the display charac-

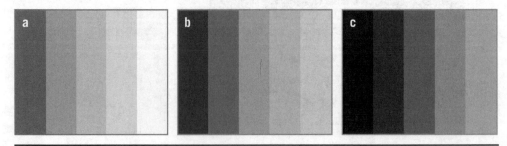

FIGURE 12-3   (*a*) Lower-contrast picture weighted toward the brighter gray areas. (*b*) Lower-contrast picture weighted toward middle gray. (*c*) Lower-contrast picture weighted toward darker gray or near-black level.

Figure 12-4 (*a*) Adding too much brightness also causes highlight blocking or loss of details in the upper gray areas. (*b*) A normal-contrast picture with the brightness control set in the middle or 0 volt DC offset. (*c*) Lowering the brightness control also lowers the contrast of the picture.

teristics (i.e., contrast and brightness) of a cathode-ray tube (CRT) TV monitor (Figure 12-5). As shown in Figure 12-5, a video signal is sent to an inverting-gain amplifier –A1. The output of amplifier –A1 is changeable in amplitude via VR1. Its output is connected to AC (alternating-current) coupling capacitor C1.

Diode D1, capacitor C1, and DC voltage amplifier +A2 form a DC restoration circuit to clamp the positive peak (e.g., synchronizing pulses' sync tips) of the video

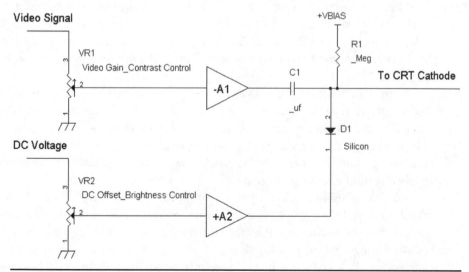

Figure 12-5   Diagram of contrast and brightness controls for a video or picture signal.

signal to a positive voltage as adjusted by brightness control VR2. The output of the DC restoration circuit at the anode of D1 is connected to the cathode of a cathode ray tube (CRT). The DC restoration circuit ensures that the black level does not shift with various bright and dark scenes in the program video source.

Many older TV sets that used vacuum tubes did not have DC restoration circuits at the cathode of the CRT. Instead, the cathode of the CRT was AC coupled from the output of the video amplifier, and the cathode was biased via a resistor to a variable voltage source for changing the brightness. When a scene faded to black, the display actually faded to gray.

Resolution or Fine Detail of a TV Picture

We now turn to the concept of resolution, which is measured by the number of lines that are displayed and sensed by the viewer. With pictures, determining resolution in general is a subjective way of placing a numerical figure on how much fine detail is observed. One criterion for defining resolution is when the fine detail information is about one-quarter the contrast of a low-spatial-frequency reference signal. Figure 12-6a shows a TV monitor with a "multiburst" signal with increasing spatial frequencies across the screen. The resolution of a video signal is primarily determined by the bandwidth or highest frequency that can be delivered to a monitor.

On the display shown in Figure 12-6, the multiburst signal starts with white and black levels followed by six packets (e.g., 0.5 MHz, 1 MHz, 2 MHz, 3 MHz, 3.58 MHz, and 4.2 MHz from the video signal) of increasingly finer resolution of vertical lines. The first three packets are clearly discernible with very good contrast between the white and black stripes. As we look at the last three packets of finer-pitched vertical lines, we see that the white and black stripes do not have as much contrast. In particular, the last packet of the finest-pitch vertical lines on the right side is discernible, but not as well as the first packet of lower-pitched vertical lines on the left side of the screen. A one-horizontal-line waveform of the multiburst signal is shown in Figure 12-7.

In the monitor itself, the resolution is often determined by the number of pixels for a flat-screen device, and in a CRT (cathode ray tube), the resolution is determined by the number of dots or stripes across the screen and the spot size of the electron beam. Also, the video bandwidth or frequency response of the video signal's processing amplifier affects the horizontal resolution (e.g., ability to resolve fine-pitched vertical lines of the display).

So what is resolution, and what does it mean to have X number lines of resolution? For digital photographs and TV, the definition of a *line* is the ability to observe *either* a dark or a white line or *either* a darker gray line or a lighter gray line. A pair of alternating light gray and dark gray lines counts as two lines, just as a pair of alternating black and white lines also counts as two lines. Figure 12-8 is a cropped

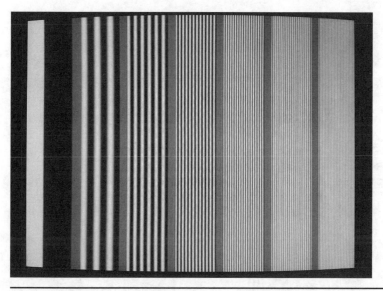

FIGURE 12-6   Multiburst signal displayed on a TV monitor.

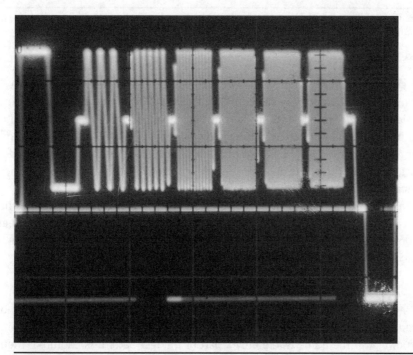

FIGURE 12-7   A portion of a multiburst waveform starting with a white and a black level followed by six packets (e.g., 0.5 MHz, 1 MHz, 2 MHz, 3 MHz, 3.58 MHz, and 4.2 MHz) of sine waves of increasing frequency and finally ending with a horizontal sync pulse.

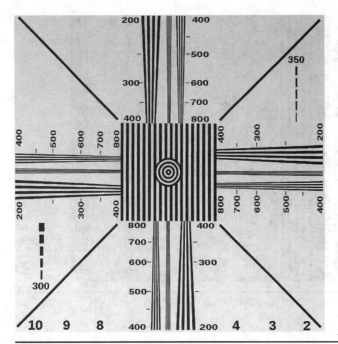

Figure 12-8 Examples of TV or photographic lines in the vertical and horizontal directions.

and close-up picture from a 1956 Electronics Industries Association (EIA) test chart. From the 12 o'clock position, we see three sets of vertical lines starting on the left side from 200 to 400 and on the right side from 400 to 800. The finer-pitched lines in the center or in between these two sets denotes 800 to 1,600. These numbers refer to the number of TV lines for horizontal resolution. Similarly, in the 9 o'clock position, we see three sets of horizontal lines denoting the number of TV lines for vertical resolution.

Now the question is what does it mean that there are X number of lines in the vertical and horizontal directions, and how do we test for this? The answer is that for a monitor or a camera, we use a test pattern with wedges of converging lines in both vertical and horizontal directions (see Figure 12-9).

NOTE The test pattern must be framed in such a manner that the two white triangles on each of the four sides coincide with the outer boundaries for the frame of the picture taken. If the test pattern is zoomed in and overflowing off the frame where only a part of the test pattern is captured, then the resolution measurement will be in error, favoring a higher resolution number. If the test pattern is zoomed out such that the test pattern is smaller within the frame, then the resolution measurement will be in error, favoring a lower resolution number.

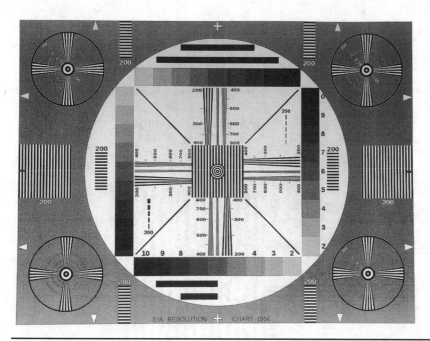

Figure 12-9 A TV test pattern for standard-definition TV from the EIA 1956 test resolution chart (http://en.wikipedia.org/wiki/File:EIA_Resolution_Chart_1956.svg).

To determine the lines of resolution in the vertical and horizontal directions, we look at the converging lines until the lines are merged into gray or the white and black lines are no longer distinguishable (see Figure 12-10). When we determine vertical resolution, we look at the wedge of lines going across the screen, and for measuring horizontal resolution, we examine the wedge of lines going from top to bottom. Where the lines are no longer clearly distinguishable is the resolution.

For vertical resolution, what is measured in the middle wedge of lines at the 9 o'clock position is about 1,400 lines, which says exactly what one would think. However, when we look at horizontal resolution in the 12 o'clock position, that number is the number of horizontal lines per picture height, which is also approximately 1,400 lines in this example. The actual number of lines across the screen is more because our viewing screens are generally not square (in shape). To determine the number of horizontal lines across the screen, we must multiply the number of horizontal lines resolved by the aspect ratio, which is 4:3 in standard-definition TV displays. For example, if the horizontal resolution from the resolution chart reads 300 lines for a 4:3 aspect ratio screen (e.g., standard-definition TV standard), the actual number of lines that can be resolved across the screen is 300 lines × 4/3 = 400 lines (across the screen for a 4:3 aspect ratio screen).

By rendering with Adobe Photoshop, blur filter has been added in Figure 12-11 to show the effects of lowered resolution. In this example (as seen in Figure 12-11),

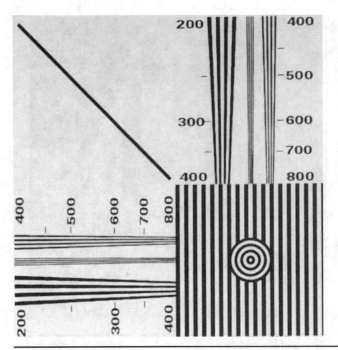

FIGURE 12-10   Example of determining the number of lines of resolution from a Nikon digital still camera.

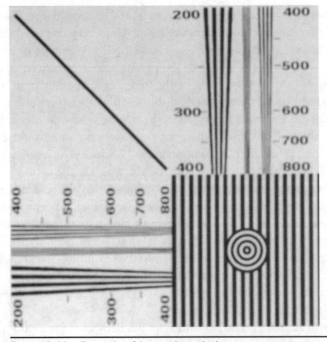

FIGURE 12-11   Example of lowered resolution.

the resolution has been reduced to about 700 lines vertically and horizontally. At 800 lines and above, all we see is an average gray level.

Aspect Ratio

Given a particular screen or display, the aspect ratio is given by the width divided by the height. For example, for standard-definition TVs (SDTV), the aspect ratio is 1.33:1 = 1.33/1.00 = 1.33, which is generally known as 4:3 = 4/3 = 1.33. This type of display is more on the square side than rectangular (in shape). High-definition TV (HDTV) displays are widescreen and more rectangular and have an aspect ratio of 16:9 = 1.78:1. Figure 12-12 shows examples of standard- and high-definition displays and their associated aspect ratios.

Another way of determining the aspect ratio is to take the number of pixels across a screen (horizontally displayed pixels) and divide it by the number of pixels up and down the screen (vertically displayed pixels). For example, for standard-definition VGA monitors, the number of horizontal pixels is 640, and the number of vertical pixels is 480. Thus the aspect ratio is 640/480 = 1.33 = 4:3. If we look at HDTV, such as the 1,080-scan-line system, there are 1,920 pixels across and 1,080 pixels up and down for an aspect ratio of 1920/1080 = 1.78 = 16:9.

The next question is, if we have a number of pixels vertically or across a screen, does that automatically determine the number of lines resolved in each of the directions? For example, for the VGA system of 640 by 480 pixels horizontally and vertically, do we actually get 640 lines of resolution across and 480 lines up and down the screen? The answer is no in almost all cases.

The only time the resolution matches the number of lines and pixels in the TV display is when the program material (e.g., TV show) has a scene that lines up perfectly with the pixels of the display. For example, 240 black lines and 240 white lines from the picture being taken are exactly lined up or matched with the pixels or lines

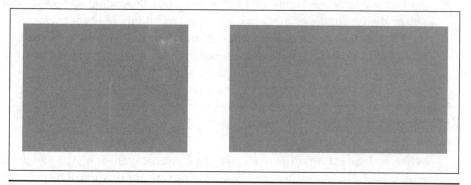

Figure 12-12 Aspect ratios from left to right for standard (4:3) and high-definition (16:9) displays.

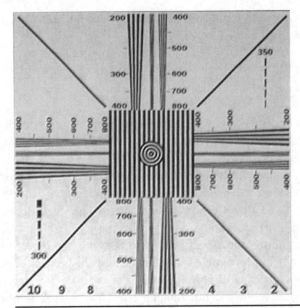

FIGURE 12-13 Test chart taken in the camera in the original position.

in the vertical direction, and similarly, 320 black and 320 white lines are lined up with the horizontal pixels. In practice, this almost never happens. Instead, generally, the lines from the TV program do not align themselves with the pixels on the screen. Figure 12-13 shows the effect of capturing fine detail in one position, while a small shift of the picture results in a different resolution.

In Figure 12-13, note at the 12 o'clock position the 400 to 800 lines; at 500 and 600 lines, the wedge of lines is almost fused into a solid gray with no lines discernible. Now observe in Figure 12-14 the result with a slight change in camera position (shifted) slightly up or down but at the same distance. Notice at the same 12 o'clock position the 400 to 800 lines; for 500 and 600 lines, about 4 wedge lines can be discerned. If one observes the wedges of lines at the 9 o'clock position, Figure 12-13 resolves 500 lines, but Figure 12-14 does not. Thus, depending on how the camera's sensor aligns with the picture it is taking, the resolution can change.

The problem we are seeing is *aliasing*, and this was especially noticeable for vertical resolution because of the discrete number of TV scanning lines that were used in camera tubes and some of the first solid-state TV camera sensors.

Enter the Kell Factor

When we look at a TV display, all pixels in the vertical direction (up and down) are represented as scanning lines or just rows of picture information. It turns out that by experimentation, in general, the maximum number of lines resolved with a discrete

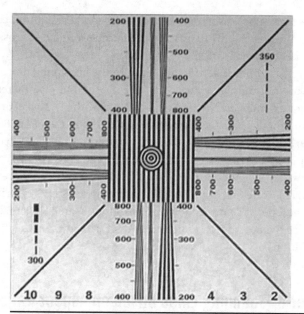

Figure 12-14 Test chart taken with the camera shifted in position but at the same distance.

number of displayed or active scanning lines scanned (or number of rows of pixels) is somewhere between 0.65 and 0.8 of the number of scanning lines (or rows of pixels). This number between 0.65 and 0.8 is known as the *Kell factor*.

In general, the resolution is determined by:

Kell Factor × number of scanning lines or rows displayed

In general, the Kell factor can be taken as 0.7.

Let's take two examples from standard-definition (480 lines) and high-definition television (1,080 lines). The actual resolutions are

480 lines × 0.7 = 336 lines vertically   For SDTV

1,080 lines × 0.7 = 756 lines vertically   For HDTV

In the horizontal direction, the number of lines for equal resolution for vertical and horizontal pixels is determined by using the same numbers for the vertical resolution in the following manner for 480 and 1,080 line TV systems, respectively:

336 lines × aspect ratio for SDTV = 336 lines × 4/3 = 448 lines across

756 lines × aspect ratio for HDTV = 756 × 16/9 = 1,344 lines across

The next question would be, do we have to have equal resolution (aka square pixels) along the vertical and horizontal directions? The answer is no. We can actually have *more* resolution in one direction than the other. This was especially true in the days of CRT (cathode ray tube) displays, where the horizontal resolution could be higher than the vertical resolution by increasing the number of pixels across the screen or by increasing the video bandwidth.

By increasing the number of pixels across the screen in the horizontal direction, a higher-resolution picture not only is perceived by the viewer, but also increased fine detail is provided. For example, in Nikon digital cameras with a 3:2 aspect ratio, the models D1 (2,000 horizontal pixels by 1,312 vertical pixels) and D1X (4,028 horizontal pixels by 1,312 vertical pixels) have the same vertical resolution. However, the D1X has twice as many pixels across, thus yielding about twice the horizontal resolution of the D1 (see www.dpreview.com/reviews/nikond1/19 and www.dpreview.com/reviews/nikond1x/25). Note that although *DP Review* rates the D1X's horizontal resolution at 1,650 lines, on close inspection, the resolution is in excess of 2,000 lines, which is about twice the horizontal resolution of the D1 (approximately 1,100 lines of horizontal resolution).

Determining the Number of Scanning Lines Based on Aspect Ratio, Kell Factor, Frame Rate, and Bandwidth

Modern motion pictures, as shown to the general public in the 1930s, projected 24 frames of pictures per second to avoid flickering. Slower frame rates were possible, but the pictures would flicker. Home movies from the 1960s, for example, were made on 8-mm photographic film, which had a frame rate of 16 frames per second. An improved Super 8-mm motion picture system had a frame rate of 18 frames per second. Both these home movie systems suffered from flicker, but they were deemed acceptable to the general public.

In the projection of film, the whole frame is displayed at once. However, in television systems that were created in the mid-twentieth century, the display devices were cathode ray tubes, CRTs. A CRT will display an image by sweeping or tracing out the picture from left to right for each horizontal line. After one horizontal line is displayed, the next line is spaced downward, and that line is traced or drawn.

The display of TV images on a CRT can be thought of as similar to the way a word processor types one line at a time, and at the end of each line, there is a carriage return (e.g., hit "Enter") to push the cursor one line space down to start the next line of typing. After typing many lines, the lines of information (words) will end on the last line at the bottom of the page, and the word-processing program will continue the document on a new page from the top. In terms of displaying a TV signal on a CRT monitor, the carriage return may be thought of as a horizontal line retrace, and the new page can be thought of as a vertical field or vertical frame retrace.

Now back to the requirement of how many lines do we need to show a TV picture? In essence, the number of lines is determined by how large the CRT is. For most CRT TV sets from the twentieth century, the diagonal dimensions varied from less than 10 inches to as much as about 27 inches. There were 36-inch CRT TV sets, but these were not that common. For small screens (about 20 inches measured diagonally), 480 scanning lines met the requirements for good detail in the TV pictures. Before we look into the 480-line TV system, let's take a look at a simpler system that uses 240 viewable lines, which some portable devices display in the twenty-first century.

To display a scanned-line system without flicker, it was found that an acceptable rate is about 50 or more frames or fields per second. A *frame* is considered to be the whole picture, whereas a *field* is one-half the picture, so it takes two fields in succession to display a frame. TV fields are used in *interlaced scanned* formats, whereas if there are no interlaced fields but just frames, we are working with a *progressively scanned* format.

The *i* in 480i or 576i represents an interlaced format such as NTSC or PAL television standards. The *p* in 720p represents a progressive scan format for a 720-line HDTV system. We will now discuss further progressive and interlace scanned TV systems.

Let's take a look at a 240p format with 60 frames per second for a 4:3 (4/3) aspect ratio. If we have a frame rate of 60 frames per second, this means that 240 lines are displayed every 1/60th of a second. This also means that one line takes about (1/60) second × 1/240 = 69.44 μs (μs = μsec = microsecond). If we calculate how many lines of vertical resolution we have, it should be the Kell factor × 240 lines = 0.7 × 240 lines = 168 lines per picture height. The number of lines in the horizontal direction would be 168 lines × 4/3 = 224 lines across.

If we look at a sine-wave signal, on one cycle, the positive peak can represent a bright line, and the negative peak can represent a dark line. Thus one cycle represents two lines. Or put another way, there are two lines to every cycle of the sine wave. Since we have 224 lines across the screen, the number of cycles is 224/2 = 112 cycles.

The question is, if we try to fit 112 cycles into one line period of 69.44 μsec, what is the frequency of the sine wave? Put another way, the relationship is 69.44 μsec/112 cycles = 0.62 μsec/cycle. The period of the sine wave is then 0.62 μsec, or the frequency is then 1/period = 1/0.62 μsec = 1.61 MHz.

In order to display a 240-line progressive scan system at 60 Hz, we need about 1.6 MHz of bandwidth. But can we actually make a TV signal where we just have video information and no other signals ... such as signals that tell the monitor when the start of a new horizontal line occurs (e.g., carriage return) or another signal to tell the monitor to scan a new frame or field (e.g., new page)? The answer is no. We need to add timing signals for the monitor to extract such that the monitor knows when a new line and field or frame starts. These signals are called *horizontal* and *vertical synchronizing signals*. Even with the synchronizing signals added to the picture information, is there another signal we can add? The answer is yes, and that signal is called the *blanking-level pulse*, which defines the darkest of the black-level reference for the TV signal.

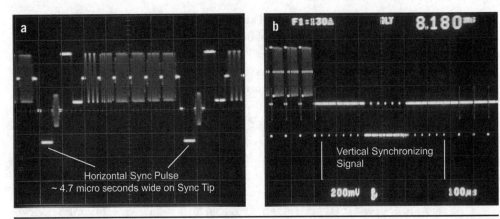

FIGURE 12-15 (*a*) TV signal with a horizontal sync pulse. (*b*) TV signal with vertical sync pulses.

We now can show what a TV signal looks like in practice. See Figure 12-15.

After factoring in the horizontal sync pulses, vertical sync pulses, and so on, the video signal can be described in the following manner: One horizontal line has about 63.6 μsec of duration, which has about 11 μsec to include a "front porch" and "back porch" blanking-level reference pulse and a 4.7 μsec horizontal sync pulse. The remaining portion (63.6 – 11) μsec = 52.6 μsec is devoted to the picture information area, such as the luminance values.

> **NOTE** The details of the front porch, back porch, horizontal sync pulse, and other reference signals will be discussed in Chapter 13.

For the total number of lines for the 240p system, we actually have about 262 lines, in which 240 lines display the image, and the other 22 lines are used for the vertical sync signals plus extra data lines that may carry test signals, time code, closed-caption data, and so on.

We now recalculate the required bandwidth to display the 224 lines across the screen, but this time with 53.6 μsec instead of 69.44 μsec. The period of one cycle is then 52.6 μsec/112 cycles = 0.469 μsec, which translates to 1/0.469 μsec = 2.12 MHz (instead of 1.6 MHz). This makes sense—we are squeezing in the same 112 cycles into a smaller period of time. Now back to the standard definition TV (SDTV) signal, which has 480 lines interlaced per frame.

Because we as humans (don't know about animals) have a retention of a flashed image, known as *persistence*, it is possible to display one field on odd lines and the succeeding field on even lines to form a TV frame. Two fields make a frame, and each field has a half-line offset (equivalent to a vertically half-pitch offset) from each other. Each field (odd field and even field) has 262.5 lines. Because two TV fields fill in to complete one TV frame, there are 2 × 262.5 lines = 525 lines—a number many in the TV field are familiar with. So when we talk about 480i, we are really relating to a 525-line TV system.

The TV fields are refreshed at a rate of roughly 1/60th of second. One TV frame is then displayed in two TV field's time = 2 × 1/60th second = 1/30th second = period of one TV frame. Although the frame rate is 1/30th of a second, one may think that there would be a flicker problem. Fortunately, because the TV fields are almost the same except for a half-vertical-pitch offset, the human eyes see the displayed image as something that is refreshing at 1/60th of a second, and thus flicker is not noticed for the most part.

With 480 displayed lines per frame, and using a Kell factor of 0.7, the number vertical of lines we can resolve is 0.7 × 480 lines = 336 lines. The number of horizontal lines across the screen with a 4:3 aspect ratio is then 336 lines × 4/3 = 448 lines. Note that 448 lines can be represented by 224 cycles—recall that a cycle of a sine wave is equivalent to two lines, a light line and a dark line.

With 525 lines within 1/30th second each, horizontal line can be calculated as:

525 lines/(1/30th second)

which, when the numerator and denominator are divided by 525, gives us:

1 line/[(1/30th second)(525)] = 1 line per 63.5 μsec.

To figure out the bandwidth required, let's assume that the front porch, back porch, and sync areas take up in total about 10.9 μsec, which leaves (63.5 μsec − 10.9 μsec) = 52.6 μsec. We need to fit 224 cycles of sine wave into 52.6 μsec. This yields 224 cycles/52.6 μsec, which equals 1 cycle/0.2348 μsec, which implies a frequency = 1/0.2348 μsec = 4.26 MHz for a 480i TV system. Note that 4.26 MHz ≈ 2 × 2.12 MHz, or twice the bandwidth requirement for 240p.

As we can see with doubling the number of lines via an interlaced system of 480 lines, we need twice the bandwidth of a progressive scan 240-line system to ensure twice the resolution both vertically and horizontally. But why not just go to a 60-frame-per-second 480-line progressive scan system and not deal with interlacing two fields? Let's see what happens in terms of the bandwidth.

We will need to have 525 lines in the period of 1/60th of a second for a progressive scan system versus 1/30th of a second for the interlaced system. Note that (1/60) second = (½) of (1/30th) second. This means that the period of one horizontal line in the progressive format is half the period of the horizontal line period in the interlaced system. That is, one horizontal line = (½)63.5 μsec = 31.75 μsec for a 525-line progressive scan TV signal. If we allocate 5.5 μsec (half of 11 μsec) for the front porch, back porch, and horizontal sync, we are left with 26.25 μsec for the picture area. We still need to fit in 448 lines across the screen or, equivalently, 224 cycles of sine wave.

Therefore, we have 224 cycles/26.25 μsec = 1 cycle/0.1172 μsec. This translates to an 8.53-MHz (1/0.1172 μsec) sine wave, or roughly twice the frequency or bandwidth of the 4.26 MHz required for the interlaced system for the same number of scanning lines.

The penalty for going with a progressive scan is that it takes more bandwidth. The advantage of progressive over interlaced scanning is that for movements in the TV program, the interlaced system can show *artifacts*, usually a zigzag rendition of a moving object from one field to another. However, the progressive scan version will show no motional artifacts. For the most part, the interlaced version works fine, and its motional artifacts are not *that* noticeable.

Sharpness and Frequency Response

Traditionally, the sharpness control enhanced the frequency response of the video signal, that is, the pixel information that is formatted in a horizontal line as a signal is processed to have a high-frequency boost. In inexpensive TV monitors, this enhancement provided an asymmetrical edge enhancement to the image via a single resistor-capacitor network. That is, there are overshoots only. Most TV sets used this type of sharpness control to provide a single-edge enhancement to a pulse waveform. However, some TV receivers did employ symmetrical-edge sharpness enhancement (e.g., a Zenith 23-inch black-and-white vacuum-tube TV set). However, there is also another way to add sharpness to the video signal via a phase linear high-frequency boost that leads to symmetrical edge enhancement.

Batman's Ears to the Rescue for Image Enhancement

If one uses two RC circuits or the equivalent of cascading two high-pass filters and combining the output with the original input, a phase linear high-frequency boost is achieved. The result to a pulse is something that looks like Batman's ears (see Figure 12-16, bottom figure).

Here you can see single-edge overshoots denoted by A and B and phase linear preshoots denoted by A" and B' along with overshoots A' and B".

In Figure 12-16, the top pulse is the input waveform that is unaltered by edge enhancement. The middle pulse has a single-edge sharpness enhancement applied to the input waveform. Areas A and B denote a white-level overshoot and black-level overshoot, respectively. Finally, the bottom waveform uses a phase linear approach to add edge enhancement to the input waveform. The phase linear edge-enhancement technique includes white and black overshoots A" and B" (just like the single-edge-enhanced signal), but additionally there are preshoot signals A' and B' to further highlight and outline the pulse. One observation: don't the pointy peaks formed by A" and B' look like "Batman's ears? Hmm, "How right you are old chum!" Figure 12-17a shows the original image and Figure 12-17b illustrates both single-edge and phase linear (double edge) enhancement techniques.

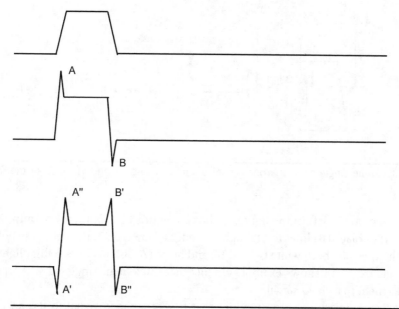

FIGURE 12-16 Effects on a pulse with single edge and phase linear double-edge enhancement. Note: A" and B' look like Batman's ears.

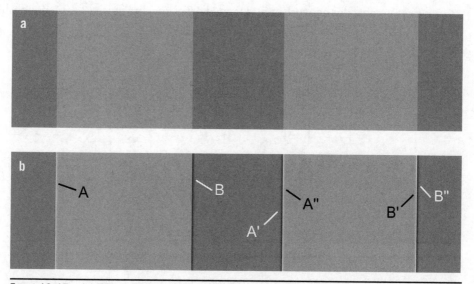

FIGURE 12-17 (*a*) Two light gray "boxes" without edge enhancement. (*b*) Single-edge enhancement shown on the "box" on the left, and phase linear edge enhancement shown on the "box" on the right.

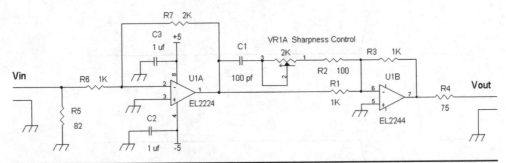

FIGURE 12-18 Example single-edge sharpness control circuit using C1 and VR1/R2 for the RC filter.

As seen on the left in the light gray box in Figure 12-17*b*, there are only either white (A) or black (B) lines to highlight the edges. However, in the light gray box on the right, there are both white (A"and B') and black (A' and B") lines to highlight the edges. Figure 12-18 shows example circuits that implement single-edge sharpness enhancement to a video signal.

U1A is an inverting-gain amplifier that feeds the video signal to R1, an input resistor for the second inverting amplifying stage U1B. A high-frequency boost is provided by having C1, VR1A, and R2 couple high-frequency signals into U1B. The combination of the two signals from C1 and R1 allows the video signal to provide a single-edge sharpness enhancement. A phase linear sharpness control is shown in Figure 12-19, which uses a more elaborate RC high-pass filter.

A phase linear sharpness control can be implemented in other ways by using an inductor-capacitor network or by a tapped delay line. In Figure 12-19, a double high-

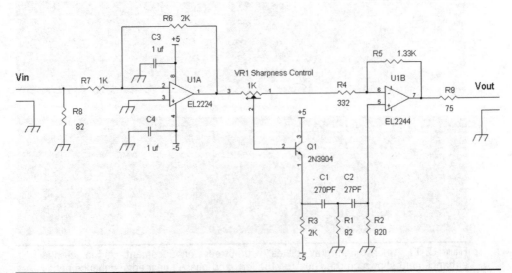

FIGURE 12-19 A phase linear sharpness control using R1, C1, R2, and C2 for an RC filter.

pass filter circuit formed by C1, R1, C2, and R2 provides a +12 dB/octave slope high-frequency signal to the (+) input of U1B instead of the (−) input and R4. The reason why is that each high-pass filter section C1 and R1 or C2 and R2 provides a 90-degree phase shift. So cascading two of these sections provides 2 × 90 degrees = 180 degrees of phase shift, and therefore, the high-frequency boosted signal has to have its phase reversed via coupling the signal from R2 into the (+) input terminal of U1B. The boosted high-frequency signal via R2 is in phase with the video signal into R4 and provides the double-edge sharpness enhancement.

So why is phase linear image enhancement better than the simple RC filter type of image enhancement? The simple RC filter (i.e., high-pass filter plus original signal) proves a pulse waveform with only overshoots. This only provides a superwhite or supergray level at the edge transition of the pulse,

However, a phase linear sharpness control (original signal − double RC high-pass filter) has both undershoots and overshoots on the pulse waveform. What this does is provide a superblack level preceding the pulse's edge at A' and a superwhite level immediately after the edge at A". Also, there is a superwhite level preceding the light gray to dark gray transition at B' followed by the superblack level at B". In essence, a phase linear sharpness control adds a halo of both white and black outlines at the transition from one gray level to another (see Figure 12-17b again).

Sharpness enhancement performs two duties. The first is that it can compensate for loss of high-frequency response (e.g., caused by the diameter of the electron beam that scans across a cathode ray tube) to restore fine-detail resolution. The second is that it can further add a boost at high frequencies for edge enhancement that leads to a crisper-looking picture. Generally, the boost starts at about one-third or one-half the maximum bandwidth of the video signal. For example, an NTSC or standard-definition signal has about a 4.2 MHz of bandwidth, so the boost will start anywhere between 1.4 MHz and 2.1 MHz with a rising frequency response up to 4.2 MHz or beyond.

Some Color TV Basics

If we take three primary colors—red, green, and blue—a combination of the three colors can give us virtually any other color, such as violet, brown, yellow, turquoise, and so on and any shade gray (neutral color). In the capture of color pictures, usually three sensors are required. The simplest may just be having each a red, a green, and a blue sensor. Other combinations are possible, such as a luminance sensor with red and blue sensors.

For color TV during the post analog TV era, there are typically three color TV signals—the luminance (Y), the red color difference (R-Y) signal Pr, and the blue color difference (B-Y) channel Pb, where R = red channel, B = blue channel, and Y = luminance or black-and-white channel. The luminance channel Y represents black-

and-white information and can be expressed as the combination of the primary colors red, green, and blue:

$$Y = 0.59G + 0.30R + 0.11B \tag{12-1}$$

where G = green channel's signal, R = red channel's signal, and B = blue channel's signal.

By algebraic "magic" or "grinding," we can eventually derive that the color-difference green channel is a linear combination of (R-Y) and (B-Y), which results in the following equation:

$$(G\text{-}Y) = -0.51(R\text{-}Y) - 0.19(B\text{-}Y) \tag{12-2}$$

Recall that the output of a Blu-ray, DVD, or computer monitor occurs in the form of Y, (R-Y) = P_r, and (B-Y) = P_b, and we can extract R, G, and B from these three signals, as shown in Figure 12-20.

In terms encoding an S-video (Y-C signals) or composite video signal (CVBS), we modulate the Pr and Pb signals at a color subcarrier frequency with a cosine wave and a sine wave. For an S-video or composite analog color TV signal, the modulated P_r and P_b signals form the chroma or C channel.

The P_r and P_b signals are modulated with an in-phase carrier signal (cosine signal) and a quadrature-phase (sine signal) carrier signal at frequencies at about 3.579545MHz (NTSC, National Television System Committee) or 4.433619 MHz

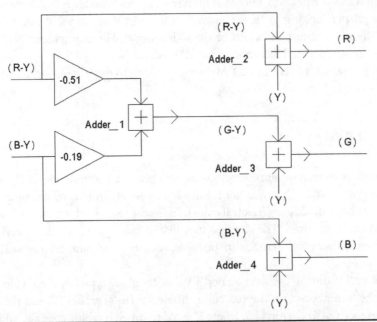

FIGURE 12-20 Recovering red (R), green (G), and blue (B) signals via Y, (R-Y) = P_r, and (B-Y) = P_b.

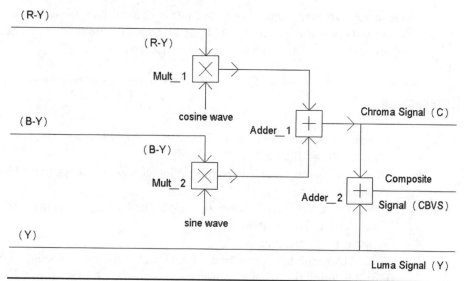

FIGURE 12-21   An example of encoding Y, P_r, and P_b signals into an S-video and a composite color TV signal.

(PAL, phase alternate line subcarrier). Other analog standards, such as SECAM (sequential color with memory) and variants of PAL such as PAL-M, PAL-N, etc. still are based on modulating the P_r and P_b channels in some manner. Figure 12-21 show encoding of Y, P_r, and P_b into S-video and composite (CVBS) color TV signals using quadrature phase and amplitude modulation.

As illustrated in Figure 12-21, the P_r = (R-Y) and P_b = (B-Y) signals are modulated or multiplied by 0- and 90-degree signals via Mult_1 and Mult_2. Cosine is defined as a 0-degree phase reference, and sine is defined as a 90-degree phase reference to perform I and Q AM modulation. In essence, the chroma signal is a quadrature AM signal at frequencies such as 3.579545 MHz (NTSC) or 4.433619 MHz (PAL). Not shown in Figure 12-21 are the low-pass filters that usually filter the outputs of each multiplier.

The choice of selecting such odd frequencies for the subcarrier modulation such as 3.79545 MHz or 4.433619 MHz is not a fluke. These frequencies were chosen specifically to display in a compatible manner with a monochrome or black-and-white-only TV monitors. When the chroma signal is present in a scene, this produces varying amplitudes and phases of the modulated subcarrier signal such that there is a *dithered* effect from one horizontal line to the next. In an NTSC color signal, the color frequency is chosen to have 227.5 cycles of subcarrier per line. This means that the adjacent TV lines will display the color subcarrier signal in a checkerboard pattern on a black-and-white monitor.

This checkerboard pattern of dots displayed on a black-and-white TV set is much less objectionable than if the color frequency is chosen to have an even number of subcarrier signal cycles per horizontal line. If the color subcarrier frequency is chosen to have an even number of cycles per horizontal line, we would see thin stripes up and down the screen. It would put all the actors and actresses from a TV show "behind bars" all the time when the scene has color.

References

1. K. Blair Benson and Jerry Whitaker, *Television Engineering Handbook*. New York: McGraw-Hill, 1992.
2. Howard W. Sams, *Color TV Training Manual, 4th ed*. Indianapolis: Howard W. Sams Co., Inc., 1977.
3. Howard W. Coleman, ed., *Color Television: The Business of Colorcasting*. New York: Hastings House, 1968.
4. *Batman*, TV series, 1966–1968.
5. "1956 EIA Resolution Test Chart"; available at: http://en.wikipedia.org/wiki/File:EIA_Resolution_Chart_1956.svg.

CHAPTER 13

Video Circuits and Systems

Video signals are a little bit more complicated than audio signals because they have to include synchronizing information (e.g., sync pulses, color burst signals, etc.). Horizontal and vertical synchronizing signals are essential to command a display when and where to display the picture information. Also, unlike audio signals that can be additively mixed together with any other audio signal, video signals have to synchronize in phase and frequency to each other or to a common clock signal before any additive mixing can be done.

This chapter will show how to control the contrast, brightness, sharpness, and horizontal picture position for a display. The circuits that implement these functions on their own belong to a video processing system that includes extra circuitry to ensure that timing (e.g., sync signals) and black or blanking levels are set properly. A four-page schematic of a video processor is provided, which requires advanced hobbyist skills and is not meant to be an actual project. However, if the reader is up to building such a circuit, by all means please try. In addition to showing how to make a practical video processor that allows adjustment for contrast, brightness, and sharpness of an external TV signal, we will illustrate how to make a video scrambler.

How Easy Is It to Make a Brightness and Contrast Video Circuit?

If the video signal is internal to a display such as a TV monitor where the vertical and horizontal sync signals have already been extracted and used within the TV monitor, then the brightness and contrast circuits include just a direct-current (DC) offset and an adjustable-gain amplifier. However, when brightness and contrast controls are implemented for video signals *outside* the TV set, the circuitry is a bit more complicated.

Let's start with the brightness control, where we want to add brightness adjustment from an external video signal such as one from a DVD player. If a DC offset

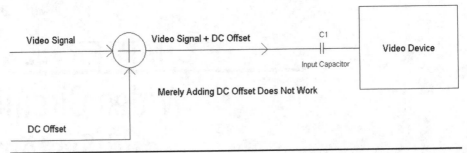

Figure 13-1   A DC offset voltage is added to the input video signal.

voltage is simply added to the video signal and sent to a TV set as seen in Figure 13-1, then there will be no effect.

As seen in Figure 13-1, most video devices such as a TV set have an AC (alternating-current) coupling capacitor (e.g., C1) that blocks any DC offset voltage. Thus there will be no change in brightness with the DC offset voltage added.

To add a brightness signal, the DC offset voltage is added selectively to the active video line portion of the video signal. This active video line portion in a standard-definition TV signal is about 52 μsec (μsec = microsecond = μs) in length and starts about 10 μsec after the start of the horizontal sync pulses (see Figures 13-2 and 13-3). In Figure 13-2, a switch allows passage of the DC offset voltage during active video line only to add a pedestal voltage to the video signal's pixel voltage. The active line gate signal is generated from a *sync separator* circuit along with a timing generator. Figures 13-4 and 13-5 provide examples of increasing and decreasing the brightness of a video signal.

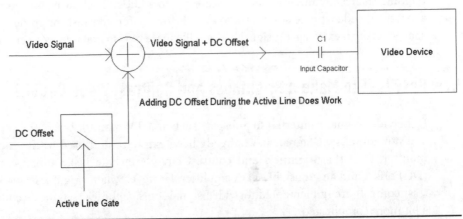

Figure 13-2   Controlling brightness to a video signal is accomplished by selectively adding the DC offset voltage.

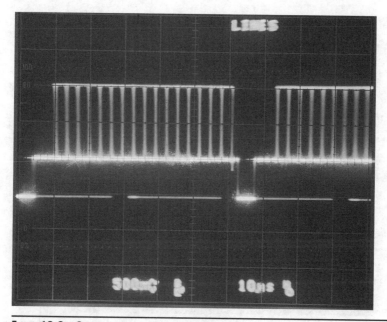

Figure 13-3 Convergence pattern of a test signal with 0 volt DC offset voltage added to the active line portion of the video signal.

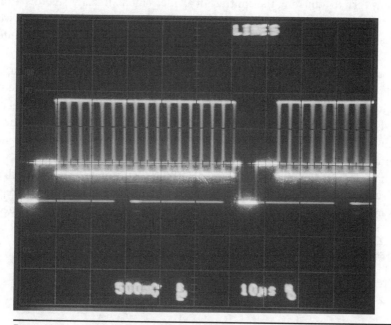

Figure 13-4 An example of adding a negative DC offset voltage during an active line portion to decrease the brightness of a video signal.

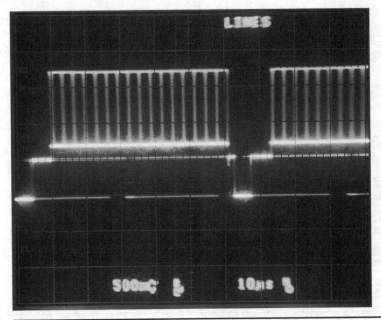

FIGURE 13-5   An example of adding a positive DC offset voltage to increase brightness.

We now turn to adjusting the contrast of a video signal. In a TV set, this is done by simply changing the amplitude of the video signal. However, if we do this when an external input video signal is provided, as shown in Figure 13-6, we encounter a problem.

In Figure 13-6, when the video signal level is varied, both the pixel information's amplitude and the synchronizing pulses' level will be changed. If VR1 is adjusted for low amplitude to provide a low-contrast picture, the horizontal and vertical synchronizing pulses will be attenuated to cause the TV set to unlock. To ensure that the TV set will always lock, a solution is to vary the pixel amplitude but keep the sync pulses at their normal amplitude (see Figure 13-7).

When we look at Figure 13-7, we see something that is more of a system than a simple circuit. There is a timing generator, a clamped black level amplifier, a (regenerated) sync generator ("New Sync"), and a multiplexer. All these circuits are neces-

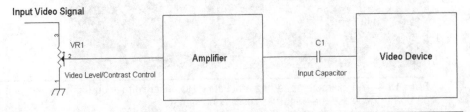

FIGURE 13-6   Varying the amplitude of a video signal.

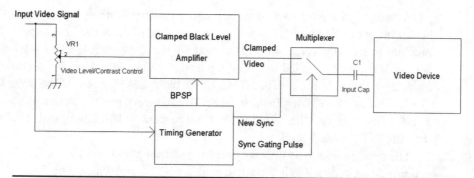

Figure 13-7 Inserting "New Sync" to ensure that a stable picture is viewed while varying the contrast.

sary to provide changes in video levels while not affecting the amplitude of the sync timing pulses.

The timing generator consists of a *sync separator* that literally extracts the horizontal and vertical synchronizing pulses from the input video signal. Once the synchronizing pulses have been extracted, they can be inserted into a new video signal that contains the active video signal whose amplitude is adjusted. The timing generator also sends a control signal (sync gating pulse) to the multiplexer such that regenerated sync pulses are inserted appropriately with the varied amplitude video signal.

A black level clamp circuit is required to establish the blanking level correctly. Many video signals are AC coupled, and with different scenes such as dark and bright ones, the blanking level will vary (see Figures 13-8 through 13-10).

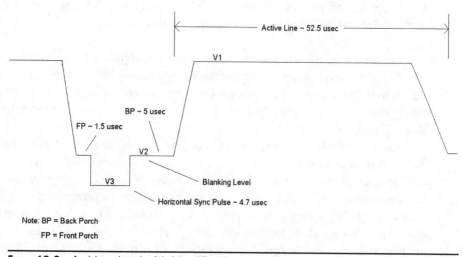

Figure 13-8 A video signal with identification of the front porch, horizontal sync pulse, and back porch.

As shown in Figure 13-8, there are four basic components to a TV line. The beginning of the "front porch" region denotes the end of the active line portion of the previous line and thus is the start of the next TV line. Its voltage level is the blanking voltage V2, which is called the *blanking level*. The blanking level is generally the black level, except in the case of NTSC, where black is 7.5 percent of the white level (7.5 IRE units, where IRE = Institute of Radio Engineers, and the peak white level is at 100 IRE units), and the blanking level is 0 percent (0 IRE units), which is lower than the (NTSC) black level.

The sync tip level is at V3, and the sync pulse amplitude is (V2 – V3). The normal sync amplitude is 286 mV peak to peak for NTSC and 300 mV for PAL.

NOTE   The number 286 is not exactly easy to remember, but if you are into vintage computers, think of the 286 processor, which was succeeded by the 386, and later by the 486 central processing units (CPUs).

The "back porch" region is normally set at the blanking level, and this region is commonly sampled by clamp circuits to set the blanking level or back porch voltage to a predetermined level. Following the back porch, the pixel information is located at the active line portion and its level is denoted by V1 and its amplitude is (V1 – V2). For NTSC, the pixel voltage value for a standard-level signal (V1 – V2) is between 0 mV and 714 mV and for a PAL signal (V1 – V2) is between 0 mV and 700 mV. Peak white is then 714 mV in the NTSC TV standard.

NOTE   714—where did this number come from? "Dum dee dum dum." The 714-mV number is a result of the fact that from peak white to sync tip, the voltage is (V1 – V3) = 1,000 mV, or 714 = (1,000 – 286). Or another way I have told people how to remember 714 is to think of Sgt. Joe Friday's badge number, 714, on the TV show *Dragnet*. After all, this is a chapter on TV.

Now we turn to the effects of AC coupling a video signal and how the back porch signal's voltage level changes with different video signals. Note in the AC-coupled stair-step signal in Figure 13-9 that the back porch level is a negative voltage that is below 0 volts at about –200 mV. The stair-step signal has an average picture level of about 50 percent.

Notice that the black-level signal in Figure 13-10, when AC coupled, has its back porch level close to 0 volts, and its average picture level is about 0 percent. The clamped black-level amplifier in Figure 13-7 can receive an AC-coupled video signal and then restore the black level (e.g., back porch level) to a predetermined voltage such as 0 volts regardless of the average picture level of the input video signal (e.g., when the video signal = 0 percent white or 100 percent white). We will later show a method to keep the back porch level set to a predetermined voltage via a feedback clamp circuit.

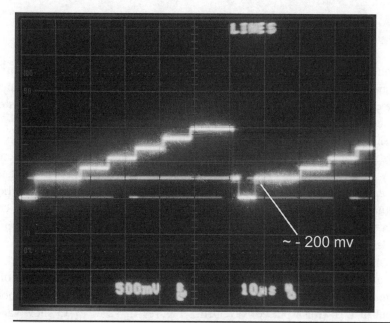

Figure 13-9   An AC-coupled video signal with a stair-step signal.

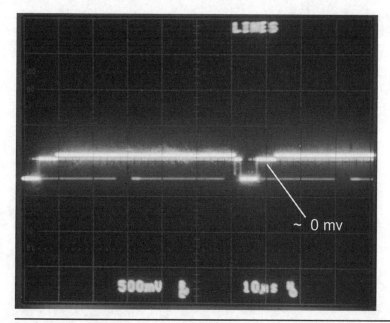

Figure 13-10   An AC-coupled video signal that is representative of a black level.

A Sharpness Circuit: Can It Be Simple?

It turns out that to make a stand-alone sharpness circuit, a sync separator and timing circuits are not necessary. Since the sharpness circuit affects the high-frequency response of the video signal, it can be treated like an audio equalizer for a stereo system (see Figure 13-11).

The sharpness circuit in Figure 13-11 is similar to the one in Figure 12-19 except that there is an analog signal delay circuit formed by R4 and C6. This delay circuit is necessary because high-pass filters C4, R8, C5, and VR1A produce a slight time delay. Even though the high-pass filters provide the double-pointed peaks when a pulse signal is present in Vin, these peaks are delayed somewhat, and when combined via the (+) input of U1B with the output signal from U1A, one of the peaks gets hidden in the rising or falling edge of the pulse signal, leaving only the overshoots visible without preshoots.

Low-pass filter and delay circuit R4 and C6 then provides the proper delay of the pulse signal such that the pulse response has both peaks (preshoot and overshoot) at the output of U1B (for a pulsed waveform). Figures 13-12 to 13-15 show various frequency responses of a multiburst signal for different sharpness settings and a pulse response at the maximum sharpness setting.

> **NOTE** It was found for the ciruit in Figure 13-11 that C6 can be in the range of 270 pF to 330 pF.

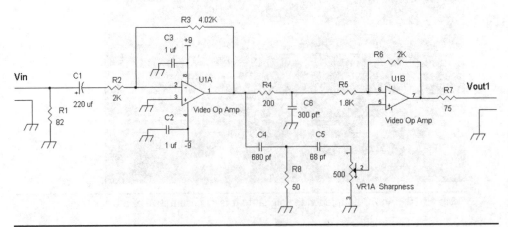

Figure 13-11 A practical phase linear sharpness control with the "Batman's ears" pulse response.

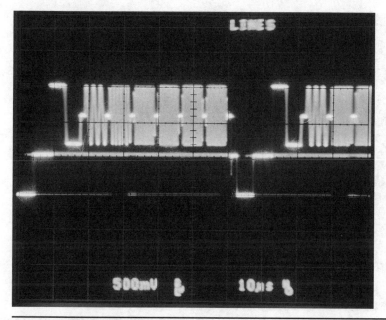

Figure 13-12   Sharpness control set for flat frequency response.

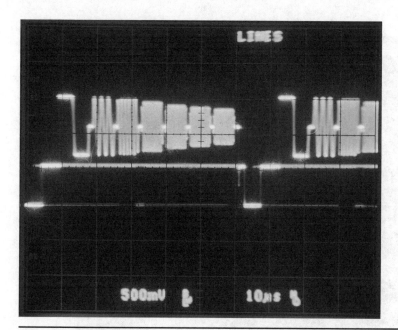

Figure 13-13   Sharpness control set to roll off the frequency response for a softer-looking picture.

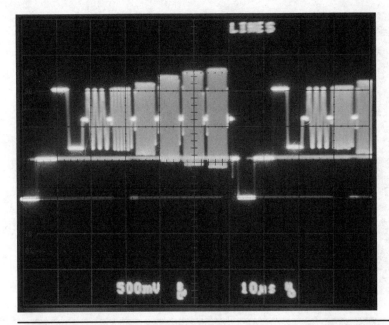

Figure 13-14   Sharpness control set to enhance the high-frequency response to provide edge enhancement.

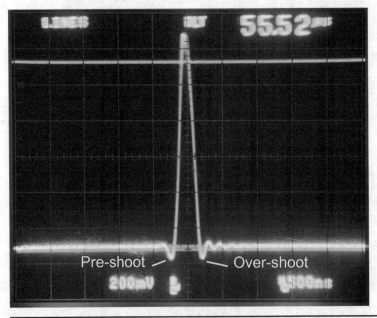

Figure 13-15   Pulse response when sharpness control is set at maximum provides preshoot and overshoot pulses.

A Video Processor to Adjust for Brightness, Contrast, and Sharpness

We will now show a video processor circuit, which requires a DVD player with the Y-channel output of the component or S-video connection connected to the video processor. The DVD player is set to standard-definition TV (SDTV), either 480i (derived from the NTSC 525-line TV standard) or 576i (derived from the PAL 625-line TV standard). A TV monitor or display will also be needed to see the adjustment effects of the video processor. Also, it is highly recommended that a two-channel oscilloscope of at least 15-MHz analog bandwidth is used in observing the various video and timing waveforms from the video processor.

This video processor circuit may be more advanced and complicated than any of the previous circuits shown so far because it is a combination of many different circuits. Figure 13-16 shows a top-level block diagram.

An input video signal source is connected to Vin, which is connected to two circuits, a sync separator and the input to the contrast-sharpness circuit. With the video signal coupled to the contrast and sharpness circuit, a video signal-amplitude control via a potentiometer allows adjustment of the contrast. Another control allows adding high-frequency signals derived from the video signal to add edge enhancement and high-frequency equalization to the video signal. The output circuit for the contrast and sharpness control also has a blanking level clamp circuit that allows the blanking level to be varied from a negative voltage to a positive voltage. Normally, the voltage blanking level is set to "about" 0 volt.

A processed video signal is provided at the output of the contrast and sharpness circuit. This circuit also has a variable or controllable blanking voltage that will later allow for changing the brightness. This processed video signal is fed to one of the two inputs of the multiplexer circuit.

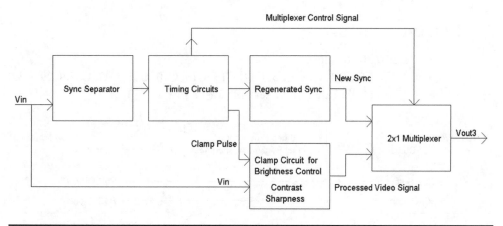

FIGURE 13-16   Block diagram of the video processor.

NOTE The blanking level is generally the black level of a video signal. In "older" 525-line TV standard sets used in the United States, blanking level = 0 percent of peak white level, whereas the black level = 7.5 percent of the peak white level. However, in most other TV standards, the blanking level = black level, which is 0 percent of the peak white level.

As previously mentioned, to change the contrast of a video signal requires more circuitry to ensure that the horizontal and vertical synchronizing signals are not affected when the contrast control changes the amplitude of the pixel information. At some point, we need to insert new sync pulses that are not affected by controlling the contrast. This requires a sync separator to extract the synchronizing signals from the input video source. The output of the sync separator circuit provides a "derived" sync signal, which is fed to timing circuits that allow "New Sync" signals to be inserted with the processed video signal.

Another job of the timing circuits is to provide a multiplexer control signal that commands the multiplexer to insert the new synchronizing, front porch, and back porch signals into the processed video signal. Finally, the timing circuit also provides a sampling pulse that is coincident within the back porch region of the input video signal. This back porch pulse serves as the clamp pulse for the contrast and sharpness output amplifier.

The video processor can be used to alter the brightness, contrast, and sharpness of a DVD program, as shown in Figure 13-17. Most DVD and Blu-ray players include a Y (luma) output signal connector, which can be used as the video source for the video processor to vary the contrast, brightness, or sharpness of the displayed picture (see Figure 13-18). The video processor circuits shown in Figures 13-19 through 13-22 will require a standard-definition Y (black and white) interlaced TV signal. Either 480i or 576i will do.

Now let's look into the circuits, which are divided among four pages. We start with the contrast, brightness, and sharpness circuits that include a feedback clamp circuit.

Parts List

- C1, 220 µF 16 volts
- C2, C3, C9, C11, C12, C13, C14, C15, C16, C20, C21, C23, C27, C29, C33, C37, and C38, 1 µF film or ceramic 5% or 10%

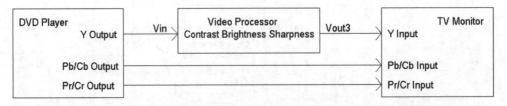

Figure 13-17   Video processor integrated with an example DVD player and TV monitor.

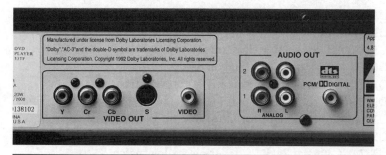

FIGURE 13-18 The Y, P_r (C_r), and P_b (C_b) video outputs with RCA connectors on the back of a DVD player.

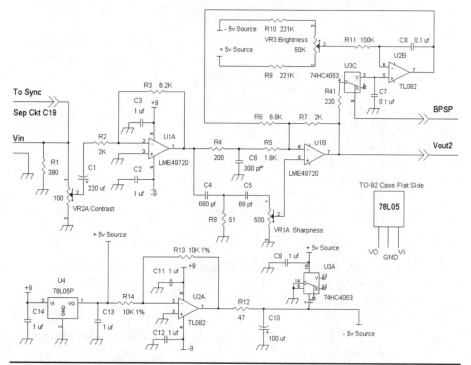

FIGURE 13-19 Schematic diagram for adjusting contrast, brightness, and sharpness.

- C4, 680 pF, 5% or 10% silver mica or ceramic
- C5, 68 pF 5% or 10% silver mica or ceramic
- C6, 300 pF or alternatively, 330 pF or 270 pF) 5% silver mica or ceramic
- C7, C8, 0.1 μF, 5% film
- C10, C17, C19, and C34, 100 μF 16 volts
- C18, 1000 μF 16 volts
- C22, 10 μF 25 volts
- C24, C26, C31, C32, C35, and C36, 0.001 μF, 5% film

- C25, 82 pF (or, alternatively, 100 pF) 5% or 10% silver mica or ceramic
- C28, 0.01 µF, 5% film
- C30, 330 pF, 5% or 10% silver mica or ceramic
- D1, D2 1N914, or 1N4148 silicon diodes
- Q1, Q2, Q3, and Q4 2N3904 or 2N4124 NPN transistors
- DIP 8 socket for U1 (socket is for experimenting with other op amps)
- U1, LM49720 (LM4562) or LM6172 eight-pin DIP dual op amp
- U2, TL082, LF353, or TL072 eight-pin DIP dual op amp
- U3, 74HC4053 or 74HCT4053 triple analog switch
- U5,74HC14 CMOS Schmitt trigger hex inverter
- U6, U8, and U9, 74HC221 dual one-shot monostable
- U7,74HC74 or 74HCT74 dual D flip-flop
- U10,74HC00 or 74HCT00 quad NAND gate
- +9 volt and –9 volt supply from batteries or a regulated ±9 volt to ±12 volt supply
- All resistors are ¼ watt, 5% unless noted
- R1, 390 Ω
- R2 and R7, 2 kΩ
- R3, 6.2 kΩ
- R4, 200 Ω
- R5, 1.8 kΩ
- R6, 6.8 kΩ
- R8, 51 Ω
- R9, R10, 2.21 kΩ 1%
- R11 and R28, 100 kΩ
- R12, 47 Ω
- R13, R14, R22, R30, R32, R35, and R40, 10 kΩ 1 percent (Note: R40 may be 10 kΩ 5%)
- R15, R23, R24, and R31, 2.7 kΩ
- R16 and R21, 47 Ω or 51 Ω
- R17, 510-Ω, ½ watt, 5% or alternatively, parallel two 1kΩ resistors for R17
- R18, 68 Ω
- R19, 9.1 kΩ
- R20, 1.2 kΩ
- R25 and R26, 1 kΩ
- R27, 220 Ω
- R29, 4.7 kΩ
- R33, 3.32 kΩ, 1%
- R34, 100 kΩ, 1%
- R36, 49.9 kΩ, 1%
- R37, 1 kΩ, 1%
- R38, 1.5 kΩ

- R39, 4.75 kΩ, 1%
- Potentiometers are ⅛ watt to ¼ watt rating
- VR1A, 500 Ω pot
- VR2A, 100 Ω pot
- VR3, 50 kΩ pot
- VR4, 5 kΩ pot

The video signal from a DVD player's Y channel is connected to Vin, which is coupled to a variable voltage divider circuit formed by VR2A (Figure 13-19). The output of VR2A is AC coupled to an inverting-gain amplifier U1A that has a fixed gain of approximately –3. Typically, Vin with a peak white signal will be on the order of 1 volt peak to peak since R1 in parallel with VR2A forms approximately 75 Ω for the standard input resistance. By adjusting VR2A, the output of U1A at pin 1 varies between 0 volts and 3 volts peak to peak of an inverted video signal.

To turn the video signal back into its noninverting form, U1B forms a unity-gain inverting amplifier via feedback resistor R7 and series input resistors to pin 6 of R4 and R5. The voltage gain from pin 1 to pin 7 of U1A and U1B is

$$-[R7/(R4 + R5)] = [2,000/(200 + 1,800)] = -1$$

To provide a range of high-frequency boost to the video signal that results in edge enhancement and an improvement in resolution, two high-pass filter sections C4, R8, C5, and VR1A add a high-frequency signal into pin 5, the (+) input terminal of U1B. The high-frequency signal's amplitude is adjustable via potentiometer VR1A. At the output of pin 7 of U1B is a video signal that has variable amplitude and sharpness via VR2A and VR1A, respectively.

We now also want to establish a predetermined or adjustable blanking level of the output signal at pin 7 of U1B. To accomplish this, a sample and hold circuit via U3C and C7 "records" the voltage of the video signal at pin 7 of U1B during a portion of the back porch period (approximately 3 μsec out of the approximately 5 μsec back porch period), which is after the trailing edge of sync pulses. Variable resistor VR3 allows adjustment of a range of positive to negative voltages via the +5v and the –5v voltage sources connected to R9 and R10. U2B serves as a feedback integrating amplifier with C8 and send a clamp error signal to R6.

The DC voltage at the slider of VR3 servos or forces the back porch voltage of the video signal to be the same voltage. For example, if the voltage is 0 volt at the slider of VR3, pin 7 at U1B will have a video signal whose back porch voltage is 0 volt. If the voltage at the slider of VR3 is +150 mV DC, then the video signal at pin 7 will have DC offset, and its back porch voltage will be +150 mV. Likewise, if VR3 slider's voltage is –100 mV DC, the resulting video signal at pin 7 will have a negative DC offset with its back porch voltage at –100 mV.

To provide a regulated +5 volt source to R9, U4, a low-power TO-92 positive regulator (78L05) integrated circuit (IC), is connected to the raw power supply of +9 volts. This regulated +5 volt source is also used to power the logic gates, analog switches, one-shot timing circuits, and sync separator circuit R28, Q4, and R29.

A regulated −5 volts is generated by using an inverting-gain op amp circuit U2A with resistors R13 and R14. The output at pin 1 of U2A is connected to a 47 Ω resistor prior to connection to capacitor C10 to avoid oscillation. An op amp in general does not load well directly into a capacitor and requires a small-value series resistance (e.g., 47 Ω) to isolate its output terminal from a capacitive load. The analog switch U3 requires a negative voltage because there will be a positive and a negative voltage range video signal at the input of one of the analog switches.

NOTE Although an LME49720 (or, equivalently, an LM4562) audio op amp is used for this video-processing circuit, it is well advised that the circuit be made with an 8 pin DIP socket. The socket allows easier changing of op amps. Higher-performance op amps suitable especially for video purposes such as the LM6172 and AD828ANZ can be used instead.

The output of U1B is Vout2, which is connected to the X input (pin 2 via R21) of the 74HC4053 analog switch (see Figure 13-20). The C Sync signal is a logic-level signal from 0 to +5 volts. Capacitor C17 blocks the DC voltage and level shifts the signal to be approximately 0 to −5 volts at R19. By dividing the C Sync signal appropriately via R20 and R19, a "New Sync" signal from about 0 to −600 mV is provided to the Y input terminal of U3B. When the command or control signal at pin 10 is logic high, the "New Sync" signal at pin 1 of U3B is transferred to the output of the switch at pin 15 of U3B and to R16.

The command signal to insert or switch in the "New Sync" is a combination of two sources, the VBI (vertical blanking interval) and the HBI (horizontal blanking interval) signals. These two signals, VBI and HBI, are logically OR'd via diodes D1, D2, and R15.

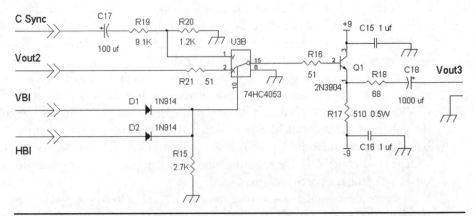

FIGURE 13-20   Output circuit with sync replacement.

When the VBI command is logic high, vertical sync signals from the original video input source are inserted. For inserting new horizontal sync signals, the HBI signal enables a high-logic level pulse to gate in the new front porch, sync pulse, and back porch signals. Also, as we will see later, the horizontal sync pulse is variable in phase or time-wise to demonstrate centering the displayed picture in a lateral manner.

Output amplifier emitter-follower Q1 buffers (amplifies with an approximate gain of 1) the processed signal that has adjustable contrast, brightness, and sharpness along with "New Sync" to the output terminal Vout3 via DC blocking capacitor C17. When Vout3 is terminated with 75 Ω, the sync level is reduced to half to provide about 300 mV peak to peak, which is a standard sync amplitude level. Note that R17, a 510 Ω, ½ watt resistor may be substituted with two 1 kΩ, ¼ watt resistors connected in parallel. Now let's look at Figure 13-21 (sync separator).

To extract the sync pulses from the input video signal (e.g., Y channel output from a DVD player or TV tuner), the input signal is amplified by about –2.7 × via common-emitter amplifier Q2 with collector load resistor R24 and emitter resistor R25. The gain is set by approximately –R24/R25 = –2.7 kΩ/1 kΩ = –2.7.

An inverting video signal with its positive-going sync pulses at about 2.7 × 0.3 volt = 0.8 volt is provided at the collector of Q2. Emitter follower Q3 provides a low-impedance drive signal into the sync separator circuit C22, Q4, R28, and R29. From the negative end of C22, the inverted video signal has positive-going sync pulses DC restored to about 0.7 volt at the sync tips. The base-emitter junction of Q4 acts like a diode with the anode connected to the – terminal of C22 and the cathode connected to ground. R28 is biased to + 5 volts to DC restore the sync tips to +0.7 volts and turn on Q4 as a switch. When the inverted sync pulse (from the inverted video signal at Q2's collector) drops below +0.7 volt, Q4 turns off. As a result, a negative-going sync pulse from about 0 volt to +5 volts is provided at the collector of Q4. Any video information or voltage levels, which are at least 0.8 volt below (inverted) sync tip at the base of Q4, are ignored.

Inverter gate U5A then inverts the sync-separated pulses to positive-going pulses. These positive-going sync pulses are sent to the data or D input of a flip-flop circuit U7A. A delayed version (approximately 0.2 μs delay) of these positive-going pulses via U5B, R31, C25, and U5C is coupled to the positive trigger input of one-shot multivibrator U6B. The output of U6B at pin 12 provides a negative-going approximately 3 μs pulse that is framed within a normal 4.7 μs wide horizontal sync pulse.

Whereas an analog switch and capacitor form a sample-and-hold circuit for an infinite number of levels in an analog signal, a D flip-flop is like a 1-bit sample-and-hold circuit. Both circuits will latch onto the value of the input signal during the sample pulse. In the analog switch U3C in Figure 13-19, the video signal provides the voltage to be sampled by the clamp pulse.

In a D flip-flop, the input is at the D input terminal, pin 2 in this case, and the sampling pulse is the output of the one-shot timing circuit at pin 12 of U6B. The

74HC74 samples the D input during the rising edge of the clock pulse at pin 3 of U7A.

So let's see why a flip-flop is being used to sense a beginning of the vertical blanking interval or, put another way, at the end of the active field. When an active field ends in the TV signal, there are pre- and post-equalizing sync pulses that precede and follow the broad vertical sync pulses. These equalizing pulses are half the width (2.3 μsec) of the 4.7 μsec wide horizontal sync pulses.

The object, then, is to detect the first pre-equalizing pulse and generate a VBI signal. When horizontal sync pulses of 4.7 μsec are at the D input, U7A will output a logic-high output signal at pin 5 because the clock pulse triggers a rising edge at 3 μsec away from the leading edge of horizontal and all sync pulses and catches a high state of the positive-going horizontal sync pulse.

However, when a 2.3 μsec pre-equalizing pulse is inputted into the flip flop, the 3 μsec clock pulse will go past the high state of this pre-equalizing pulse and catch a low state instead. The output at pin 5 of U7A then goes from logic high to logic low when the first pre-equalizing pulse is sensed. The complementary output of U7A is pin 6, which goes high at detection of the first pre-equalizing pulse, and this positive-going pulse triggers one-shot timing circuit U6A to generate a pulse of approximately 100 kΩ × 0.01 μF = 1 msec (msec = millisecond) to provide the VBI pulse at pin 13 (at U6A).

For illustration purposes, Figure 13-21 also shows the traditional method of sensing vertical sync pulses that are approximately 27 μsec wide versus the 4.7 μsec pulsewidth of a horizontal sync pulse. This method includes a simple RC low-pass filter R30 and C24 that filters the horizontal and vertical sync pulses. Negative-going sync pulses are fed to R30, and when the low-state vertical sync pulses are present, the voltage across C24 will drop sufficiently to allow inverter gate U5E at pin 10 to flip to a positive-going vertical pulse with a frequency of about 60 Hz or 59.94 Hz. Although the signal at pin 10 of U5E is not used for other parts of the video processor, it can be used as a vertical rate-triggering pulse signal for an oscilloscope.

We will now explain how the back porch sampling pulse (BPSP) is provided by U5D pin 8. The slightly delayed positive-going sync pulses from U5C pin 6 are coupled to a high-pass filter C30 and R32 to generate a negative going spike on the falling edge of the positive-going sync pulses. This negative spike is on the trailing edge of the sync pulse and starts in the beginning of the back porch area of the video signal. U5D also serves as a half-shot timing generator, which provides a positive-going pulse of approximately 3 μs within the approximately 5 μs-wide back porch of the video signal. This pulse then is used for sampling the back porch or blanking-level voltage of the incoming video signal for U3C in Figure 13-19.

For generating the HBI (horizontal blanking interval) signal, which is a pulse about 11 μs long, we start with a negative-going composite sync signal from pin 4 of U5B (Figure 13-22). First, we trigger the leading edge of sync via U8B to generate about a 10 μs pulse via pin 5 that ends at the beginning of the active video line. The

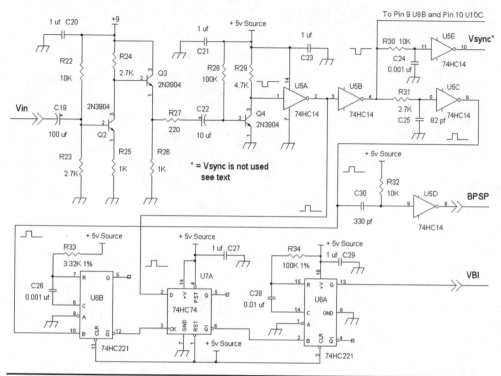

Figure 13-21   Input circuit with sync separator circuit.

output pulse at pin 5 of U8B is then sent to the negative or falling edge trigger input of the next timing generator U8A, which generates a pulse at pin 13 indicative of the active video line, about 52 μsec. The inverted output at pin 4 of U8A then is low during the active horizontal line portion of the video signal and high during the horizontal blanking interval (HBI) that includes the front porch, horizontal sync pulse, and back porch areas.

To regenerate the vertical and horizontal sync pulses, we could just insert sync pulses from pin 4 of U5B for the C-Sync signal and be done and go home. However, wouldn't we like a little adventure? Why not show the effects of phasing the picture laterally and add wiggle as well? That is, we can shift the picture on the display from side to side manually with a potentiometer or, alternatively, couple a signal generator or audio signal source to U8A to electronically shake the picture up to provide scrambling.

Nominally, the output pulse at pin 13 of U8A finishes at the end of the active TV line. At this point, U9B provides about a 1.5 μsec pulse to start the newly regenerated horizontal sync pulse of 1.5 μsec (the duration of the front porch) into the horizontal blanking interval. U9A generates horizontal sync pulses of the standard 4.7 μsec pulsewidth via pin 13 of U9A. These regenerated horizontal pulses can be

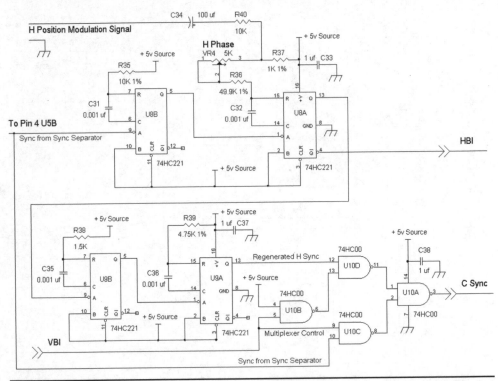

FIGURE 13-22   Timing circuits to provide horizontal blanking interval and composite sync signals.

shifted to lead or to lag, which causes the TV monitor to display the picture to the side in either direction or centered correctly via H Phase variable resistor VR4.

If a sine wave or any other waveform of frequency less than 15 kHz is inputted to C34, the regenerated horizontal sync pulses will be position modulated that results in a wiggle. Typical amplitudes would be on the order of 200 mV to 1.2 volts peak to peak depending on the strength of the horizontal shift. The modulated waveform's amplitude is attenuated approximately tenfold via R40 and R37. By modulating the voltage at VR4 and R36, the charging time of C32 is varied to provide a pulsewidth-modulated signal from pin 13 of U8A.

To insert regular vertical sync pulses and the regenerated horizontal sync pulses, we multiplex or switch between the types of sync pulses to provide the regenerated composite sync signal, C Sync. By using the VBI (vertical blanking interval) signal, the derived vertical sync pulses from the input video signal is spliced in with the regenerated horizontal sync pulses that are capable of position modulation.

When the VBI signal is logic high, the vertical sync signals from the sync separator are switched to the output of the multiplexer, U10A pin 3. The regenerated horizontal sync signals that can be position modulated are switched to pin 3 of U10A when the VBI signal is logic low, which happens during the active video field. The vertical and horizontal sync signals are then sent to the (+) terminal of C17 (Figure 13-20), which couples these sync signals to U3B for insertion into the processed video signal that allows adjustment for contrast, brightness, and sharpness.

To experiment with modulating the phase or position of the new horizontal sync pulses, a function generator is connected to C34 (negative side) and ground in Figure 13-22. Frequencies from 100 Hz to 100 kHz can be tried. The sensitivity of the positional shift is that about 1.2 volts peak to peak from the (function) generator results in about 2.3 μsec peak to peak of horizontal sync displacement. At frequencies near the horizontal line frequency, f_H (e.g., approximately 15.734 kHz for 480i and approximately 15.625 kHz for 576i), the scrambling effect has approximately a frequency = $|f_H - f_{osc}|$, where f_{osc} is the oscillator frequency. One will find that when the oscillator is set close to multiples of f_H, there will also be picture-displacement effects. For a reference, Figure 13-23a has no modulation.

Figure 13-23b–f shows the effects of connecting a 1.2 volt peak-to-peak signal at approximately 240 Hz with differently shaped waveforms. Figure 13-23g shows what happens when the oscillator frequency is set to an odd multiple of the frame rate

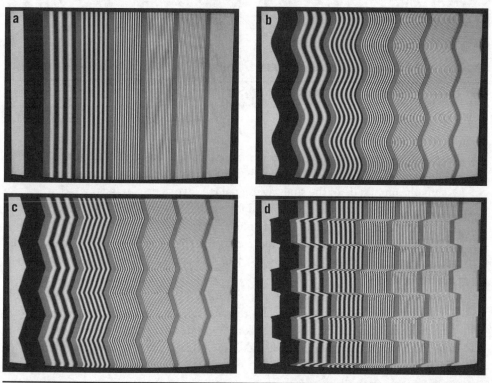

FIGURE 13-23 (a) A multiburst test signal with the video processor's modulation signal at C34 turned off. (b) Sine-wave modulation. (c) Triangle-wave modulation. (d) Square-wave modulation.

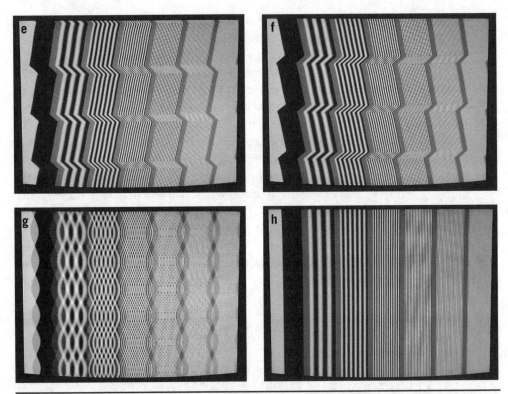

Figure 13-23 (*e*) Sawtooth-wave modulation. (*f*) Second sawtooth-wave modulation. (*g*) Interleaved sine-wave pattern. (*h*) No modulation.

(e.g., 29.97 Hz), and an interleaved scrambling pattern is provided. Approximately, a frequency of 270 Hz or 210 Hz may be applied to result in an interleaved pattern. Again, the amplitude of the sinewave signal is 1.2 volts peak to peak.

Other types of signals may be used for modulating or scrambling the TV signal. For example, an audio signal from a radio or an audio player may be used. Random-noise generators may also be connected to the C34. A general *block* layout of the video processor is shown in Figure 13-24, and a picture of its breadboard is shown in Figure 13-25.

As mentioned previously, this video processor project is probably more suitable for the advanced hobbyist. Generally, the signal leads are kept short. Alternatively, from a low-resistance drive such as from the slider of VR2A, the lead can be long when it is kept away from other signal lines and kept close to the ground plane for shielding.

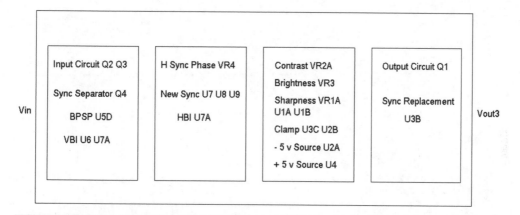

Input Circuit Q2 Q3	H Sync Phase VR4	Contrast VR2A	Output Circuit Q1
Sync Separator Q4	New Sync U7 U8 U9	Brightness VR3	Sync Replacement
BPSP U5D	HBI U7A	Sharpness VR1A U1A U1B	U3B
VBI U6 U7A		Clamp U3C U2B	
		- 5 v Source U2A	
		+ 5 v Source U4	

FIGURE 13-24 Block layout of the video processor.

FIGURE 13-25 Copper-clad breadboard of the video processor.

References

1. K. Blair Benson and Jerry Whitaker, *Television Engineering Handbook*. New York: McGraw-Hill, 1992.

2. Howard W. Sams, *Color TV Training Manual, 4th ed*. Indianapolis: Howard W. Sams Co., Inc., 1977.

3. Howard W. Coleman, ed., *Color Television: The Business of Colorcasting*. New York: Hastings House, 1968.

4. *Dragnet*, the TV series, 1966–1970.

5. John O. Ryan, Ronald Quan, James R. Holzgrafe, and Peter J. Wonfor, "Method and Apparatus for Scrambling and Descrambling of Video Signal with Edge Fill," United States Patent Number 5,438,620, filed February 28, 1994.

CHAPTER 14

High School Mathematics with Electronics

Many electronic circuits and systems can be described with high school mathematics and, in particular, algebra. For example, mathematical topics covered in high school algebra include equations of one or two unknown variables, polynomials, linear equations describing lines, parabolic functions, and graphing techniques. In this chapter, we will show applications selected from some of these mathematical topics that pertain to electronic circuits and systems.

> **NOTE** The math presented in this chapter will be written in an informal manner. What we want to show here is how the math relates to electronics and not necessarily to present the math in a very formal manner that would be found in a pure mathematics book. For example, this chapter will not necessarily show the mathematics in the same order of an algebra book, such as starting with numbers, operations of numbers, equalities and inequalities, and so on.

Equation of a Line

A line is generally described by the following equation:

$$y = mx + b \tag{14-1}$$

where the constant m is the slope of the line. The constant b is the y intercept, or in other words, when $x \to 0$:

$$y = b$$

In Equation (14-1), the variable x is allowed to have a multitude of values, but the constants m and b have predetermined values.

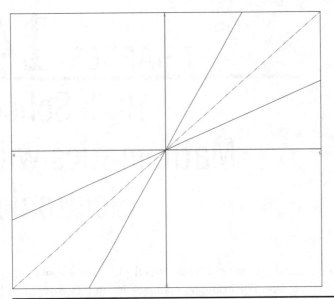

Figure 14-1 Equations of three lines with slopes m = 0.5 for the shallow slope, 1 for the slope in the middle, and 2 for the steepest sloped line.

In a simplest example, for now, let's remove or ignore the constant b and just look at the equation:

$$y = mx \qquad (14\text{-}2)$$

Equation (14-2) can be graphed with various values of m as shown in Figure 14-1.

In Equations (14-1) and (14-2), m is the slope of the line, or better yet, m is defined as:

$$m = \Delta y / \Delta x \qquad (14\text{-}3)$$

where $\Delta y / \Delta x$ is sometimes also known as the rise over the run, or $\Delta y / \Delta x$ = rise/run (see Figure 14-2).

So now let's take a look at a few examples of how Equation (14-2) relates to electronic circuits, starting with a resistive divider in Figure 14-3.

For a resistive divider circuit such as that shown in Figure 14-3, we have the familiar equation:

$$Vout = Vin[R2/(R1 + R2)] \qquad (14\text{-}4)$$

Let's substitute in the following manner:

$$Vout = y$$

$$Vin = x$$

$$m = [R2/(R1 + R2)] = slope$$

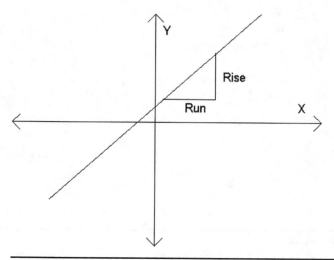

FIGURE 14-2   An example of the rise and run for determining a slope.

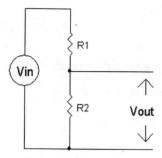

FIGURE 14-3   A two-resistor circuit forming a voltage divider.

Now we see that Equation (14-4) is just a variant of Equation (14-2).

By dividing by Vin on both sides of the equation, we have:

$$[Vout/Vin] = [R2/(R1 + R2)] = m = \text{gain of the voltage divider} \qquad (14\text{-}5)$$

Now we see that Equation (14-5) is just a variant of Equation (14-3). That is, the slope m actually tells us the gain or attenuation of the voltage divider. In practice, of course, m ≤ 1 for a voltage divider since a passive voltage divider made of the same parts (e.g., either all resistors, all capacitors, or all inductors) cannot provide voltage gain, that is, provide more voltage at Vout than at Vin.

In terms of amplifiers, let's take a look at two amplifiers, a noninverting and an inverting op amp amplifier, and their associated gains that can be graphed (see Figure 14-4).

For the noninverting op amp, we have:

$$Vout = [(RF + R1)/R1]Vin \qquad (14\text{-}6)$$

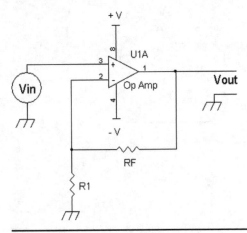

Figure 14-4   Noninverting op amp.

Thus, equivalently:

Vout = y

Vin = x

m = slope = [(RF + R1)/R1] = [Vout/Vin] = voltage gain of the noninverting op amp

And for an inverting op amp as shown in Figure 14-5, we have:

Vout = −[RF/R1]Vin   (14-7)

Again, this can be characterized equivalently as:

Vout = y

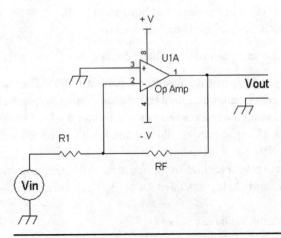

Figure 14-5   Inverting op amp.

Vin = x

m = slope = –[RF/R1] = [Vout/Vin] = voltage gain of the inverting op amp

For the direct-coupled op amp noninverting and inverting amplifiers, the input signal Vin can include both direct-current (DC) and alternating-current (AC) signals.

Offset Voltages Described by the Y Intercept b

For the following examples, the input signal Vin will be restricted to AC signals, and all capacitors will be large enough to hold their charges to provide a constant DC voltage. Consider a single-supply noninverting op amp circuit as shown in Figure 14-6.

The DC offset voltage at the output is at (½) of the supply voltage when RB1 = RB2, and Vout can be described as:

Vout = [(RF + R1)/R1]Vin + (½)(+V)   (14-8)

For example, if (+V) = 9 volts, (½)(+V) = 4.5 volts. Equation (14-8), which describes a single-supply noninverting amplifier, is an example of Equation (14-1):

Vout = y

Vin = x

m = slope = AC signal gain = [(RF + R1)/R1] = [Vout/Vin]

and:

b = (½)(+V) = DC offset = the y intercept

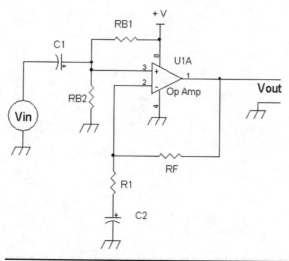

Figure 14-6   Single-supply noninverting amplifier.

If you now observe the noninverting amplifier with a single supply, pertaining to the gain or slope of the line set up by Vout and Vin, it does not matter what the value of the DC offset is. If we take the slope by its definition with y = Vout and x = Vin, then:

$$m = \Delta y/\Delta x = \Delta Vout/\Delta Vin = gain \qquad (14\text{-}9)$$

When we take a look at the gain, especially the AC gain, the DC offset voltage or y-intercept value is removed because we are only interested in the slope.

Similarly for an inverting amplifier, the output is Vout, as shown in Figure 14-7. RB1 = RB2, so the DC offset voltage sits at one-half the supply voltage +V:

$$Vout = -\,[RF/R1]Vin + (\tfrac{1}{2})(+V)$$

where:

$$Vout = y$$

$$Vin = x$$

$$m = slope = AC\ signal\ gain = -[RF/R1] = [Vout/Vin]$$

and:

$$b = (\tfrac{1}{2})(+V) = DC\ offset = the\ y\ intercept$$

The question then arises why the DC offset does not contribute to the AC gain mathematically. Intuitively, we know that adding any DC voltage to an AC signal does nothing in terms of reducing or increasing the amplitude of the AC signal; thus the DC voltage does not affect the gain of a *linear amplifier.*

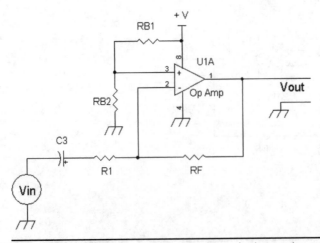

Figure 14-7   Inverting op amp circuit with a single supply.

| NOTE | We will find that in *nonlinear* amplifiers, the gain does vary with the DC offset of the amplifier. This was illustrated in Figure 8-23 |

pertaining to intermodulation distortion in amplifiers.

Suppose that we take two equations like Equation (14-1) and do the following:

$$y = m_1x + b_1 + m_2x + b_2 \qquad (14\text{-}10)$$

which can be rewritten in a more familiar form:

$$y = (m_1 + m_2)x + (b_1 + b_2)$$

The new slope is $\Delta y/\Delta x = (m_1 + m_2)$, and the new y intercept is $(b_1 + b_2)$. This also means that $\Delta y/\Delta x$ = slope by definition, so $\Delta y/\Delta x = (m_1 + m_2)$. Now let's split Equation (14-10) into two parts and confirm that we can take the slope of each individual line and sum the slopes together to come up with the answer $(m_1 + m_2)$.

$$y_1 = m_1x + b_1$$

$$y_2 = m_2x + b_2$$

$$\Delta y1/\Delta x = m1$$

$$\Delta y_2/\Delta x = m_2$$

The sum of the slopes is:

$$(\Delta y_1/\Delta x) + (\Delta y_2/\Delta x) = m_1 + m_2 = \Delta y/\Delta x$$

or, more importantly:

$$(\Delta y_1/\Delta x) + (\Delta y_2/\Delta x) = \Delta y/\Delta x \qquad (14\text{-}11)$$

Equation (14-11) is true for all types of linear equations for y_1 and y_2.

Now let's look at another specific case. Let :

$$y_1 = mx$$

and:

$$y_2 = b$$

for $y = y_1 + y_2 = mx + b$ or $y = mx + b$, where y is the sum of the two lines, y_1 and y_2. This leads to:

$$(\Delta y_1/\Delta x) = m$$

and:

$$(\Delta y_2/\Delta x) = 0$$

$$(\Delta y_1/\Delta x) + (\Delta y_2/\Delta x) = m + 0 = m = \Delta y/\Delta x$$

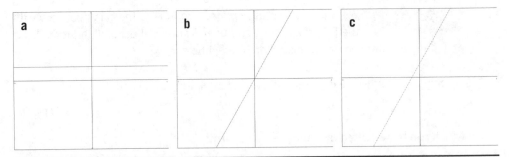

Figure 14-8   (*a*) y = 1, where the slope is = 0. (*b*) y = 2x, where the slope is 2. (*c*) y = 2x + 1, where the slope is still 2, but the line from Figure 14-8*b* is shifted up by 1, which is the y intercept or b = 1.

But we know that y = mx + b, so $\Delta y / \Delta x$ = m.

The slope of a constant is 0 since the slope is defined as the ratio rise Δy over run Δx. Because a constant has no rise for Δy, as x is changing, then Δy = 0 for a line described as a constant (see Figure 14-8).

Intuitively, this is correct in that the slope of a flat horizontal line, which is generally described as:

y = c = constant

has no slope for the line that has any level (e.g., c = 0, 1, –1, 10, etc.). For example, in a stair step, within each step there is no slope since each step is flat.

A line equation with a y intercept b can be the superposition of two lines, a line that is associated with a slope m, (y_1 = mx) and another line associated with a flat horizontal line (y_2 = b), and the "total" slope of line y = mx + b is just the sum of the slopes of the two individual lines y_1 and y_2 since $y_1 + y_2$ = y. The concept of summing two or more amplifiers' outputs is analogous to the superposition of lines.

Systems of Linear Equations Used in FM Stereo

When FM (frequency modulation) broadcast transmissions first started, there was only one main audio channel sent to the receiver. In order to achieve backward compatibility for transmitting two independent audio channels, a trick was needed. In an FM signal, recovered bandwidth in a radio is about 70 kHz.

Within this 70 kHz bandwidth, the main audio channel occupies from about 50 Hz to about 15 kHz. Another audio channel can be frequency translated up to a frequency beyond 15 kHz that is centered around 38 kHz. This 38 kHz channel is called the *stereo subcarrier channel*. It is an amplitude-modulation (AM) signal that is double sideband with suppressed carrier. The AM modulator is a multiplier.

Transmitting only one of the stereo channels, such as the left channel (L) with a bandwidth from 50 Hz to 15 kHz or the right channel (R) at around 38 kHz, would not be compatible with older radios that only receive the main channel. The older receivers would only demodulate the left channel, and the right channel information would be missing.

To maintain compatibility, the main channel transmitted the left and right channels summed together to provide a monaural channel (L + R). In this way, older radios would still receive both channels mixed together. To make use of the stereo subcarrier at 38 kHz, a difference channel (L − R) holds the left minus right audio information that is amplitude modulating the 38 kHz carrier such that the 38 kHz carrier signal is suppressed (see Figure 14-9).

The receiver has two tasks to perform. One is to demodulate the monaural channel that contains the (L + R) audio signal. The other is to demodulate the 38 kHz subcarrier channel to retrieve the (L − R) audio signal. To extract separate left and right audio channels, the (L + R) and (L − R) audio channels are summed and subtracted (see Figure 14-10).

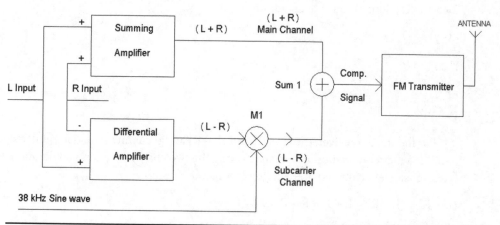

FIGURE 14-9 Transmitting a compatible stereo signal for FM radios.

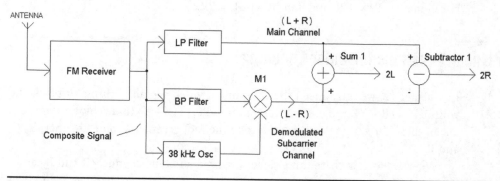

FIGURE 14-10 Stereo receiver with a summer and subtractor to recover the L and R audio channels.

In algebra, we often also use summing and subtraction of equations to solve for the unknown variables. Here is an example of summing and subtracting for solving two variables with two equations:

$x + y = 10$

$x - y = 28$

Summing the two equations, we have on both sides:

$x + y + x - y = 10 + 28$ or $2x = 38$

$x = 19$

Subtracting the two equations, we have:

$(x + y) - (x - y) = 10 - 28$

$x + y - x - (-y) = -18$

Two subtractions amount to an addition, so:

$-(-y) = +y$

$y + y = -18$

$2y = -18$

$y = -9$

For the stereo receiver, the $(L + R)$ and $(L - R)$ audio channels are demodulated, and it is a matter of summing and subtracting the two channels (main and demodulated subcarrier channels) in the following manner:

$(L + R) + (L - R) = L + L + R - R = 2L$

$(L + R) - (L - R) = L - L + R - (-R) = 2R$

Recall that $-(-R) = +R$, and thus, $R - (-R) = 2R$.

Linear Equations for Color Encoding

In Chapter 12, we introduced the component video signals Y (luma or black and white), $P_r = (R - Y)$ (red color difference), and $P_b = (B - Y)$ (blue color difference). Given these three signals, we can decode the red, green, and blue channels R, G, and B.

When we look at these three signals, the red and blue channels R and B can be extracted via summing the Y channel with the color difference channels P_r and P_b as follows:

$$P_r + Y = (R - Y) + Y = R - Y + Y = R$$

$$P_b + Y = (B - Y) + Y = B - Y + Y = B$$

What we would want is an expression for the green color difference channel $P_g = (G - Y)$ in terms of P_r and P_b. That is:

$$P_g = K_1 P_r + K_2 P_b \quad \text{or} \quad (G - Y) = K_1(R - Y) + K_2(B - Y) \tag{14-12}$$

where K_1 and K_2 are the equivalent x and y variables we want to solve for. We have one equation via Equation (14-12) and two variables K_1 and K_2 we need to solve for. So another "independent" equation with Equation (14-12) is required for solving K_1 and K_2. Also, from Chapter 12, we know that the Y channel can be expressed as:

$$Y = 0.59G + 0.30R + 0.11B \tag{14-13}$$

If we substitute $(0.59G + 0.30R + 0.11R)$ for all the Y terms in Equation (14-12), we get:

$$G - (0.59G + 0.30R + 0.11B) = K_1[R - (0.59G + 0.30R + 0.11B)] + K_2[B - (0.59G + 0.30R + 0.11B)] \tag{14-14}$$

This leads to:

$$G - 0.59G - 0.30R - 0.11B = K_1(R - 0.59G - 0.30R - 0.11B) + K_2(B - 0.59G - 0.30R - 0.11B) \tag{14-15}$$

Equation (14-15) simplifies to:

$$0.41G - 0.30R - 0.11B = K_1(-0.59G + 0.70R - 0.11B) + K_2(-0.59G - 0.30R + 0.89B) \tag{14-16}$$

From Equation (14-16), we can set up two different equations involving K_1 and K_2 by picking any two colors signals from R, G, and B. Let's take the first two colors, green (G) and red (R). For green, G, we have:

$$0.41G = K_1(-0.59G) + K_2(-0.59G) \tag{14-17}$$

We can divide G from both sides to get:

$$0.41 = K_1(-0.59) + K_2(-0.59)$$

which is the same as:

$$-0.59K_1 + (-0.59K_2) = 0.41 \tag{14-18}$$

Now let's set up equations involving the red channel (R).

$$-0.30R = K_1(0.70R) + K_2(-0.30\,R) \tag{14-19}$$

Let's divide by R on both sides to get:

$$-0.30 = K_1(0.70) + K_2(-0.30) \tag{14-20}$$

This is the same as:

$$0.70K_1 + (-0.30K_2) = -0.30 \tag{14-21}$$

With Equations (14-18) and (14-21), we now have two independent equations to solve for K_1 and K_2:

$$-0.59K_1 + (-0.59K_2) = 0.41 \tag{14-18}$$

$$0.70K_1 + (-0.30K_2) = -0.30 \tag{14-21}$$

Multiplying Equation (14-21) by 0.59/0.70 yields:

$$0.70(0.59)/(0.70)K_1 + (-0.30)(0.59)/(0.70)K_2 = -0.30(0.59)/(0.70)$$

This reduces to Equation (14-22) that will be added to Equation (14-18) to form Equation (14-23):

$$0.59K_1 + (-0.253K_2) = -0.253 \tag{14-22}$$

$$-0.59K_1 + (-0.59K_2) = 0.41 \tag{14-18}$$

Now we will add equations (14-22) to (14-18), which leads to:

$$-0.84K_2 = 0.157 \tag{14-23}$$

Divide by -0.84 on both sides of Equation (14-23) and:

$$K_2 = 0.157/(-0.84) = -0.187 \approx -0.19 = K_2$$

At this point, we can just substitute (more exactly) $K_2 = -0.187$ into Equation (14-18) or Equation (14-22) to solve for K_1. Let's just pick Equation (14-18):

$$-0.59K_1 + (-0.59)(-0.187) = 0.41$$

$$-0.59K_1 + 0.11 = 0.41$$

$$-0.59K_1 = 0.41 - 0.11 = 0.30$$

$$-0.59K_1 = 0.30 \tag{14-24}$$

Divide by -0.59 on each side of Equation (14-24) and:

$$K_1 = 0.30/(-0.59) = -0.508 \approx -0.51 = K_1$$

Now we substitute -0.51 for K_1 and -0.19 for K_2 in Equation (14-12) to provide Equation (14-25):

$$(G - Y) = K_1(R - Y) + K_2(B - Y) \tag{14-12}$$

$$(G - Y) = (-0.51)(R - Y) + (-0.19)(B - Y) \tag{14-25}$$

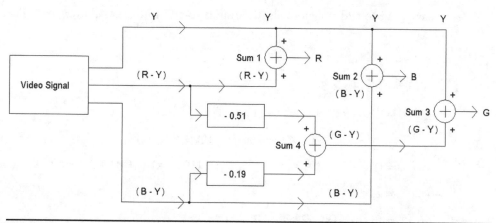

FIGURE 14-11   Color decoder to provide an R, B, and G signal via generating the (G – Y) signal from the (R – Y) and (B – Y) signals and summing with the Y signal.

Figure 14-11 shows an R, G, B decoder based on using the (G – Y) derived from the (R – Y) and (B – Y) signals.

Polynomials

When we look at the electrical voltage-to-current relationship of a resistor, it is very linear. That is:

V = IR

If we let V = y, I = x, and R = m, we get the familiar equation of a line:

y = mx + b

where b = 0 because resistors do not have built-in DC offset voltages, so:

y = mx

Since y is a function of x, we can often more aptly describe equations that depend on the variable x as f(x) = y. For example, for a line equation:

f(x) = mx + b

which denotes a first-order polynominal equation because the highest power of x is 1.

Now let's take a look at a polynominal using the variable x:

$f(x) = 5x^3 - 6x^2 + 11x - 15$

which is a third-order polynomial written in descending order starting with the highest power first, which is 3, followed by the next highest power, 2, and so on.

If we want to add polynomials, the rule is to add to the like powers of x. For example:

$f_1(x) = 7x^3 - 9x^2 + 10x - 1$

$f_2(x) = 15x^3 - 66x^2 + x + 5$

$f_1(x) + f_2(x) = (7x^3 - 9x^2 + 10x - 1) + (15x^3 - 66x^2 + x + 5)$

$$= 7x^3 + 15x^3 + (-9x^2) + (-66x^2) + 10x + x + (-1) + 5$$

$$= (7x^3 + 15x3) + [-9x^2 + (-66x^2)] + (10x + x) + [(-1) + 5]$$

$$= 22x^3 + (-75x^2) + 11x + 4$$

$$= f_1(x) + f_2(x) = 22x^3 - 75x + 11x + 4$$

Polynomials also can be expressed in ascending order, which is often the case when distortion analysis is done. Here the polynomial is written starting with the lower-power term and then moving to the higher-power terms.

The same polynomial:

$f(x) = 5x^3 - 6x^2 + 11x - 15$

can be expressed as:

$f(x) = -15 + 11x - 6x^2 + 5x^3$

Polynomials with constants or coefficients a_0 to a_n generally can be expressed as:

$$f(x) = a_0 + a_1x + a_2x^2 + a_3x^3 + \ldots a_nx^n \qquad (14\text{-}26)$$

Now let's take a look at a practical example of an electronic circuit. See an example parabolic curve of a depletion mode MOSFET in Figure 14-12.

Suppose that we have a depletion-mode MOSFET that is approximated by the following equation for drain current:

$$I_D = I_{DSS}[1 - 2(V_{GS}/V_p) + (V_{GS}/V_p)^2] \qquad (14\text{-}27)$$

where I_D = the drain current that is a function of V_{GS} = gate-to-source voltage; and V_p = pinch-off voltage, usually a negative voltage; and I_{DSS} = drain current when V_{GS} = 0 volts. Or equivalently, we can express I_D as:

$$I_D = I_{DSS} - I_{DSS}(2I_{DSS}/V_p)V_{GS} + I_{DSS}[(1/V_p)^2](V_{GS})^2 \qquad (14\text{-}28)$$

In terms of a general-form polynomial as shown in Equation (14-26):

$$f(x) = a_0 + a_1x + a_2x^2 + a_3x^3 + \ldots a_nx^n \qquad (14\text{-}26)$$

Equation (14-28) can be also expressed as the drain current that is a function of the gate-to-source voltage:

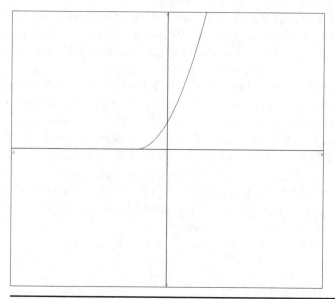

Figure 14-12 Curve of a depletion-mode MOSFET with a pinch-off voltage of −1 volt. The Y axis is the drain current, and the X axis is the gate-to-source voltage.

$$I_D(V_{GS}) = a_0 + a_1 V_{GS} + a_2(V_{GS})^2 \qquad (14\text{-}29)$$

where $a_0 = I_{DSS}$, $a_1 = -(2I_{DSS}/V_p)$, and $a_2 = I_{DSS}[(1/V_p)^2]$.

Let's take a look at an example. Suppose that for a certain depletion-mode MOSFET we have $I_{DSS} = 10$ mA and $V_p = -1$ volt and note that $[(1/-1 \text{ volt})^2] = [(1/1 \text{ volt})^2]$; then we will have:

$$I_D = 10 \text{ mA} - (20 \text{ mA}/-1 \text{ volt})V_{GS} + 10 \text{ mA}[(1/-1 \text{ volt})^2](V_{GS})^2$$

$$= 10 \text{ mA} + (20 \text{ mA}/1 \text{ volt})V_{GS} + 10 \text{ mA}[(1/1 \text{ volt})^2](V_{GS})^2 \qquad (14\text{-}30)$$

Let's take a look at the three terms 10 mA, $(20 \text{ mA}/1 \text{ volt})V_{GS}$, and $10 \text{ mA}[(1/1 \text{ volt})^2](V_{GS})^2$: 10 mA = DC quiescent drain current for $V_{GS} = 0$ volt. The 10 mA is also a DC offset term analogous to the y intercept. $(20 \text{ mA}/1 \text{ volt})V_{GS} = $ a linear term with the 20 mA/1 volt = 20 mA/volt as the transconductance coefficient. What this says is that at a DC bias for $V_{GS} \approx 0$ volt, the transconductance is 20 mA per volt or, put another way, if V_{GS} is an AC signal of approximately 0.1 volt peak, the resulting drain current from *this linear term* will be:

$(20 \text{ mA}/1 \text{ volt})0.1 \text{ volt peak} = 2 \text{ mA peak}$

Finally, the last term has a squaring function and therefore will cause harmonic and other types of distortions: $10 \text{ mA}[(1/1 \text{ volt})^2](V_{GS})^2 = $ a squaring or quadratic term, which results in distorting the input signal at V_{GS}.

For example, if the depletion-mode MOSFET is biased to about 0 volt and there is an input sine-wave signal at frequency f_1, the output current I_D will provide not only a sine-wave signal at f_1 but also a smaller-amplitude signal whose frequency is $2f_1$. This signal at $2f_1$ is a second harmonic distortion signal.

The next question is whether this second-order distortion can be reduced in some manner. It can, but the circuitry is modified to include two FETs. The cancellation, as you will see, involves adding like terms of two polynomials. Consider a system as shown in Figure 14-13. In this system, we will be using two-depletion mode MOSFETs, each driven with balanced signals or signals that are opposite in phase. That is, for Q1, $V_{GS1} = $ Vin, and for Q2, $V_{GS2} = -$Vin. The outputs of the depletion-mode MOSFETs will be connected to a differential current device (e.g., a transformer) to provide an output current $I_{OUT} = (I_{D1} - I_{D2})$. The devices are matched, and thus I_{DSS} and V_p are the same for both Q1 and Q2. Let's see what happens!

The depletion-mode MOSFETs have the following relationships.

For Q1:

$$I_{D1} = I_{DSS} - I_{DSS}(2I_{DSS}/V_p)V_{GS1} + I_{DSS}[(1/V_p)^2](V_{GS1})^2$$

$$V_{GS1} = \text{Vin}$$

$$I_{D1} = I_{DSS} - I_{DSS}(2I_{DSS}/V_p)\text{Vin} + I_{DSS}[(1/V_p)^2](\text{Vin})^2$$

For the second device, Q2:

$$I_{D2} = I_{DSS} - I_{DSS}(2I_{DSS}/V_p)V_{GS2} + I_{DSS}[(1/V_p)^2](V_{GS2})^2$$

and:

$$V_{GS2} = -\text{Vin}$$

so:

$$I_{D2} = I_{DSS} - I_{DSS}(2I_{DSS}/V_p)(-\text{Vin}) + I_{DSS}[(1/V_p)^2](-\text{Vin})^2$$

However, two negative values, $-I_{DSS}(2I_{DSS}/V_p)$ and $-$Vin, multiplied together result in a positive value. Also, a negative value multiplied by itself results in a positive value, so:

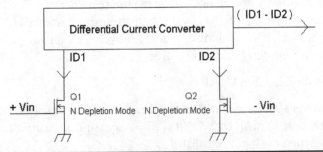

FIGURE 14-13   A push-pull amplifier system using two depletion-mode N-channel MOSFETs.

$(-Vin)^2 = (Vin)^2$

Thus:

$I_{D2} = I_{DSS} + I_{DSS}(2I_{DSS}/V_p)(Vin) + I_{DSS}[(1/V_p)^2](Vin)^2$

The output current $I_{OUT} = (I_{D1} - I_{D2})$, which is equal to subtracting the second current I_{D2} from the first one I_{D1}:

$I_{OUT} = (I_{D1} - I_{D2})$

$\quad = I_{DSS} - I_{DSS}(2I_{DSS}/V_p)Vin + I_{DSS}[(1/V_p)^2](Vin)^2 - I_{DSS} - I_{DSS}(2I_{DSS}/Vp)(Vin) - I_{DSS}[(1/V_p)^2](Vin)^2$

$\quad = I_{DSS} - I_{DSS} - I_{DSS}(2I_{DSS}/V_p)Vin - I_{DSS}(2I_{DSS}/V_p)(Vin) + I_{DSS}[(1/V_p)^2](Vin)^2 - I_{DSS}[(1/V_p)^2](Vin)^2 = I_{OUT}$

$I_{OUT} = (I_{D1} - I_{D2})$

$\quad = I_{DSS} - I_{DSS} + [-I_{DSS}(2I_{DSS}/V_p)Vin] - I_{DSS}(2I_{DSS}/V_p)(Vin) + I_{DSS}[(1/V_p)^2](Vin)^2 - I_{DSS}[(1/V_p)^2](Vin)^2$

Note that $I_{DSS} - I_{DSS} = 0$ and that also $I_{DSS}[(1/V_p)^2](Vin)^2 - I_{DSS}[(1/V_p)^2](Vin)^2 = 0$, so all we are left is:

$I_{OUT} = (I_{D1} - I_{D2})$

$\quad = -I_{DSS}(2I_{DSS}/V_p)Vin - I_{DSS}(2I_{DSS}/V_p)(Vin)$

$\quad = -2I_{DSS}(2I_{DSS}/V_p)Vin$

$I_{OUT} = (I_{D1} - I_{D2}) = -2I_{DSS}(2I_{DSS}/V_p)Vin$

What we see is that there is a distortion cancellation effect in a push-pull amplifier. If the device has a linear term and a square or quadratic term, by building a push-pull amplifier with perfectly matched devices, all distortion (at least in theory) vanishes!

However, in practice, all devices have second-order plus higher-order terms. If we look at a device in general with nth-order terms, we have:

$$f_1(Vin) = a_0 + a_1 Vin + a_2(Vin)^2 + a_3(Vin)^3 + \ldots a_n(Vin)^n \qquad (14\text{-}31)$$

And if we set up a second function in a push-pull manner having:

$$f_2(-Vin) = a_0 + a_1(-Vin) + a_2(-Vin)^2 + a_3(-Vin)^3 + \ldots a_n(-Vin)^n \qquad (14\text{-}32)$$

we know that $-Vin$ to an odd power retains its minus sign, but $-Vin$ to an even power does not and has a positive value.

Thus:

$$f_1(\text{Vin}) - f_2(-\text{Vin}) = a_0 + a_1\text{Vin} + a_2(\text{Vin})^2 + a_3(\text{Vin})^3 + \ldots a_n(\text{Vin})^n -$$
$$[a_0 + a_1(-\text{Vin}) + a_2(-\text{Vin})^2 + a_3(-\text{Vin})^3 + \ldots a_n(-\text{Vin})^n]$$

$$= a_0 + a_1\text{Vin} + a_2(\text{Vin})^2 + a_3(\text{Vin})^3 + \ldots a_n(\text{Vin})^n - [a_0 + a_1(-\text{Vin}) + a_2(\text{Vin})^2 +$$
$$a_3(-\text{Vin})^3 + \ldots a_n(-\text{Vin})^n]$$

$$= f_1(\text{Vin}) - f_2(-\text{Vin})$$

$$= 0 + 2a_1\text{Vin} + 0 + 2a_3(\text{Vin})^3 + 0 + 2a_5(\text{Vin})^5 + 0 + 2a_7(\text{Vin})^7 + 0 + 2a_9(\text{Vin})^9 + \ldots$$

$$f_1(\text{Vin}) - f_2(-\text{Vin}) = 2a_1\text{Vin} + 2a_3(\text{Vin})^3 + 2a_5(\text{Vin})^5 + 2a_7(\text{Vin})^7 + 2a_9(\text{Vin})^9 + \ldots$$
$$(14\text{-}33)$$

Equation (14-33) shows that when the push-pull configuration is compared with a single device or single-ended amplifier, the even-order distortion (2nd, 4th, 6th, 8th, etc.) products are canceled. However, there is a twofold increase in gain from the linear term and a twofold increase in odd-order distortion (3rd, 5th, 7th, 9th, etc.) for a push-pull amplifier when compared with a single-ended amplifier with one device.

For some illustrations (see Figure 14-14), let's take a look at some waveforms where $\text{Vin} = \sin(2\pi ft)$ for where:

$$\text{Vin} \to (\text{Vin})^2, \quad \text{Vin} \to (\text{Vin})^3, \quad \text{Vin} \to (\text{Vin})^4$$

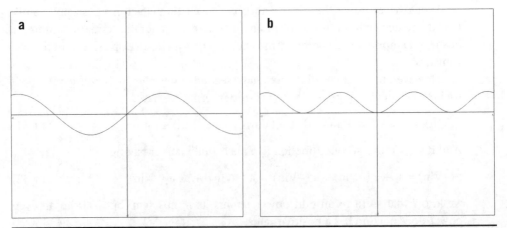

Figure 14-14 (*a*) A sinusoidal waveform Vin. (*b*) Sinusoidal waveform Vin squared, which contains a DC offset and a second harmonic sine wave.

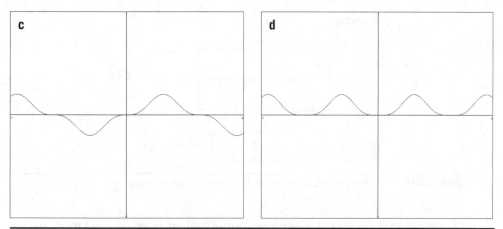

FIGURE 14-14   (*c*) Sinusoidal waveform Vin cubed, which contains a third harmonic and the fundamental frequency signal. (*d*) Sinusoidal waveform Vin raised to the fourth power, which includes a DC offset, a second harmonic, and a fourth harmonic signal. Note that the fourth harmonic signal is not obviously shown but it is there.

Negative-Feedback Systems with a (Linear) First-Order-Term Polynomial

Let's consider an amplifier such as an op amp with a relationship of:

$$y = a_1 x$$

where y = Vout, the output voltage, and x = V_{Diff}, the difference of the input voltages across the (+) and (−) terminals of the op amp, and further, a_1 is the gain of the amplifier. We can substitute the y and x variables in $y = a_1 x$ with Vout for y and V_{Diff} for x, which results in:

$$Vout = a_1 V_{Diff}$$

Figure 14-15 shows the op amp with negative feedback.

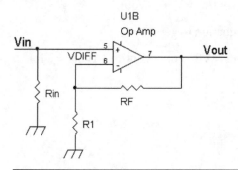

FIGURE 14-15   A feedback circuit with an op amp having Vout = $a_1 V_{Diff}$.

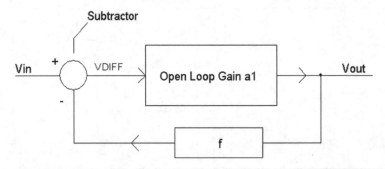

FIGURE 14-16   Op amp circuit modeled as a negative-feedback system.

When we apply feedback to the op amp circuit via resistors that form a voltage divider, we can also model this amplifier as a negative-feedback system, as illustrated in Figure 14-16. The voltage divider circuit RF and R1 provides attenuation to the output voltage Vout. This attenuation can be expressed as R1/(RF + R1) = f, where f is the feedback factor.

For a given amplifier with an open-loop gain of a_1 and a feedback factor of f, the output of the negative-feedback system in Figure 14-16 can be determined. First, we know that:

$V_{Diff} = (Vin - fVout)$

Second, we also know that:

$Vout = a_1 V_{Diff}$

and by dividing by a_1 on both sides of the equation, we get:

$Vout/a_1 = V_{Diff}$

We will want to express the input and output relationship of the negative-feedback system in terms of Vin and Vout, so we must remove V_{Diff} somehow from the equations. From the first equation, we exchange V_{Diff} with $Vout/a_1$ to get:

$V_{Diff} = Vout/a_1 = Vin - fVout$ or $Vout/a_1 = Vin - fVout$.

Let's add fVout to both sides of the equation to get:

$Vout/a_1 + fVout = Vin - fVout + fVout = Vin$ or $Vout/a_1 + fVout = Vin$

We then factor Vout such that:

$Vout(1/a_1 + f) = Vin$

or by dividing by $(1/a_1 + f)$ on both sides, we get:

$Vout = Vin[1/(1/a_1 + f)]$

We do not change the value of $[1/(1/a_1 + f)]$ if we multiply it by 1, which is also $[a_1/a_1]$. Therefore, $[a_1/a_1] \times [1/(1/a_1 + f)]$ results in:

$[a_1/a_1] \times [1/(1/a_1 + f)] = [(1)a_1/(a_1)(1/a_1 + f)] = [a_1/(1 + a_1f)]$

This results in:

$[1/(1/a_1 + f)] = [a_1/(1 + a_1f)]$ and $Vout = Vin[1/(1/a_1 + f)]$

so by substitution of $[1/(1/a_1 + f)] = [a_1/(1 + a_1f)]$, we have:

$Vout = Vin[a_1/(1 + a_1f)]$

or the closed-loop gain is:

$Vout/Vin = [a_1/(1 + a_1f)]$ (14-34)

A TL082 op amp has a DC open-loop gain range of $25{,}000 < a_1 < 200{,}000$, which is about an 8:1 ratio. Because the open-loop gains of op amps are not guaranteed within 10 percent, op amps are rarely used to amplify in the open-loop configuration. Instead, they are more useful when negative feedback is applied, and the gain of the system is $Vout/Vin = [a_1/(1 + a_1f)]$. For large values of open-loop gain a_1 where $a_1 \gg 1$ and where (a_1f) is also $\gg 1$, we have:

$(1 + a_1f) \approx a_1f$

This leads to:

$Vout/Vin \approx [a_1/(a_1f)] = 1/f = Vout/Vin$

where $R1/(RF + R1) = f$. This then equates to:

$Vout/Vin \approx 1/[R1/(RF + R1)] = (RF + R1)/R1 = Vout/Vin$

which is the familiar equation for the gain of a noninverting op amp.

The next section relates to describing a negative-feedback system that is imperfect and includes nonlinearities or distortions. This section includes how like powers of polynomials are grouped together to determine a new polynomial that describes the nonlinearities. Note that this section may contain advanced topics in distortion analysis.

Polynomials Used in Determining Distortion Effects of Negative Feedback in Amplifiers

This section relates to describing a negative feedback system that is imperfect and includes nonlinearities or distortions that are described with polynomials. Two polynomials are presented and then equated to each other by equating each of the ordered terms. For example, the linear, square, and cubic terms of both polynomials are equated.

> **NOTE** This section pertaining to determining the distortion effects from feedback in amplifiers is referenced from Professor Robert G. Meyer's EE240 notes of 1976 at the University of California, Berkeley. I am extremely grateful for his EE240 class on nonlinear integrated circuits.

Examine Figure 14-17, and suppose that we have a nonlinear system where V_e and Polynomial 1 are described as follows:

$$Vo = a_1 V_e + a_2(V_e)^2 + a_3(V_e)^3 + \ldots a_n(V_e)^n \tag{14-35}$$

where V_e = input voltage to the nonlinear system.

Now let's construct a negative-feedback system around this nonlinear system with an attenuator via a feedback factor f (see Figure 14-18). As shown in Figure 14-18, $V_e = (Vin - fV_o)$, and Equation (14-35) becomes:

$$V_{o1} = a_1 (Vin - fV_{o1}) + a_2(Vin - fV_{o1})^2 + a_3(Vin - fV_{o1})^3 + \ldots a_n(Vin - fV_{o1})^n \tag{14-36}$$

Equation (14-36) represents a polynomial equation for the feedback system shown in Figure 14-18.

Now suppose that we want to express the negative-feedback system of Figure 14-18 as an equivalent nonlinear system, as shown in Figure 14-19. We can characterize a new polynomial equation for this equivalent system using new constants b_1 to b_n in Polynomial 2 described as follows:

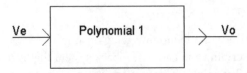

FIGURE 14-17   A nonlinear system with output voltage V_o and input voltage V_e.

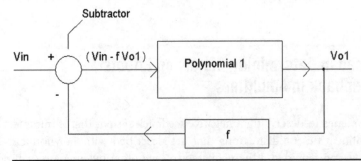

FIGURE 14-18   A negative-feedback system that includes a nonlinear open-loop amplifier.

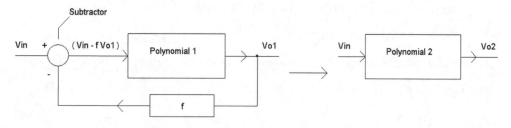

Figure 14-19 An equivalent nonlinear system to that in Figure 14-18.

$$V_{o2} = b_1 Vin + b_2(Vin)^2 + b_3(Vin)^3 + \dots b_n(Vin)^n \qquad (14\text{-}37)$$

What we want, then, is that $V_{o1} = V_{o2}$ or, equivalently:

$$a_1(Vin - fV_{o2}) + a_2(Vin - fV_{o2})^2 + a_3(Vin - fV_{o2})^3 + \dots a_n(Vin - fV_{o2})^n =$$
$$[b_1 Vin + b_2(Vin)^2 + b_3(Vin)^3 + \dots b_n(Vin)^n] = V_{o2} \qquad (14\text{-}38)$$

In particular, we would like to express each coefficient of b_1 to b_n as a function of a_1 to a_n and the feedback factor f.

In Equation (14-38), we need to substitute every fV_{o2} in $a_1(Vin - fV_{o2}) + a_2(Vin - fV_{o2})^2 + a_3(Vin - fV_{o2})^3 + \dots a_n(Vin - fV_{o2})^n$ with:

$$f[b_1 Vin + b_2(Vin)^2 + b_3(Vin)^3 + \dots b_n(Vin)^n]$$

It's pretty messy, but it's not that bad if we just take the first three terms, which bring us up to the third power. That is:

$$V_{o1} = a_1 (Vin - f[b_1 Vin + b_2(Vin)^2 + b_3(Vin)^3 + \dots]) + a_2(Vin - f[b_1 Vin + b_2(Vin)^2 + b_3(Vin)^3 + \dots])^2 + a_3(Vin - f[b_1 Vin + b_2(Vin)^2 + b_3(Vin)^3 + \dots])^3 + \dots = b_1 Vin + b_2(Vin)^2 + b_3(Vin)^3 = V_{o2} = V_{o1}$$

Let's first equate the linear term that is associated with Vin raised to the power of 1:

$$a_1(Vin - fb_1 Vin) = b_1 Vin$$

or via the distributive law of algebra:

$$a_1 Vin - a_1 fb_1 Vin = b_1 Vin \qquad (14\text{-}39)$$

Note that the other terms in the series, $b_2(Vin)^2 + b_3(Vin)^3 + \dots$, are ignored because they have powers for Vin of 2, 3, and so on that are greater than 1. Recall that we are equating to like terms of the polynomials.

Let's move $-a_1 f b_1 Vin$ out of the way on the left side by adding $a_1 fb_1 Vin$ on both sides, which results in:

$$a_1 Vin = a_1 fb_1 Vin + b_1 Vin = b_1 Vin[a_1 f + 1],$$

$$a_1 Vin = b_1 Vin[a_1 f + 1]$$

Now divide both sides by $\text{Vin}[a_1 f + 1]$ to solve for b_1:

$a_1/[a_1 f + 1] = b_1 = [a_1/(1 + a_1 f)]$

which is what we expect for the gain in a linear system, as shown in Equation (14-34):

$\text{Vout}/\text{Vin} = [a_1/(1 + a_1 f)]$ (14-34)

$b_1 = a_1/(1 + a_1 f)$ (14-40)

Now let's equate the second-order terms containing $(\text{Vin})^2$, which give rise to second-harmonic and second-order intermodulation distortions. These terms are highlighted in boldface, and the other terms that do not match second-order terms have a strike-through.

$a_1(\cancel{\text{Vin}} - f[b_1 \cancel{\text{Vin}} + \mathbf{b_2(\text{Vin})^2} + b_3\cancel{(\text{Vin})^3} + ...]) + a_2(\text{Vin} - \mathbf{f[b_1 \text{Vin}} + b_2\cancel{(\text{Vin})^2} + b_3\cancel{(\text{Vin})^3} + ...])^2 = b_2(\text{Vin})^2$ (14-41)

Let's see if we can tidy this up a little:

$a_1[-fb_2(\text{Vin})^2] + a_2(\text{Vin} - f[b_1 \text{Vin}])^2 = b_2(\text{Vin})^2$

$a_1(-fb_2(\text{Vin})^2) + a_2[(\text{Vin})^2 - 2fb_1(\text{Vin})^2 + (f[b_1 \text{Vin}])^2] = b_2(\text{Vin})^2$

and divide by $(\text{Vin})^2$ on both sides, giving:

$a_1(-fb_2) + a_2[1 - 2fb_1 + (fb_1)^2] = b_2$

Let's add $a_1(f b_2)$ to both sides to solve for b_2:

$a_2[1 - 2fb_1 + (fb_1)^2] = b_2 + a_1(fb_2) = b_2(1 + a_1 f) = a_2[1 - 2fb_1 + (fb_1)^2] = b_2(1 + a_1 f)$

$a_2[1 - 2fb_1 + (fb_1)^2]/(1 + a_1 f) = b_2$

But we can factor the polynomial $[1 - 2fb_1 + (fb_1)^2]$ as $(1 - fb_1)^2$.

NOTE   In many algebra books, the squared term $(p - q)^2 = (p^2 - 2pq + q^2)$, and let $p = 1$ and $q = fb_1$.

$a_2(1 - fb_1)2/(1 + a_1 f) = b_2$

or:

$b_2 = a_2(1 - fb_1)^2/(1 + a_1 f)$ (14-42)

However, we know from Equation (14-40) that $b_1 = [a_1/(1 + a_1 f)]$, which leads to:

$b_2 = a_2(1 - f[a_1/(1 + a_1 f)])^2/(1 + a_1 f)$ (14-43)

This equation can be further simplified if we multiply $[a_2(1 - f[a_1/(1 + a_1 f)])^2/(1 + a_1 f)]$ by 1 via:

$(1 + a_1f)^2/(1 + a_1f)^2 = 1$, which leads to b_2 = (a very long numerator)$/(1 + a_1f)^3$

This results in a $(1 + a_1f)^3$ term in the denominator and a very long numerator that has the many terms that will eventually simplify to:

$$a_2[1^2 + (a_1f)^2 + 2a_1f + (-2a_1f) - 2(a_1f)^2 + (a_1f)^2] = a_2(1) = a_2$$

which cancels every term except the a_2 term. Therefore, Equation (14-43) is also equal to:

$$b_2 = a_2/(1 + a_1f)^3 \tag{14-44}$$

We will now solve for the third-order term b_3 in terms of a_1, a_2, and f. The third-order term is especially interesting. Unlike the second-order term b_2, which is decreasing in value as the feedback f is increased, b_3 can have a null where the distortion vanishes as a function of f, but the distortion increases when f is increased further.

Let's take another look at the following and highlight the relevant cubic terms involving $(Vin)^3$ with boldface and cross out the other terms that are not related:

$a_1(\overline{Vin - f[b_1Vin + b_2(Vin)^2} + \mathbf{b_3(Vin)^3}]) + a_2(Vin - f[\mathbf{b_1Vin} + b_2(Vin)^2 + b_3(Vin)^3])^2 + a_3(Vin - \mathbf{f[b_1Vin} + \overline{b_2(Vin)^2 + b_3(Vin)^3 + ...}])^3 + ... =$
$\overline{b_1Vin + b_2(Vin)^2} + \mathbf{b_3(Vin)^3} = V_{o1} = V_{o2}$

This reduces the terms to:

$a_1(- f[b_3(Vin)^3]) + a_2(Vin - f[b_1Vin + b_2(Vin)^2])^2 + a_3(Vin - f[b_1Vin])^3 =$
$b_3(Vin)^3 = V_{o1} = V_{o2} \tag{14-45}$

For the a_1 term, $a_1(-f[b_3(Vin)^3])$ can be expressed as:

$$-a_1fb_3(Vin)^3 \tag{14-46}$$

The a_2 term, $a_2(Vin - f[b_1Vin + b_2(Vin)^2])^2$, actually has at least one term not related to the cubic power, and we need to expand this to:

$a_2(Vin - f[b_1Vin + b_2(Vin)^2])^2 = a^2[(Vin)^2 - 2Vin\, f\{b_1Vin + b_2(Vin)^2\}] + (f)^2[b_1Vin + b_2(Vin)^2]^2$

At this point, the a_2 is looking pretty messy. We can further remove those terms not related to the cubic power:

$a_2(Vin - f[b_1Vin + b_2(Vin)^2])^2 = a_2[\overline{(Vin)^2} - 2Vinf\{\overline{b_1Vin} + b_2(Vin)^2\}] + (f)^2[b_1Vin + b_2(Vin)^2]^2$

$a_2(Vin - f[b_1Vin + b_2(Vin)^2])^2 = \mathbf{a_2[-2Vinfb_2(Vin)^2]} + a_2(f)^2[b_1Vin + b_2(Vin)^2]^2$

and we need to one more round of elimination of unrelated terms:

$a_2(f)^2[b_1Vin + b_2(Vin)^2]^2 = a_2(f)^2[(b_1Vin)^2 + 2b_1Vinb_2(Vin)^2 + \{b_2(Vin)^2\}^2]$

which reduces to the following when only $(Vin)^3$ is concerned because $(b_1Vin)^2$ and $[b_2(Vin)^2]^2$ are eliminated given that they are not of the third power:

$$a_2(f)^2[b_1Vin + b_2(Vin)^2]^2 \rightarrow a_2(f)^2[2b_1Vin\ b_2(Vin)^2]$$

Getting back, we now have the a_2 terms as:

$$a_2[-2Vinfb_2(Vin)^2] + a_2(f)^2[2b_1Vinb_2(Vin)^2] = 2a_2fb_2(Vin)^3\ [-1 + b_1f] =$$
$$-2a_2fb_2(Vin)^3[1 - b_1f] \tag{14-47}$$

The a_3 term from Equation (14-45) is

$$a_3(Vin - f[b_1Vin]\ )^3 = a_3[(Vin)(1 - f\,b_1)]^3$$
$$= a_3(Vin)^3(1 - f\,b_1)^3$$
$$= a_3(Vin)^3(1 - b_1f)^3 \tag{14-48}$$

Combining the terms from Equations (14-46), (14-47), and (14-48) with Equation (14-45) gives

$$b_3(Vin)^3 = -a_1fb_3(Vin)^3 - 2a_2fb_2(Vin)^3[1 - b_1f] + a_3(Vin)^3(1 - b_1f)^3 \tag{14-49}$$

Let's move the $-a_1fb_3(Vin)^3$ to the left side of the equation, which results in:

$$b_3(Vin)^3 + a_1fb^3(Vin)^3 = -2a_2fb_2(Vin)^3[1 - b_1f] + a_3(Vin)^3(1 - b_1f)^3 \tag{14-50}$$

Let's divide both sides of the equation by $(Vin)^3$ to get

$$b_3 + a_1fb_3 = -2a_2fb_2[1 - b_1f] + a_3(1 - b_1f)^3 \tag{14-51}$$

And now we use the distributive property, which results in $[b_3 + a_1fb_3] = b_3(1 + a_1f)$, so

$$b_3(1 + a_1f) = -2a_2fb_2(1 - b_1f) + a_3(1 - b_1f)^3 \tag{14-52}$$

Divide by $(1 + a_1f)$ on both sides to get

$$b_3 = [-2a_2fb_2(1 - b_1f) + a_3(1 - b_1f)^3]/(1 + a_1f) \tag{14-53}$$

We know from Equations (14-40) and (14-44) that:

$$b_1 = a_1/(1 + a_1f) \quad \text{and} \quad b_2 = a_2/(1 + a_1f)^3$$

By substituting $a_1/(1 + a_1f)$ for b_1 and $a_2/(1 + a_1f)^3$ for b_2 into Equation (14-53), we get from Professor Meyer's EE240 notes that :

$$b_3 = [a_3(1 + a_1f) - 2(a_2)^2f]/(1 + a_1f)^5 \tag{14-54}$$

I would have shown the steps in deriving Equation (14-54), but I think everyone, including me, would just like to see the answer for b_3. The new polynomial that is equivalent to the feedback system is:

$$V_{o2} = b_1 Vin + b_2 (Vin)^2 + b_3 (Vin)^3 + \ldots b_n (Vin)^n \tag{14-37}$$

$$V_{o2} = [a_1/(1 + a_1 f)]Vin + [a_2/(1 + a_1 f)^3] (Vin)^2 + \{[a_3(1 + a_1 f) - 2(a_2)^2 f]/(1 + a_1 f)^5\}(Vin)^3 + \ldots \tag{14-55}$$

The third-order coefficient, $\{[a_3(1 + a_1 f) - 2(a_2)^2 f]/(1 + a_1 f)^5\}$, shows two interesting properties. This term is sometimes known as the *second-order interaction term*. The first interesting property is that if a pure square-law device (e.g., a perfect FET) that only has coefficients of a_1 (linear term) and a_2 (square-law term) has any type of negative feedback, the negative-feedback factor f will reduce the second-order harmonic via $[a_2/(1 + a_1 f)^3]$ but add third-order distortion via the third-order term, $\{[a_3(1 + a_1 f) - 2(a_2)^2 f]/(1 + a_1 f)^5\}$, with $a^3 = 0$, to yield:

$$[a_3(1 + a_1 f) - 2(a_2)^2 f]/(1 + a_1 f)^5 \rightarrow [-2(a_2)^2 f]/(1 + a_1 f)^5 > 0$$

For a pure square-law device, then:

$$V_{o2} = [a_1/(1 + a_1 f)]Vin + [a_2/(1 + a_1 f)^3](Vin)^2 + [-2(a_2)^2 f]/(1 + a_1 f)^5](Vin)^3 + \ldots \tag{14-56}$$

The addition of third-order harmonic distortion to a square-law device with negative feedback makes sense. Any linearization or straightening out of the original parabolic curve inherent in a square-law device will cause higher-order distortion terms to pop up because a bent parabola is no longer a parabola.

A second observation of the third-order term $[a_3(1 + a_1 f) - 2(a_2)^2 f]/(1 + a_1 f)^5$, is that one can null out the third-order distortion by adjusting the feedback factor f. That is, if the numerator $[a_3(1 + a_1 f) - 2(a_2)^2 f] = 0$, the third-order distortion vanishes. Put another way, to null out the third-order distortion:

$$a_3(1 + a_1 f) = 2(a_2)^2 f$$

The requirement is that the device have both second- and third-order distortions. This means that a_2 and a_3 have nonzero values.

In practice, a bipolar transistor with an unbypassed emitter resistor RE forms negative feedback. If $RE = (\frac{1}{2})(1/g_m)$, where $g_m = ICQ/0.026$ volt and $ICQ = DC$ quiescent collector current, then third-order distortion is canceled. For example, at $1 mA = ICQ$, $(1/g_m) = 26\ \Omega$, so $RE = (\frac{1}{2})26\ \Omega = 13\ \Omega$ to null out third-order distortion in a common-emitter or common-base amplifier when the input signal voltage has close to a $0\ \Omega$ source resistance.

Summary

We have shown some applications of algebra, such as how linear equations of lines can characterize amplifiers and how a system of two or more linear equations is used in FM stereo radios and in color TV signals. Also, we touched in a small manner on how polynomials characterize FET amplifiers. Finally, in the last section of this chapter, we showed how polynomials are used for characterizing nonlinear systems. Chapter 15 will discuss some circuit theory and will further show applications in algebra relating to electronic circuits.

References

1. Robert G. Meyer, "EE240 Notes," University of California, Berkeley, CA, 1976.
2. Edward Burger, David Chard, Paul Kennedy, et al., *Algebra 1*. New York: Holt, Rinehart and Winston, 2008.

CHAPTER 15

Some Basic Circuit Analysis Techniques

Some basic circuit analysis tools will be presented to allow further understanding of analog electronics. We will start with Kirchoff's loop and node equations and dive into Thevenin equivalent circuits to further simplify circuit analysis. Basic direct-current (DC) circuits will be covered first, followed by the more complicated or complex alternating-current (AC) circuit analysis, which includes complex numbers.

One should keep in mind that circuit analysis is a tool that is useful for some deeper understanding on how circuits work. However, unlike the type of circuit theory normally taught in much greater detail in college courses, this chapter will concentrate more on practical circuit analysis. For example, we will not be covering brain-twisting problems such as circuits with many nodes that require ≥ 3 equations to set up and ≥ 3 unknown variables to solve. In practice, if the equation requires more than two equations and two unknown variables to describe the circuit, then it will be better to somehow reduce the problem to preferably one equation and one unknown variable.

Limitations of Kirchoff's Laws for Analyzing Circuits Via Loop and Node Equations

Before we embark on analyzing circuits with Kirchoff's laws, their limitations must be known. Generally, as a rule of thumb, for circuits whose wiring of components is less than 1 percent of a wavelength of the highest frequency encountered, Kirchoff's laws hold. For example, if one is working on an RF (radio-frequency) circuit in the 70-cm band, which is approximately 440 MHz, the leads from one component to another should be less than (70 cm)/100 = 0.7 cm, which is approximately half an inch.

Wavelength = (300 million meters/second)/(Frequency in cycles per second or Hz)

For example, a 150-MHz signal has a wavelength of (300 million meters/second)/(150 million Hz) = 2 meters.

If the wiring is greater than 1 percent of a wavelength of the frequency used, Kirchoff's laws may not hold. This is especially true if the circuit has wiring approaching an eighth to a quarter wavelength of the highest frequency used in the circuit. For example, for 150-MHz circuit, the wavelength is 2 meters if the wiring is on the order ≈19 inches, which is a quarter wavelength, and Kirchoff's law will not hold because there will be different voltages along the 19 inches of wiring.

In the circuits we have worked with so far, the highest frequency is about 120 MHz (≈ 2.5 meters of wavelength) for the FM (frequency modulation) radios. At 1 percent of 2.5 meters, we have about 2.5 cm or about 1 inch lead length for the components. Normally, for such a radio, and to be on the safe side, we would have built the circuits with typically less than ½-inch leads.

In audio circuits with a 20-kHz top frequency, the wavelength is in 15-kilometer range, and using the 1 percent factor, we have leads that will obey Kirchoff's law at (15 k meter)/100, or about 150 meters. Of course, generally, we build audio circuits with less than 2 feet of wiring in the system and less than 6 inches from component to component within a circuit. In the circuits to which we do apply Kirchoff's law, there will be a reasonable assumption that the circuit elements such as resistors, inductors, capacitors, transistor, diodes, and ICs (integrated circuits) have been built with sufficiently short leads.

Loop Equations

One of Kirchoff's laws involves summing voltages around a loop to equal zero (see Figure 15-1). In Figure 15-1, the voltage source Vs1 is summed head to toe (minus of one element to a positive of the next element, or vice versa) with each of the voltages formed across the elements (e.g., resistors, lamps, and/or forward-biased semiconductors) Element 1 to Element 3. In this example, we have Vs1 + V1 + V2 + V3 = 0. At first glance, this does not seem to make sense.

To bear more light on how the sum of voltages equals zero with the polarities of each element adding head to toe, we first look at another example that is more intuitive. The example will show that one of the voltages is equal to the sum of the other voltages along the loop. See Figure 15-2, with voltage source Vs2 and elements, Element 1a to Element 3a. We see intuitively that Vs2 = Va1 + Va2 + Va3, especially if we make Vs2 = 6 volts DC and Va1, Va2, and Va3 all 2 volt lead-acid batteries where they are added in series. From Figure 15-2, we are more used to seeing one voltage equal to the sum of all the other voltages in a loop.

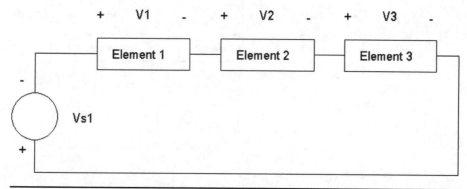

FIGURE 15-1   A voltage source with circuit elements that form voltage drops.

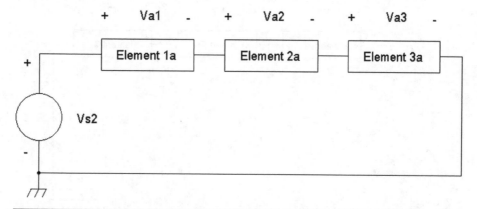

FIGURE 15-2   A practical example showing a version of Kirchoff's law with voltage source Vs2 having its negative terminal grounded.

Enter the Zero-Voltage Source

If we modify Figure 15-1 to include a series zero-voltage source V0 = 0 with its negative terminal as shown in Figure 15-3, we can now see that V0 = Vs1 + V1 + V2 + V3, and by substituting 0 for V0, we have 0 = Vs1 + V1 + V2 + V3 (see Figure 15-3). However, in most cases of human circuit analysis, it is more intuitive to set up an equation where one voltage source is equal to the series-connected voltages of the rest of the circuit.

Defining the Direction of the Current and Voltage Drops

The definition of electric current flow is the number of electron charges per second or the amount of change in charge Q per change in time t, which is $\Delta Q/\Delta t$. Electric current is then the flow of electrons or negative charges. While one can analyze a circuit in terms of electron flow, which requires that the electric currents have nega-

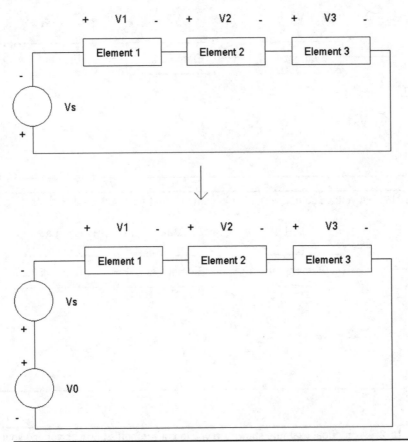

FIGURE 15-3   A zero-voltage source V0 to clarify how the loop equations work.

tive values, an equivalent and correct analysis can be made of the same circuit if we say that we can define a positive-charged current (Figure 15-4).

Intuitively, this works out easier when using voltmeters, ammeters, and other test equipment, and we do not have to keep track of negative-value currents as much. In this example, the voltage source is –Vsg, and the positive current I flows into the resistors to provide a positive voltage across each of the resistors, as shown in Figure 15-4.

Using the Kirchoff loop equation:

$$-Vsg = VR\_1 + VR\_2$$

And by using one of Ohm's laws, which states that the voltage across a resistor is equal to current multiplied by its resistance, we get:

$$-Vsg = I(R1) + I(R2)$$

Note that –Vsg may be a positive or negative voltage referenced to ground.

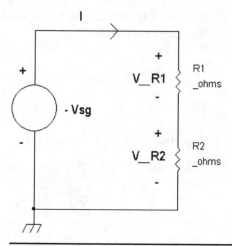

FIGURE 15-4   Defining current as a flow of positive charges.

Deriving the Voltage Divider Formula Using Positive-Charged Currents

Consider the circuit in Figure 15-5. The current I is defined as a positive current flowing out of the positive terminal of the voltage source V:

$$V = I(R1) + I(R2) = I(R1 + R2) = V$$

Solving for the current I:

$$I = V/(R1 + R2)$$

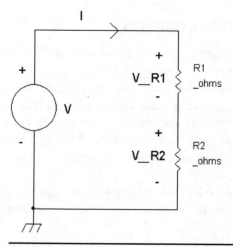

FIGURE 15-5   A voltage divider circuit with voltage source, V, and resistors R1 and R2.

To determine the voltage across R2, which is the output of the circuit, as shown in Figure 15-5, we multiply the current I by the resistance R2, or:

V_R2 = I(R2)

But

I = V/(R1 + R2)   (15-1)

so we substitute [V/(R1 + R2)] for I in the equation, V_R2 = I (R2) and get:

V_R2 = [V/(R1 + R2)](R2) = R2[V/(R1 + R2)] = [(R2)V/(R1 + R2)] = [VR2/(R1 + R2)]

or:

V_R2 = V[R2/(R1 + R2)]   (15-2)

Equation (15-2) is the familiar equation for a voltage divider circuit. For example, if V = 1.5 volts and R1 = 10 Ω and R2 = 20 Ω:

Vout = 1.5 volts[20/(10 +20)] = 1.5(20/30) = 1 volt

Note that Equation (15-1) shows that if we do not wish to build a voltage divider and just drain the same amount of current I from the voltage source V with another resistor R, we just have:

R = (R1 + R2)

Or equivalently, a series of connected resistors is just the sum of the resistances. For n resistors in series, the total resistance is:

R = R1 + R2 + ... + Rn   (15-3)

Node Equations to Determine Parallel Resistances

Now let's take a look at currents flowing through parallel resistances. This time we will apply Kirchoff's law for node equations (see Figure 15-6a). Kirchoff's current law states that all the currents summed into a node equal zero. Again, this is not always the easiest concept to understand. In general, we like to think that electric currents are like water streams that split out and that the main stream's current is equal to all the other minor streams that have branched out. We can sometimes think of a zero-current branch Izero that is analogous to the zero-voltage source in Figure 15-6b.

An example of a circuit that can be solved for the unknown voltage V is shown in Figure 15-7. To set up an equation for the circuit in Figure 15-7, we can start with:

I1 + I2 = I3

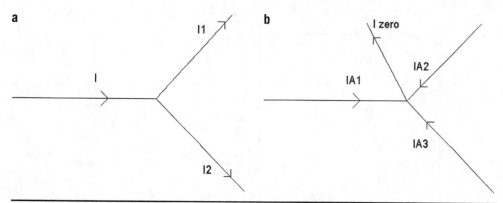

FIGURE 15-6 (*a*) A current source I with other currents branching out to I1 and I2 on the left. (*b*) A zero-current branch, Izero on the right.

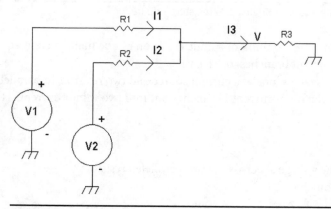

FIGURE 15-7 A circuit with a node denoting the unknown voltage V to be solved.

or, equivalently, by Ohm's law, a resistor current is equal to the voltage across the resistor divided by its resistance. This leads to:

$(V1 - V)/R1 + (V2 - V)/R2 = V/R3$

$V1/R1 - V/R1 + V2/R2 - V/R2 = V/R3$

$V1/R1 + V2/R2 = V/R3 + V/R1 + V/R2 = V[(1/R1) + (1/R2) + (1/R3)]$

$([(V1/R1) + (V2/R2)]/ [(1/R1) + (1/R2) + (1/R3)]) = V$

We now can solve for V given the voltages and resistances. Once V is determined, we can go back and solve the current flow through each resistor via the following equations:

$I1 = (V1 - V)/R1$

$I2 = (V2 - V)/R2$

$I3 = V/R3$

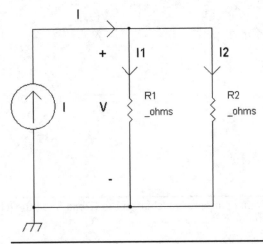

FIGURE 15-8   A current source I driving two resistors in parallel.

Intuitively, it's often better to think of a node equation as one main current equal to the sum of minor current branches, or vice versa.

For example, let's take a look at a current source and two resistors in parallel (see Figure 15-8). Here the main current I branches out into two minor currents I1 and I2. Thus:

$$I = I1 + I2$$

The unknown is the voltage V across the two paralleled resistors.

We know this much:

$$I1 = V/R1 \quad \text{and} \quad I2 = V/R2$$

Let's substitute V/R1 for I1 and V/R2 for I2. With:

$$I = I1 + I2$$

we get:

$$I = V/R1 + V/R2 = V(1/R1 + 1/R2) = I \qquad (15\text{-}4)$$

If we divide by (1/R1 + 1/R2) on both sides of Equation (15-4), we get:

$$V = I/(1/R1 + 1/R2)$$

or, equivalently:

$$I = I \times 1$$

$$V = I[1/(1/R1 + 1/R2)] \qquad (15\text{-}5)$$

If we recall one of Ohm's laws, we know that voltage is equal to current times resistance. Therefore, the equivalent resistance to get the same voltage V is:

R = 1/(1/R1 + 1/R2)

or, equivalently, if we multiply 1/(1/R1 + 1/R2) by 1 = (R1R2)/(R1R2) =1, we get:

R = (R1R2)/(R1R2)(1/R1 + 1/R2) = (R1R2)/(R1 + R2) = R

or:

R = (R1R2)/(R1 + R2)   (15-6)

Equation (15-6) is a familiar equation for two resistors connected in parallel.

If we had n resistors in parallel, we would have an expanded Equation (15-4) that solves for V as:

V = I[1/(1/R1 + 1/R2 + ... + 1/Rn)]   (15-7)

And the equivalent resistance for n resistors in parallel is:

R = 1/(1/R1 + 1/R2 + ... + 1/Rn)   (15-8)

Thevenin Equivalent Circuits

While it may be an interesting exercise to solve for currents and voltages in a particular circuit with multiple variables and equations, in practice, if we are going to solve by hand (e.g., without computers), it would be better to somehow reduce complex circuits into simpler ones. Not only will the calculations be easier, but the reduced-parts circuit will be easier to understand intuitively.

A Thevenin equivalent circuit for the voltage divider shown in Figure 15-9 reduces the circuit from two resistors to one resistor. To calculate the Thevenin voltage source Vth, it is simply the output voltage:

Vout = Vsource[R2/(R1 +R2)] = Vth   (15-9)

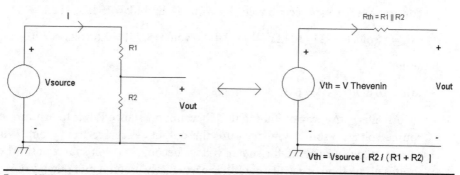

FIGURE 15-9   A voltage divider circuit on the left and an equivalent Thevenin equivalent circuit on the right.

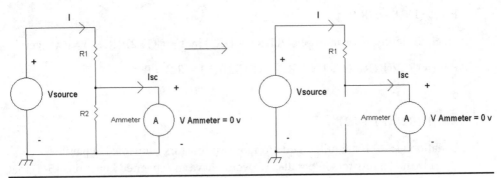

FIGURE 15-10   Measuring the short-circuit current to find Isc.

The Thevenin resistance Rth can be calculated two ways. The first method is:

Vth/Isc

where Isc is the short-circuit current shown in Figure 15-10. Isc = Vsource/R1 since placing an ideal ammeter across R2 shorts R1 to ground.

NOTE   The concept of measuring the short-circuit current can be dangerous in many circuits. For simulations or hand calculations without actually building the circuit, finding the short-circuit current Isc is safe. It is not recommended to ever try to measure the short-circuit current of a working circuit. Damage to the circuit and/or injury to the person involved can happen.

Rth = Vth/Isc = Vsource [R2/(R1 +R2)]/ [Vsource /R1] = [R1(R2)]/(R1 +R2) = R1∥R2

which is the parallel resistances of R1 and R2 equaling Rth. To reiterate, then:

Rth = R1∥R2   (15-10)

Vth = Vsource[R2/(R1 +R2)]

For example, let Vsource = 6 volts, R1 = 10 kΩ, and R2 = 15 kΩ. Then:

Vth = 6 volts [15 kΩ/(10 kΩ + 15 kΩ)] = 6 volts(15/25) = 3.6 volts = Vth

and:

Rth = 10 kΩ∥15 kΩ = 6kΩ

An alternative way of finding the Thevenin resistance Rth is to just turn off the source voltage Vsource and measure the output resistance from Vout. With the source voltage turned off, the source voltage becomes a 0 volt source of short circuit itself. This method of calculating Rth is particularly useful if more than two resistors are in the original circuit (see Figures 15-11 to 15-13).

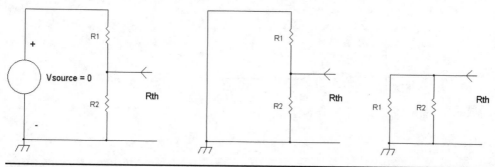

FIGUURE 15-11 Example of the alternate method of determining Rth for a voltage divider circuit where Rth = R1∥R2.

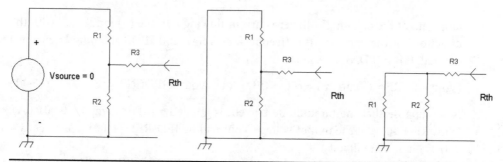

FIGURE 15-12 An example of determining Rth with a three-resistor circuit where Rth = (R1∥R2) + R3.

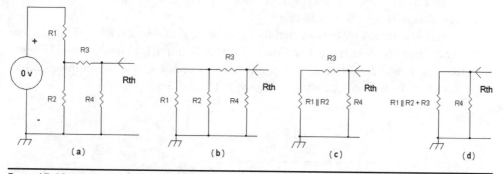

FIGURE 15-13 An example of determining Rth with a four-resistor circuits where Rth = [(R1∥R2) + R3]∥R4.

A Practical Example of Analyzing Via the Thevenin Equivalent Circuit

Suppose that we wish to make a transresistance amplifier that provides a voltage output for a current generator input source such as a photodiode (see Figure 15-14). In the circuit shown in this figure, we can use node equations to set up and solve for Vout. The (+) input of the op amp is grounded to 0 volt, so the (–) input V(–) of the op amp is servoed or adjusted to 0 volt by negative feedback and the very large open-

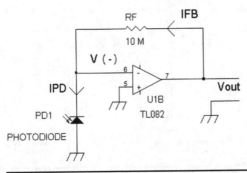

FIGURE 15-14   Photodiode preamp circuit with feedback resistor RF and feedback current IFB.

loop gain of the op amp. With zero current flowing into the (−) input, and with the directions of the currents, IFB (feedback current), and IPD (photodiode current) defined, IFB = IPD. But we know that:

$$(Vout - 0)/RF = (Vout/RF) = IFB = IPD \quad or \quad Vout = IPD(RF) \qquad (15\text{-}11)$$

Now suppose that the photodiode current is about 1 nA (10^{-9} amp or 0.001 μA). With RF = 10 MΩ, the output voltage Vout will be IPD(RF) = 0.001 μA × 10^7 Ω = 0.010 volt or 10 millivolts.

What if we want to increase the photodiode current amplification? We can have RF = 100 MΩ, but these high-value resistors are less common. Can we still use standard-value resistors and provide the equivalent of a 100 MΩ feedback resistor? The answer is yes. See Figure 15-15.

At first glance, we can see that there are three resistors, R1, R2, and RF, and we *could* assign three currents that flow to them as I1 into R1, I2 into R2, and IFB into RF, but we will not. Instead, we can simplify this circuit with a Thevenin equivalent circuit in the feedback network, as shown in Figure 15-16.

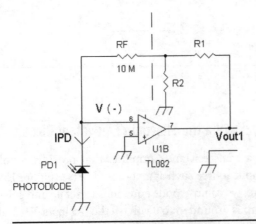

FIGURE 15-15   A more complicated feedback network for the photodiode preamp.

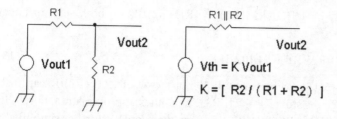

FIGURE 15-16 A Thevenin equivalent circuit for Figure 15-15.

The output Vout1 from the output of the op amp is voltage divided by R1 and R2 such that:

$$\text{Vout1}[R2/(R1 + R2)] = \text{Vout 2} \tag{15-12}$$

the Thevenin voltage, and R1 and R2 now form a Thevenin resistance or R1||R2, as shown in Figure 15-16. An equivalent circuit can be modeled as shown in Figure 15-17.

We now have two resistors in series, RF and R1||R2 in the feedback loop, and we will use Vout2 to set up the node equation:

$$(\text{Vout2} - 0)/(RF + R1||R2) = IPD$$

$$\text{Vout2}/(RF + R1||R2) = IPD \quad \text{or} \quad \text{Vout2} = (RF + R1||R2)IPD \tag{15-13}$$

However, we want to express the output voltage in reference to Vout1, the output terminal of the op amp, which we have already calculated as:

$$\text{Vout1}[R2/(R1 + R2)] = \text{Vout 2}$$

If we substitute Vout1[R2/(R1 +R2)] for Vout2, we get:

$$\text{Vout1}[R2/(R1 + R2)] = (RF + R1||R2)IPD \tag{15-14}$$

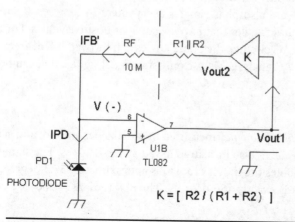

FIGURE 15-17 Thevenin equivalent circuit model.

And if we multiply by $1/[R2/(R1+R2)] = (R1+R2)/R2$ on both sides of Equation (15-14), the result is:

$$Vout1 = [(R1+R2)/R2](RF + R1||R2)IPD \qquad (15\text{-}15)$$

From Equation (15-15), we get a multiplying factor of $(R1+R2)/R2$ against the equivalent feedback resistors $RF + R1||R2$. For a quick approximation, normally, we can set $R1||R2 << RF$ or $(RF + R1||R2) \approx RF$ so that Vout1 is further simplified to:

$$Vout1 = [(R1+R2)/R2](RF)IPD \qquad (15\text{-}16)$$

For example, if $R2 = 1\ k\Omega$, $R1 = 10\ k\Omega$, and $RF = 10\ M\Omega$, then:

$$R1||R2 = 10\ k\Omega || 1\ k\Omega \approx 900\ \Omega << 10\ M\Omega$$

Also $[(R1+R2)/R2 = (10\ k\Omega + 1)/1 = 11$, so:

$$Vout1 = [(R1+R2)/R2](RF)IPD = 11(10\ M\Omega)IPD = (110\ M\Omega)IPD = Vout1$$

and now we have the equivalent of a 110 MΩ feedback resistor.

Some AC Circuit Analysis

In DC circuits, there are only two signal phases of the DC voltage or current, 0 degrees and 180 degrees. That is, do the DC voltages add or subtract? For example, in a two-cell flashlight, if the two batteries are inserted correctly, the voltages add in an in-phase manner of 0 degrees. If one of the batteries is reversed such that the (+) and (+) or (–) and (–) are connected, then the voltages of the batteries subtract, or combine in a 180-degree manner, and the total voltage will be less. In this example with two batteries, each of equal voltage, the total voltage to the lamp will be 0 volt when connected back to back.

For an AC circuit, not only are there 0 and 180 degrees for addition or subtraction of signals, but there are also phase angles everywhere in between 0 and 360 degrees and sometimes phase angles that have negative or positive values. For more complicated or cascaded AC circuits, the phase angles can be over 360 degrees.

Normally, for simple AC analysis, we use only sinusoidal signals that can be represented as:

$$V1(t) = A \sin(\omega t + \varphi)$$

where $\omega = 2\pi f$ is the *angular frequency* measured in radians per second, and φ is the *phase shift* or *phase-angle delay*, also measured in radians or degrees. The signal V(t) is time varying, and t denotes the time. A radian is approximately 57.3 degrees, and 1 radian multiplied by π is exactly 180 degrees. Thus 2π radians = 360 degrees, or one cycle.

However, we are more familiar with cycles per second or hertz, denoted by f. So an AC signal is often described as:

$$V1(t) = A \sin(2\pi ft + \varphi 1) \tag{15-17}$$

One should note that the quantity inside the sine function $(2\pi ft + \varphi 1)$ is an angle measured in radians, and more important, the two terms ωt and $2\pi ft$ both have dimensions of an angle measured in radians or, equivalently, in degrees. The peak amplitudes of the sinusoids are represented by A and B in Equations (15-17) and (15-18).

Alternatively, the AC signal can be described as a cosine function:

$$V2(t) = B \cos(2\pi ft + \varphi 2) \tag{15-18}$$

Most AC circuits are analyzed with sinusoidal signals, but they are also analyzed for transient response with nonsinusoidal waveforms such as squarewaves or pulses.

For sinusoidal signal analysis, this is sometimes called the *steady-state AC analysis*, and it requires the use of complex numbers such as:

$$c + di$$

where $i = \sqrt{-1}$, where c and d are real numbers, and i is an *imaginary number*.

Because the description of electronic circuits already denote the letter i as current, another letter j is used instead to equivalently state the complex number:

$$c + dj$$

where $j = \sqrt{-1}$.

Generally, whenever you see the imaginary number j being used in AC analysis, we are working with only sinusoidal signals.

The number j actually denotes a 90-degrees phase shift on a sinusoidal signal. We express the amplitude and phase of a signal in terms of A and φ. Because j is 90 degrees in relationship to the number c, one can construct a right-triangle relationship as shown in Figure 15-18.

By using the Pythagorean theorem and trigonometry, there are the magnitude and phase related to the complex number c + dj. The magnitude is characterized as:

$$M = (c^2 + d^2)^{1/2}$$

The magnitude of 4 + 2j in Figure 15-19 is shown by the length of the diagonal vector, which is:

$$M = (4^2 + 2^2)^{1/2} = (20)^{1/2} \approx 4.47$$

and the phase:

$$\psi = \arc \tan(d/c) = \tan^{-1}(d/c)$$

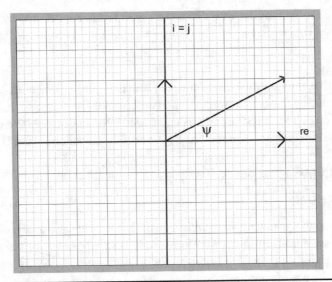

Figure 15-18   A right-angle relationship using the complex number c + dj, where c = 4 and d = 2 for a complex number of 4 + 2j.

that relates to a sinusoidal signal. Again, from Figure 15-18, c = 4 and d = 2, so ψ = $\tan^{-1}(d/c) = \tan^{-1}(2/4) \approx 26.67$ degrees.

In AC circuits, when a sinusoidal signal is used as the input signal, amplitude and/or phase at the output of a circuit will be a function of the frequency and can be more generally expressed as:

$M \rightarrow M(\omega)$

$\psi \rightarrow \psi(\omega)$

Let's take a look at a couple of practical examples. We will start with a simple RC low-pass filter, as shown in Figure 15-19. The low-pass filter circuit can be looked at as another voltage divider circuit, but this time we replace one of the components

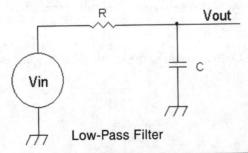

Figure 15-19   An RC low-pass filter with sinusoidal waveform input.

with a capacitor. The capacitor's *impedance* is $Z_c = 1/j\omega C$, where C is the capacitance value in farads.

> **NOTE**   FYI, the impedance of an inductor is $Z_L = j\omega L$.

The RC circuit is then characterized like a resistive voltage divider except we use Z_c:

$$\text{Vout}(\omega) = \text{Vin}(\omega)[Z_c/(R + Z_c)] \tag{15-19}$$

where $\text{Vout}(\omega)$ is an output sinusoid waveform that is a function of the angular frequency ω, and $\text{Vin}(\omega)$ is the input sine wave that also depends on the angular frequency ω.

Let's substitute $1/j\omega C$ for Z_c in Equation (15-19), which results in:

$$\text{Vout}(\omega) = \text{Vin}(\omega)(1/j\omega C)/(R + 1/j\omega C) \tag{15-20}$$

We can multiply by $j\omega C/j\omega C = 1$ on $(1/j\omega C)/(R + 1/j\omega C)$ to result in:

$$\text{Vout}(\omega) = \text{Vin}(\omega)[1/(j\omega CR + 1)] \tag{15-21}$$

It is common to express the low-pass filter in terms of its frequency-dependent gain function by dividing $\text{Vin}(\omega)$ on both sides of Equation (15-21) to provide an output-to-input relationship:

$$\text{Vout}(\omega)/\text{Vin}(\omega) = 1/(j\omega CR + 1) \tag{15-22}$$

Or equivalently, we can just change the order of $j\omega CR + 1$ that still equals $1 + j\omega CR$:

$$\text{Vout}(\omega)/\text{Vin}(\omega) = 1/(1 + j\omega CR) \tag{15-23}$$

We wish to express $1/(1 + j\omega CR)$ in terms of a complex number in the form of $c + dj$, and this can be accomplished by multiplying the numerator and denominator of Equation (15-23) by $(1 - j\omega CR)/(1 - j\omega CR)$:

$$\text{Vout}(\omega)/\text{Vin}(\omega) = (1 - j\omega CR)/(1 + j\omega CR)(1 - j\omega CR) = (1 - j\omega CR)/(1^2 - [j\omega CR]^2)$$

$$\text{Vout}(\omega)/\text{Vin}(\omega) = (1 - j\omega CR)/(1^2 - j^2\omega^2 C^2 R^2)$$

Note that $j^2 = -1$, and $-j^2 = 1$ that results in $-j^2\omega^2 C^2 R^2 = \omega^2 C^2 R^2$, so:

$$\text{Vout}(\omega)/\text{Vin}(\omega) = (1 - j\omega CR)/(1 + \omega^2 C^2 R^2)$$

The phase shift:

$$\psi(\omega) = \tan^{-1}(-\omega CR/1) = -\tan^{-1}(\omega CR)$$

$$\psi(\omega) = -\tan^{-1}(\omega CR) \tag{15-24}$$

and the maginitude of the gain function is:

$$M(\omega) = [(1 + \omega^2 C^2 R^2)^{1/2}]/(1 + \omega^2 C^2 R^2) = 1/[(1 + \omega^2 C^2 R^2)^{1/2}$$

or:

$$M(\omega) = 1/[(1 + \omega^2 C^2 R^2)^{1/2}] \qquad (15\text{-}25)$$

For a sinusoidal signal, $V1(t) = A \sin(2\pi f t + \varphi_1) = Vin$ as the input signal into the simple RC low-pass filter characterized in phase and magnitude respectively by Equations (15-24) and (15-25), the output of the low-pass filter at Vout(t) is then:

$$Vout(t) = M(\omega)A \sin[2\pi f\, t + \varphi_1 + \psi(\omega)]$$

$$Vout(t) = 1/[(1 + \omega^2 C^2 R^2)^{1/2}]A \sin[2\pi ft + \varphi_1 + (-\tan^{-1})\omega CR]$$

or:

$$Vout(t) = [1/(1 + \omega^2 C^2 R^2)^{1/2}]A \sin[2\pi ft + \varphi_1 - \tan^{-1}(\omega CR)] \qquad (15\text{-}26)$$

A more convenient way of expressing the magnitude function includes finding the break or cut-off frequency known as the –3-dB frequency when:

$$\omega^2 C^2 R^2 = 1 \qquad (15\text{-}27)$$

Substituting Equation (15-27) into Equation (15-26) results in:

$$Vout(t) = [1/(1 + 1)^{1/2}]A \sin[2\pi ft + \varphi_1 - \tan^{-1}(\omega CR)]$$

$$= Vout(t) = [1/\sqrt{2}]A \sin[2\pi ft + \varphi1 - \tan^{-1}(\omega CR)]$$

The $1/\sqrt{2}$ factor causes the original input amplitude $A \to A/\sqrt{2}$ or $A \to 0.707\, A$, which is equivalently causing the input signal's amplitude to be reduced by –3 dB. This also leads to $\omega CR = 1$ by taking the square root of both sides of Equation (15-27).

We can solve for the particular cut-off frequency ω in the equation $\omega CR = 1$, but let's assign a new omega ω_{-3dB} because this a single fixed frequency, and ω has a range of frequencies. Thus we have:

$$\omega_{-3dB} CR = 1$$

and by dividing by CR on both sides, we get:

$$\omega_{-3dB} = 1/CR = 1/RC = \omega_{-3dB} \quad \text{or} \quad \omega_{-3dB} = 1/RC \qquad (15\text{-}28)$$

Also note that to express the cut-off frequency in terms of hertz instead of radians per second as in ω_{-3dB}:

$$\omega_{-3dB} = 2\pi f_{-3dB}$$

which equivalently brings Equation (15-28) to:

$$2\pi f_{-3dB} = 1/RC \qquad (15\text{-}29)$$

Dividing by 2π on both sides of Equation (15-29) gives:

$$f_{-3dB} = 1/[2\pi RC] = 1/2\pi RC = f_{-3dB} \qquad (15\text{-}30)$$

For example, suppose that we want to make a 16 kHz low-pass filter. We can start with R = 10 kΩ and determine the capacitor value:

$$16 \text{ kHz} = 1/[2\pi(10 \text{ k}\Omega)C]$$

$$C = 1/[2\pi(10 \text{ k}\Omega)(16 \text{ kHz})] \approx 0.001 \text{ }\mu F = C$$

We can express $\omega_{-3dB} = 1/RC$ equivalently in terms of RC by multiplying by RC and dividing by ω_{-3dB} on both sides of Equation (15-28), which leads to:

$$RC = 1/\omega_{-3dB} \qquad (15\text{-}31)$$

And by squaring both sides of Equation (15-31), we get:

$$(RC)^2 = R^2C^2 = [1/\omega_{-3dB}]^2 \qquad (15\text{-}32)$$

Then we can substitute $[1/\omega_{-3dB}]^2$ for R^2C^2 in Equation (15-25) so that we can express the magnitude function in terms of the cut-off frequency.

Note that:

$$M(\omega) = |Vout(\omega)/Vin(\omega)|$$

$$M(\omega) = 1/[(1 + \omega^2 C^2 R^2)^{1/2} \qquad (15\text{-}25)$$

$$M(\omega) = 1/(1 + \omega^2[1/\omega_{-3dB}]^2)^{1/2} = 1/(1 + [\omega/\omega_{-3dB}]^2)^{1/2} \qquad (15\text{-}33)$$

Because we commonly work with cycles per second or Hz (hertz) instead of radians per second, we can factor out the 2π since $\omega/\omega_{-3dB} = 2\pi f/2\pi f_{-3dB} = f/f_{-3dB} = \omega/\omega_{-3dB}$, which leads to the RC low-pass filter's magnitude as a function of f (frequency):

$$M(f) = 1/(1 + [f/f_{-3dB}]^2)^{1/2} \qquad (15\text{-}34)$$

With $RC = CR = 1/\omega_{-3dB}$ substituted into Equation (15-24) for the phase function that describes the phase shift at the output of the filter in reference to the phase of the input sine wave signal:

$$\psi(\omega) = -\tan^{-1}(\omega CR) \qquad (15\text{-}24)$$

leads to:

$$\psi(\omega) = -\tan^{-1}(\omega/\omega_{-3dB})$$

And again we can use $f/f_{-3dB} = \omega/\omega_{-3dB}$, which leads to the phase as a function of f (frequency in Hz or hertz):

$$\psi(f) = -\tan^{-1}(f/f_{-3dB}) \qquad (15\text{-}35)$$

At the cut-off frequency $f \rightarrow f_{-3dB}$, the phase shift of the RC low-pass filter is:

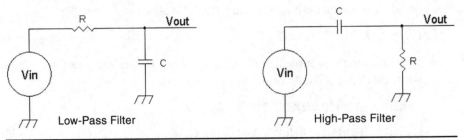

FIGURE 15-20   Low- and high-pass filter circuits.

$$\psi(f_{-3dB}) = -\tan^{-1}(f_{-3dB}/f_{-3dB}) = -\tan^{-1}(1) = -45 \text{ degrees} = \psi(\omega_{-3dB})$$

We can look at an RC high-pass filter, which will be summarized in Table 15-1 (see Figure 15-20). For a RC high-pass filter, the output-input relationship is:

$$Vout(\omega)/Vin(\omega) = [R/(R + Zc)] = R/(R + 1/j\omega C)$$

$$Vout(\omega)/Vin(\omega) = R/(R + 1/j\omega C) \tag{15-36}$$

And the results of the magnitude and phase of both low- and high-pass filters are shown in Table 15-1.

TABLE 15-1   Summary of Magnitude and Phase Characteristics of Low- and High-Pass Filters

Filter Type	M(f) Magnitude	ψ(f) Phase	Notes
Low-pass RC	$1/(1 + [f/f_{-3dB}]^2)^{1/2}$	$-\tan^{-1}(f/f_{-3dB})$	$f_{-3dB} = 1/2\pi RC$
High-pass RC	$[f/f_{-3dB}]/[(1 + [f/f_{-3dB}]^2)^{1/2}$	$90° - \tan^{-1}(f/f_{-3dB})$	$f_{-3dB} = 1/2\pi RC$

An example of a high-pass filter is an audio filter for a simple two-way high-fidelity speaker. The woofer speaker may be connected directly to the audio power amplifier, but the tweeter requires high-pass filtering. Suppose that the 8 Ω tweeter operates from 1,200 Hz and above. A small-value series capacitor between the amplifier and the tweeter is required to satisfy the following:

1,200 Hz = $1/2\pi RC$

where R = 8 Ω. Thus:

1,200 Hz = $1/[2\pi(8 \ \Omega)C]$

C = $1/2\pi(8 \ \Omega)(1,200 \text{ Hz})$ = 16.6 μF

See Figure 15-21.

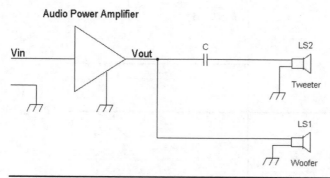

FIGURE 15-21   A simple capacitor crossover network for a tweeter loudspeaker.

Simple Transient or Pulse Response to RC Low- and High-Pass Filters

Although sinusoidal analysis is a very powerful tool for determining performance of circuits such as amplifiers and filters, the transient response of these circuits can be just as important. Using square wave or pulsed waveforms can aid in determining, for instance, the stability of a system or amplifier. In this section, we will briefly go over some basics of transient response. The subject itself normally requires much higher mathematics that includes *differential equations* and *Laplace transforms*, which are normally taught in colleges or universities.

Fortunately, we can build RC circuits and input a squarewave or pulse waveform to determine the relative shape of the pulse at the output of the circuit (see Figures 15-22 and 15-23). If we take a peek at Laplace transforms, we find that they are similar to sinusoidal analysis, except that we replace the $j\omega$ term with s.

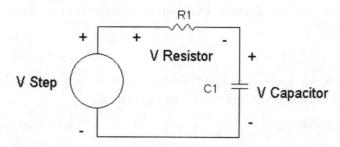

FIGURE 15-22   An RC low-pass filter circuit with a step-function input signal V Step and output at V Capacitor.

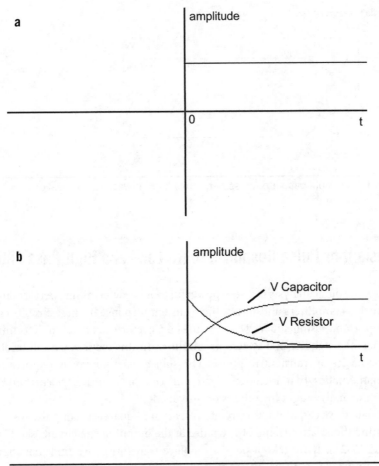

a

amplitude

0

t

b

amplitude

V Capacitor

V Resistor

0

t

Figure 15-23 (*a*) Pulsed waveform (step function) at the input of an RC low-pass filter. (*b*) The relative output signals across the capacitor and the resistor.

For an RC high-pass filter with the output as Vout1(ω) and input signal Vin(ω), the sinusoidal analysis is:

$$\text{Vout1}(\omega)/\text{Vin}(\omega) = R/(R + 1/j\omega C) \tag{15-36}$$

For transient analysis, we modify Equation (15-36) to replace jω term with s and also write Vout/Vin in terms of s (see Figure 15-24, left-side circuit).

$$\text{Vout1}(s)/\text{Vin}(s) = R/(R + 1/sC) \tag{15-37}$$

Because we want to look at the output response, we multiply by Vin(s) on both sides:

$$\text{Vout1}(s) = \text{Vin}(s)R/(R + 1/sC) \tag{15-38}$$

We can further multiply by sC/sC = 1 on the right side of Equation (15-38) to get:

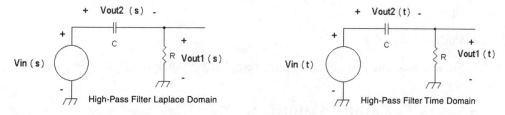

Figure 15-24 A high-pass filter shown with the variable s for the Laplace domain on the left side and its time domain equivalent circuit on the right side.

$$\text{Vout1}(s) = \text{Vin}(s)RCs/(RCs + 1) \tag{15-39}$$

Also, we can divide by RC in the numerator and denominator to result in:

$$\text{Vout1}(s) = \text{Vin}(s)[s/(s + 1/RC)] \tag{15-40}$$

A *unit* step function for Vin(s) is 1/s in the Laplace domain, which results in:

$$\text{Vout1}(s) = (1/s)[s/(s + 1/RC)] = 1/(s + 1/RC)$$

$$\text{Vout1}(s) = 1/(s + 1/RC) \tag{15-41}$$

So why did I do all this algebraic manipulation? The object was to get Vout(s) in the form of $1/(s + a)$ so that the inverse Laplace transform will give us a transient time waveform in the following manner with $a = 1/RC$:

$$\text{Vout1}(s) [1/(s + a)] \rightarrow e^{-at} = e^{-t/RC} = \text{Vout}(t) \tag{15-42}$$

We also have to make one very important initial condition with the circuit, which is that the capacitor C had no electrical charge or initially charged voltage just before the step-function input signal was applied. If there was a charge in the capacitor, Vout(t) will include yet another output signal.

If Vin(t) is a step function from 0 to V_o the output Vout1(t), the voltage across the resistor will scale accordingly and will be:

$$\text{Vout1}(t) = V_o e^{-t/RC} \tag{15-43}$$

Intuitively, Vout1(t) makes sense because the capacitor is initially at 0 volt, and as the step-function signal is enabled, the capacitor charges up to V_o and blocks DC or current from flowing into resistor R, causing the voltage across the resistor to trend toward 0 volt over time.

From the time response across the resistor, we can derive what the pulse response across the capacitor is. From Figure 15-24 on the right side circuit, we know by Kirchoff's loop equation law that:

$$\text{Vin}(t) = \text{Vout2}(t) + \text{Vout1}(t)$$

where Vout2(t) is the voltage across the capacitor C.

$Vin(t) = V_o$ for $t \geqslant 0$ and:

$Vin(t) = 0$ for $t < 0$

We will also state that the capacitor is completely discharged and its voltage is 0 volt at $t = 0$:

$Vin(t) = V_o = Vout2(t) + Vout1(t)$

By means of Equation (15-43), we substitute $V_o e^{-t/RC}$ for Vout1(t):

$V_o = Vout2(t) + V_o e^{-t/RC}$

$V_o - V_o e^{-t/RC} = Vout2(t) = V_o(1 - e^{-t/RC})$

$$Vout2(t) = V_o(1 - e^{-t/RC}) \tag{15-44}$$

where Vout2(t) is the voltage across the capacitor. Note that the voltage across the capacitor C is also the voltage from an RC low-pass filter. Thus Equation (15-44) is the step-pulse response of an RC low-pass filter (Figure 15-25).

Actually, in practice, to solve for both transient and steady-state sinusoidal responses, normally, we start with the impedance of a capacitor as $Z_{capacitor} = 1/sC$ or the impedance of an inductor as $Z_{inductor} = sL$, crank out the equations, and take the final equation in s for determining the transient response and then replace s with jω (i.e., $s = j\omega$) to determine the frequency and phase response to a sinusoidal signal. To show the effects of low- and high-pass filtering (see Figure 15-26).

The multiburst signal has one cycle of a squarewave followed by six packets of sine-wave signals starting at 500 kHz and ending at 4.2 MHz. The resistor value is R = 1 kΩ, and the capacitor's value is C = 330 pF for providing the low-pass and high-pass filtering effects. As can be seen in the center trace, the low-pass filtering effect shows a rolled-off corner on the squarewave signal that is described as Vout2(t) = $V_o(1 - e^{-t/RC})$ for the positive half of the squarewave. Also, note that the sine-wave packets are more attenuated as the frequency of the sine-wave packets increases.

The bottom trace shows a high-pass filtering effect, and the original squarewave signal has turned into a positive and negative spike. The positive spike can be described as Vout2(t) = $V_o e^{-t/RC}$ for the transient response. Note that the first sine-

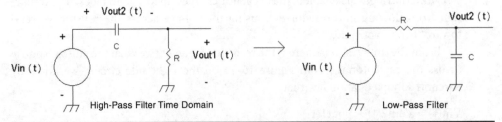

Figure 15-25   Vout2(t) on the left has a low-pass filter output and is equivalent in terms of a low-pass filter shown in the circuit on the right.

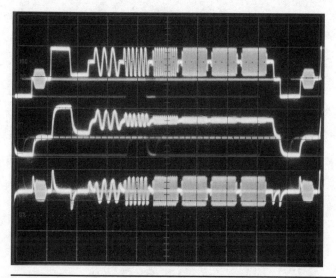

Figure 15-26   (*Top trace*) Input signal. (*Center trace*) Low-pass filtered signal. (*Bottom trace*) High-pass filtering effect.

wave packet has more attenuation than those at the higher frequencies, which shows the effect of the high-pass filter. In Chapter 16, we will continue with more circuit analysis.

References

1.   Charles A. Desoer and Ernest S. Kuh, *Basic Circuit Theory*. New York: McGraw-Hill, 1969.

2.   Allan R. Hambley, *Electrical Engineering: Principles and Applications, 2nd ed.* Upper Saddle River, NJ: Prentice-Hall, 2002.

3.   B. P. Lathi, *Linear Systems and Signals*. Carmichael, CA: Berkeley-Cambridge Press, 2002.

16

A Review and Analysis of What We Have Built So Far

Some circuits that were shown in previous chapters will be reviewed in this chapter with a more detailed analysis. We will start with frequency-response analysis of some of the amplifying circuits such as Wein oscillators and audio amplifiers. Also, we will look into how amplifying systems work. Most amplifiers include an input stage and an output stage. The characteristics of gain, frequency response, input and output resistances, and feedback will be analyzed. We will take a second look at Wein oscillators and radios with alternative circuit designs that generally lead to an improvement in performance. For example, in the Wein oscillator, it was found that larger-wattage bulbs caused the output signal to produce an unstable amplitude. A direct-current (DC) biasing current to the bulb resulted in a more stable amplitude. Some large-signal behaviors of amplifiers will be examined, including calculating for harmonic distortion in simple amplifiers.

Wein Oscillator RC Feedback Network

A Wein oscillator is shown in Figure 16-1 (shown earlier as Figure 9-7). The oscillator includes a noninverting gain of $[(VR1/R_{Lamp1}) + 1]$ from pin 7 (output) to pin 6 (inverting input) of U1B. A positive-feedback network that includes R1, C1, R2, and C2 is coupled from the output of U1B and the noninverting input at pin 5.

The key to oscillation is that there is a 0-degree phase shift at a particular frequency from this positive-feedback network. That is, at the resonant frequency, the network has no j term or reactive component that would cause a non-0-degree phase shift.

A simplified block diagram of the Wein oscillator is shown in Figure 16-2. For a top-level view, let's assign the following:

$$Z_1 = R1 + Z_{C1}$$

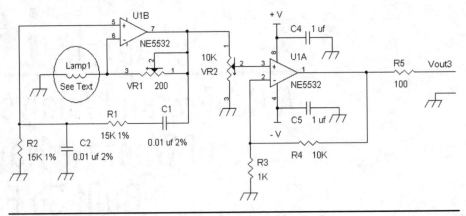

FIGURE 16-1   A Wein sine-wave oscillator circuit from Chapter 9.

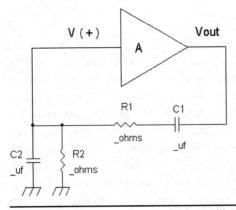

FIGURE 16-2   Block diagram of a Wein oscillator.

$$Z_2 = R2 || Z_{C2}$$

where $Z_{C1} = 1/j\omega C1$ and $Z_{C2} = 1/j\omega C2$. By way of a voltage-divider formula:

$$\text{Vout}\ [Z_2/(Z_1 + Z_2)] = V(+)$$

$$\text{Vout}[(R2||Z_{C2})/(\ R1 + Z_{C1} + R2||Z_{C2})] = V(+) \tag{16-1}$$

The algebra will be getting messy, so it will be better to show Equation (16-1) in terms of a fraction:

$$\text{Vout}\ \frac{R2||Z_{C2}}{R1 + Z_{C1} + R2||Z_{C2}} = V(+) \tag{16-2}$$

$$R2 \| Z_{C_2} = \frac{\left[R2 \frac{1}{j\omega C2} \right]}{R2 + \frac{1}{j\omega C2}}$$

so we get:

$$Vout \frac{\dfrac{\left[R2 \dfrac{1}{j\omega C2} \right]}{R2 + \dfrac{1}{j\omega C2}}}{R1 + \dfrac{1}{j\omega C2} + \dfrac{\left[R2 \dfrac{1}{j\omega C2} \right]}{R2 + \dfrac{1}{j\omega C2}}} = V(+)$$

(16-3)

If we multiply by:

$$1 = \frac{\left[R2 + \dfrac{1}{j\omega C2} \right]}{R2 + \dfrac{1}{j\omega C2}}$$

we get:

$$Vout \frac{R2 \dfrac{1}{j\omega C2}}{\left(R1 + \dfrac{1}{j\omega C2} \right)\left(R2 + \dfrac{1}{j\omega C2} \right) + R2 \dfrac{1}{j\omega C2}} = V(+)$$

(16-4)

In Figure 16-1, we made the two resistors and two capacitors equal in value, which leads to:

$$R = R1 = R2 \quad \text{and} \quad C = C1 = C2$$

$$Vout \frac{R \dfrac{1}{j\omega C}}{\left(R + \dfrac{1}{j\omega C} \right)\left(R + \dfrac{1}{j\omega C} \right) + R \dfrac{1}{j\omega C}} = V(+)$$

(16-5)

Now, by multiplying by:

$$1 = \frac{j\omega C}{j\omega C}$$

$$Vout \frac{R}{\left(Rj\omega C + 1 \right)\left(R + \dfrac{1}{j\omega C} \right) + R} = V(+)$$

(16-6)

$$Vout \frac{R}{\left(R^2 j\omega C + 2R + \dfrac{1}{j\omega C} \right) + R} = V(+)$$

(16-7)

Let's add the 2R and R terms in the denominator and group the imaginary and real numbers as follows:

$$Vout \frac{R}{\left(R^2j\omega C + \frac{1}{j\omega C}\right) + 3R} = V(+)$$

(16-8)

Also note that:

$$\frac{j}{j} \frac{1}{j\omega C} = \frac{-j}{\omega C} = \frac{1}{j\omega C}$$

$$Vout \frac{R}{\left(R^2j\omega C + \frac{-j}{\omega C}\right) + 3R} = V(+)$$

(16-9)

We can now solve for a frequency where the imaginary number goes to zero for a particular frequency ω, which will be the resonant or oscillation frequency. That is:

$$R^2j\omega C - \frac{j}{\omega C} = 0$$

$$R^2\omega C = \frac{j}{\omega C}$$

and we can divide by j on both sides to get:

$$R^2j\omega C = \frac{1}{\omega C}$$

and if we multiply by ωC on both sides, the result is:

$$R^2\omega^2C^2 = 1 \quad \text{or} \quad \omega^2 = 1/R^2C^2$$

Taking the square root of both sides and having only a positive frequency, we have the oscillation frequency as:

$$\omega = 1/RC$$

(16-10)

Or in terms of the oscillation frequency in hertz, Hz, we know that:

$$\omega = 2\pi f$$

This leads via Equation (16-10) to:

$$2\pi f = 1/RC$$

and by dividing by 2π on both sides, the oscillation frequency in hertz, Hz, is:

$$f = 1/2\pi RC$$

(16-11)

At the oscillation frequency, the reactive component that is associated with the j terms disappears, and the positive-feedback network is described as a purely resistive voltage divider:

$$\text{Vout}\ \frac{R}{3R} = \text{Vout}\ \frac{1}{3} = V(+)$$

This represents a gain factor of one-third, which is a loss, and to provide an oscillation, we need to make up the loss and have:

$$\frac{1}{3} \times A \geqslant 1$$

Or put another way, an oscillation occurs when the gain A is $\geqslant 3$ to make up for the losses or attenuation factor of (1/3) due to the positive-feedback network.

Thus $A \geqslant 3$ for a reliable oscillation. For the op amp circuit in Figure 16-1, the gain is:

$$A = ([VR1/R_{Lamp1}] + 1)$$

where VR1 is adjusted until there is an oscillation, and R_{Lamp} is the alternating-current (AC) resistance as the lamp's filament is being heated. Thus:

$$A = ([VR1/R_{Lamp1}] + 1) \geqslant 3$$

which leads to:

$$([VR1/R_{Lamp1}] + 1) \geqslant 3$$

By subtracting 1 from both sides of $([VR1/R_{Lamp1}] + 1) \geqslant 3$, we get:

$$[VR1/R_{Lamp1}] \geqslant 2$$

Or put another way, the resistance of VR1 has to be at least twice the resistance of the lamp to ensure oscillation.

Before leaving the Wein oscillator circuit, let's take a look at the magnitude of the positive-feedback voltage divider network from Equation (16-9), which we will call $M(\omega)$:

$$M(\omega) = \sqrt{\frac{R^2}{\left(R^2\omega C - \frac{1}{\omega C}\right)^2 + 9R^2}}$$

(16-12)

We know at the oscillation frequency $\omega = 1/RC$ that $M(\omega) = M(1/RC) = (1/3)$. At frequencies where $\omega \ll 1/RC$, the term $1/\omega C$ dominates by becoming very large in comparison with any of the other terms in the denominator and causes $M(\omega) \to 0$. If we look at Figure 16-1 at frequencies below the oscillation frequency, such as close to DC, C1 will block out the signal from the output of pin 7 of U1B.

Also, at frequencies where $\omega \gg 1/RC$, the term $R^2\omega C$ becomes dominant and very large in comparison with any of the other terms in the denominator and again causes $M(\omega) \rightarrow 0$. Again, intuitively, this makes sense if we look at Figure 16-1, where C2 at high frequencies will behave as an AC short circuit to ground that zaps out any signals from the output of pin 7 of U1B.

From looking at the *boundary conditions* or the limits of the extremes, as I like to call them, the positive-feedback network has the characteristic of a band-pass filter, which is generally desirable in an amplifier feedback oscillator. The maximum output of the network described by Equation (16-12) occurs at the oscillation frequency, and any other frequency above and below the oscillation frequency will have a smaller output.

A Phono Preamp Revisited

Let's take a look at the two-stage RIAA (Radio Industries Association of America) phono equalization circuit that was shown in Chapter 8 [see Figure 16-3 (shown previously as Figure 8-12)]. The two equalization networks are implemented by R5 and C5 and by R6, R7, R8, and C9.

From U1A's output pin 1 to U1B's input pin 5, R5 and C5 form a frequency-dependent voltage divider that can be described as:

$$\text{Vout (pin 1)} \frac{Z_{C5}}{R5 + Z_{C5}} = \text{Vin (pin5)} \tag{16-13}$$

where $Z_{C5} = 1/j\omega C5$.

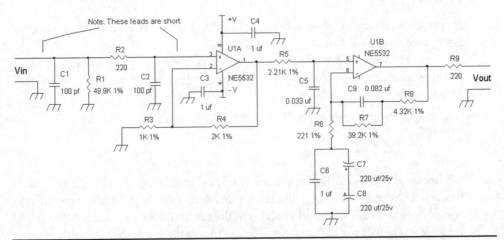

FIGURE 16-3   Two-stage phono preamp from Chapter 8.

$$\text{Vout (pin 1)} \frac{\frac{1}{j\omega C5}}{\left(R5 + \frac{1}{j\omega C5}\right)} = \text{Vin (pin5)}$$

and by multiplying the numerator and denominator by $j\omega C5$, we have:

$$\text{Vout (pin 1)} \frac{1}{(R5j\omega C5 + 1)} = \text{Vin (pin 5)}$$

or, equivalently:

$$\text{Vout (pin 1)} \frac{1}{(j\omega R5C5 + 1)} = \text{Vin (pin 5)} \qquad (16\text{-}14)$$

The magnitude of:

$$\frac{1}{(j\omega R5C5 + 1)} = \sqrt{\frac{1^2}{(\omega R5C5)^2 + 1^2}} = M_1(\omega)$$

which means that when $\omega = 1/R5C5$:

$$\sqrt{\frac{1}{(1)^2 + 1^2}} = \sqrt{\frac{1}{2}} \approx 0.707 = M_1(\omega)$$

The −3 dB cut-off frequency yields a magnitude of 0.707 referenced to a maximum magnitude of 1.0, and its frequency in hertz, Hz, is:

$$\omega = 2\pi f = 1/R5C5 \quad \text{or} \quad f = 1/2\pi R5C5 \qquad (16\text{-}16)$$

For this particular example, $f \approx 2.12$ kHz, given that R5 = 2.26 kΩ and C5 = 0.033 μF. There is a slight error with R5 = 2.21 kΩ, resulting in $f \approx 2.18$ kHz, but it is still close to the 2.12-kHz spec.

At very low frequencies $<< 1/2\pi R5C5$, $\omega \to 0$, and $M1(\omega) \to 1$. This makes sense because at DC, capacitor C5 is an open circuit allowing the signal from pin 1 to pass unscathed via R5 to pin 5 in Figure 16-3.

However, at very high frequencies $>> 1/2\pi R5C5$, we see that the term gets very large in the denominator of :

$$\sqrt{\frac{1^2}{(\omega R5C5)^2 + 1^2}} = M_1(\omega)$$

and $M1(\omega) \to 0$. Again, this makes sense because at high frequencies, C5 becomes an AC short circuit that grounds pin 5.

For the next network (from Figure 16-3), that is, R6, R7, R8, and C9, we can calculate the gain of the second section with U1B from output pin 7 in reference to input pin 5 as:

$$\text{Vout (pin 7)} = \text{Vin (pin 5)} \; \frac{(R7 || Z_{C9} + R8) + R6}{R6} \tag{16-17}$$

$$\text{Vout (pin 7)} = \text{Vin (pin 5)} \; \frac{(\dfrac{\left[R7\dfrac{1}{j\omega C9}\right]}{R7 + \dfrac{1}{j\omega C9}} + R8) + R6}{R6} \tag{16-18}$$

At this point, we need to simplify somewhat by bringing out the 1/R6 term, which is a scaler of sorts, and note that R6/R6 = 1. Thus:

$$\text{Vout (pin 7)} = \text{Vin (pin 5)} \; [\frac{1}{R6} \left(\frac{\left[R7\dfrac{1}{j\omega C9}\right]}{R7 + \dfrac{1}{j\omega C9}} + R8 \right) + 1]$$

And now the algebra gets a little messy, as if it is not a already a mess. We will multiply R8 by:

$$\frac{\left[R7 + \dfrac{1}{j\omega C9}\right]}{R7 + \dfrac{1}{j\omega C9}} = 1$$

to form a common denominator:

$$\text{Vout (pin 7)} = \text{Vin (pin 5)} \; [\frac{1}{R6} \left(\frac{\left[R7\dfrac{1}{j\omega C9}\right]}{R7 + \dfrac{1}{j\omega C9}} + R8 \frac{\left[R7 + \dfrac{1}{j\omega C9}\right]}{R7 + \dfrac{1}{j\omega C9}} \right) + 1]$$

$$\text{Vout (pin 7)} = \text{Vin (pin 5)} \; [\frac{1}{R6} \left(\frac{\left[R7\dfrac{1}{j\omega C9}\right]}{R7 + \dfrac{1}{j\omega C9}} + \frac{\left[R8R7 + \dfrac{R8}{j\omega C9}\right]}{R7 + \dfrac{1}{j\omega C9}} \right) + 1]$$

$$\text{Vout (pin 7)} = \text{Vin (pin 5)} \; [\frac{1}{R6} \left(\frac{\left[R7\dfrac{1}{j\omega C9}\right] + \left[R8R7 + \dfrac{R8}{j\omega C9}\right]}{R7 + \dfrac{1}{j\omega C9}} \right) + 1] \tag{16-19}$$

We are almost there—just a few of more steps. We will multiply the numerator and:

denominator of $\left(\dfrac{\left[R7\dfrac{1}{j\omega C9}\right] + \left[R8R7 + \dfrac{R8}{j\omega C9}\right]}{R7 + \dfrac{1}{j\omega C9}}\right)$ by $j\omega C9$

$$\text{Vout (pin 7)} = \text{Vin (pin 5)}\ [\tfrac{1}{R6}\left(\dfrac{[R7] + [R8R7j\omega C9 + R8]}{R7j\omega C9 + 1}\right) + 1]$$

$$\text{Vout (pin 7)} = \text{Vin (pin 5)}\ [\tfrac{1}{R6}\left(\dfrac{[R7 + R8] + [R7R8]j\omega C9}{R7j\omega C9 + 1}\right) + 1]$$

(16-20)

If we factor out the term [R7 + R8], we will see that there is an equivalent to two resistors in parallel, R8 and R9, in the numerator:

$$\text{Vout (pin 7)} = \text{Vin (pin 5)}\ [\tfrac{1}{R6}[R7 + R8]\left(\dfrac{1 + \dfrac{R7R8}{R7 + R8}j\omega C9}{R7j\omega C9 + 1}\right) + 1]$$

But $\dfrac{R7R8}{R7 + R8} = R7\|R8$, so:

$$\text{Vout (pin 7)} = \text{Vin (pin 5)}\ [\tfrac{1}{R6}[R7 + R8]\left(\dfrac{1 + (R7\|R8)j\omega C9}{R7j\omega C9 + 1}\right) + 1]$$

(16-21)

The magnitude of $\left(\dfrac{1 + (R7\|R8)j\omega C9}{R7j\omega C9 + 1}\right)$ is:

$$M_2(\omega) = \sqrt{\dfrac{(\omega[R7\|R8]C9)^2 + 1^2}{(\omega R7C9)^2 + 1^2}}$$

(16-22)

The roll-off frequency associated with R7 is:

$f = 1/2\pi R7C9$

and with R7 = 39.2 kΩ and C9 = 0.082 μF, $f \approx 49.5$ Hz , which is close to the 50-Hz spec. The leveling-off frequency after a roll-off at 49.5 Hz is:

$f = 1/2\pi(R7\|R8)C9$

and given that R7 = 39.2 kΩ, R8 = 4.32 kΩ, and C9 = 0.082 μF, $f \approx 499$ Hz, which is close to the 500-Hz spec.

A Look at an Emitter-Follower Circuit

We now turn our attention to amplifiers, which were covered in Chapter 6. However, I would like to now finally cover them in more detail, especially in terms of input and output resistances, gain, and transconductance.

An amplifier normally includes one or more stages of amplification, where we define amplification as an increase in power gain. This power gain is more easily understood as a voltage or a current gain. In terms of current gain, we can generally associate this with the capability of loading into lower resistances or impedances at the output. For now, we will look at two types of amplifiers—the common emitter and the emitter follower.

In any amplifier, we usually would like to know the AC characteristics in terms of input resistance, gain, and output resistance. The bipolar transistors are described in the following manner: the NPN and PNP transistors are defined by the collector, emitter, and base currents shown in Figure 16-4.

First, let's take a look at bipolar transistors in terms of their input (VBE) and output (IC) relationships starting with the following equations for NPN and PNP collector currents.

The collector currents are described as:

NPN: $IC_{npn} = I_s e^{(VBE/V_T)}$ (16-23)

PNP: $IC_{pnp} = I_s e^{(VEB/V_T)}$ (16-24)

where $V_T = kT/q = 0.026$ mV at room temperature (approximately +25°C), and the number $e \approx 2.71821828$. [If you ever want to remember the e approximation, then one way is based on some U.S. history. Andrew Jackson (or Mr. $20 bill) served "2" terms as the "7th" president of the United States and won his first election in "1828." All this is courtesy of one of my high school math teachers.] I_s is the saturation current, which is typically 0.01 pA to 0.0001 pA. VBE is the total voltage, which usually

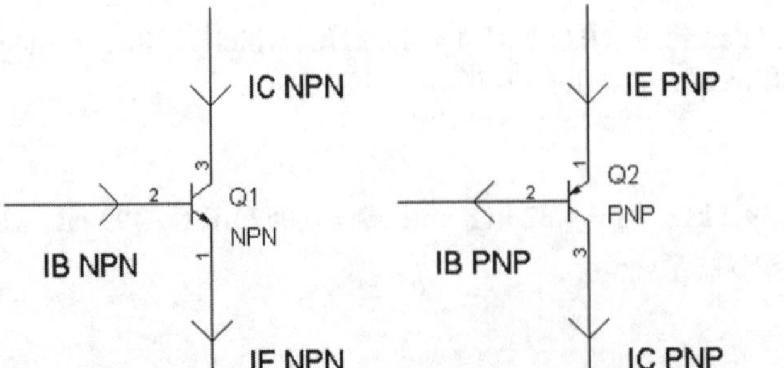

FIGURE 16-4 NPN and PNP transistors with their respective currents.

is the sum of two voltages, a quiescent DC bias voltage VBEQ and a small-signal AC signal Vbe:

$$VBE = [VBEQ + Vbe]$$

for the NPN transistor, and similarly, for the PNP transistor:

$$VEB = [VEBQ + Veb]$$

For now, let's just look at the NPN transistor. Thus:

$$IC_{npn} = I_s e^{([VBEQ + Vbe]/V_T)} \tag{16-25}$$

Using the exponential law of $a^{(b+c)} = a^b a^c$, we have:

$$IC_{npn} = I_s [e^{([VBEQ + Vbe]/V_T)}] = IC_{npn} = I_s [e^{(VBEQ/V_T)} e^{(Vbe/V_T)}] \tag{16-26}$$

When Vbe = 0, we have a quiescent collector current IC_{npnQ} that is a no-AC-signal condition, resulting in $e^{(Vbe/V_T)} = e^{(0/V_T)} = 1$ and:

$$IC_{npnQ} = I_s [e^{(VBEQ/V_T)}] = ICQ$$

where ICQ is the quiescent DC collector current. With the AC signal Vbe, Equation (16-26) is equivalent to:

$$IC_{npn} = ICQ\, e^{(Vbe/V_T)} \tag{16-27}$$

We can make a linear approximation of $e^{(Vbe/V_T)}$ because $e^x \approx 1 + x$, and let x = Vbe/V_T. Then:

$$IC_{npn} \approx ICQ(1 + Vbe/V_T) \tag{16-28}$$

$$IC_{npn} \approx ICQ + (ICQ/V_T)Vbe \tag{16-29}$$

$$IC_{npn} \approx (ICQ/V_T)Vbe + ICQ \tag{16-30}$$

Let's relate this to the familiar line equation:

$$y = mx + b$$

Let $y = IC_{npn}$, $m = (ICQ/V_T)$, x = Vbe, and b = ICQ
    The slope or gain of the line is:

$$\Delta y/\Delta x = \frac{\Delta y}{\Delta x} = \frac{\Delta IC_{npn}}{\Delta Vbe} = (ICQ/V_T) \tag{16-31}$$

where:

$$\frac{\Delta IC_{npn}}{\Delta Vbe}$$

is defined as the transconductance g_m.

However, let's look at total base-emitter voltage VBE and the small signal across the base emitter Vbe. What is the relationship between ΔVBE and ΔVbe?

VBE = (VBEQ + Vbe) → ΔVBE = Δ(VBEQ + Vbe) = ΔVBEQ + ΔVbe

ΔVBE = ΔVBEQ + ΔVbe

We know that when the "whole" base-emitter voltage VBE changes due to the small signal added to the DC bias voltage VBEQ, VBEQ stays constant. Thus the change in VBEQ is zero. For example, suppose that the transistor is biased to VBEQ = 0.600 volt DC, and the small-signal Vbe is an AC voltage that varies ±10 mV. VBE = 0.600 volt ± 10 mV. The change in VBE is from 0.610 volt to 0.590 volt, or put another way, the change in VBE is ±10 mV, which is exactly the same change in the small-signal voltage Vbe. This leads to:

ΔVBE = ΔVbe

which says that:

$$\frac{\Delta IC_{npn}}{\Delta Vbe} = \frac{\Delta IC_{npn}}{\Delta VBE}$$

$$g_m = \frac{\Delta IC_{npn}}{\Delta Vbe} = \frac{\Delta IC_{npn}}{\Delta VBE} = (ICQ/V_T) = \frac{ICQ}{V_T} \tag{16-32}$$

$$g_m = \frac{ICQ}{V_T} = \frac{\Delta IC_{npn}}{\Delta VBE} \tag{16-33}$$

For example, if the quiescent collector current ICQ is 1 mA:

$$g_m = \frac{ICQ}{V_T} = \frac{1\ mA}{0.026\ V} \approx 38.4\ mA/volt$$

Because g_m is directly proportional to ICQ, we can say that at 0.1 mA = ICQ, $g_m \approx$ 3.84 mA/volt, or one-tenth the transconductance for 1 mA = ICQ.

When we look at the curve of the function $I_s e^{(VBE/V_T)}$ in Figure 16-5, we can see that the slope increases as the collector current increases, thereby increasing gain or transconductance g_m.

When the small-signal transconductance g_m is known, we can then calculate the output resistance for an emitter follower. For simplicity, we will use a PNP emitter-follower circuit, as shown in Figure 16-6, but the answer will be the same for an NPN emitter follower.

For a PNP transistor, note that we use Veb instead of Vbe. Note from the NPN equation (16-32) that we can also say for a PNP transistor that ΔVeb = ΔVBE. Thus:

$$\frac{\Delta IC_{pnp}}{\Delta Veb} = g_m = \frac{ICQ}{V_T} = \frac{\Delta IC_{pnp}}{\Delta VEB} \tag{16-34}$$

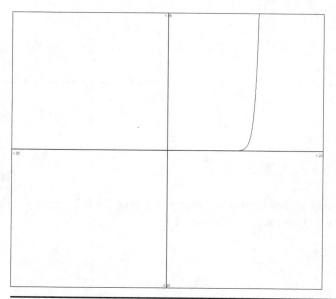

Figure 16-5   Graph of the collector current $I_s e^{(VBE/V_T)}$ as a function of VBE, where the vertical axis is the collector current and the horizontal axis is the base-emitter voltage.

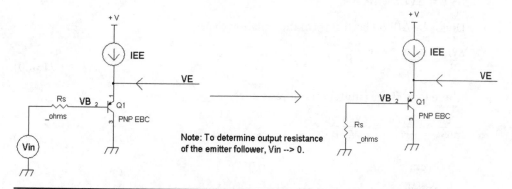

Figure 16-6   PNP emitter-follower circuit.

Also, the current gain $\beta = IC/IB$, or the ratio of collector current to base current, and $IB + IC = IE$, the emitter current. We can relate the following:

$$\beta IB = IC \qquad\qquad (16\text{-}35)$$

$$IB + IC = IE = IB + \beta IB = IE$$

$$IE = (\beta + 1)IB \qquad\qquad (16\text{-}36)$$

This also leads to a relationship between the collector and emitter currents as:

$$IE/IC = [(\beta + 1)IB]/[\beta IB] = (\beta + 1)/\beta = IE/IC$$

$$IE = IC[(\beta + 1)/\beta]$$

or, for AC signals:

$$\Delta IE = \Delta IC[(\beta + 1)/\beta] \tag{16-37}$$

This also says that:

$$IC = IE[\beta/(\beta + 1)]$$

or, for AC signals:

$$\Delta IC = \Delta IE[\beta/(\beta + 1)] \tag{16-38}$$

We are now ready to measure the output resistance of the PNP emitter-follower circuit in Figure 16-6. What we want to know is:

$$\frac{\Delta VE}{\Delta IE} = ?$$

$$VE = VEB + VB$$

where VEB is the emitter-to-base voltage of the PNP transistor.

$$\Delta VE = \Delta VEB + \Delta VB \tag{16-39}$$

Divide by ΔIE on both sides of the equation to get:

$$\frac{\Delta VE}{\Delta IE} = \frac{\Delta VEB}{\Delta IE} + \frac{\Delta VB}{\Delta IE} = Rout \tag{16-40}$$

where Rout is the emitter output resistance.

$$\frac{\Delta VEB}{\Delta IE} = \frac{\Delta VEB}{\Delta IC[\frac{\beta+1}{\beta}]} = \frac{\Delta VEB}{\Delta IC} \frac{\beta}{\beta+1} \tag{16-41}$$

However:

$$\frac{\Delta IC}{\Delta VEB} = \frac{\Delta IC}{\Delta Veb} = g_m \quad \text{and} \quad \frac{\Delta VEB}{\Delta IC} = \frac{1}{g_m} \tag{16-42}$$

because VEB represents the DC bias (VEBQ) plus the small-signal voltage Veb, and when we are looking at ΔVEB, the DC bias subtracts out, leaving only ΔVeb. This leads to:

$$\frac{\Delta VEB}{\Delta IE} = \frac{\Delta VEB}{\Delta IC} \frac{\beta}{\beta+1} = \frac{1}{g_m} \frac{\beta}{\beta+1}$$

$$\frac{\Delta VEB}{\Delta IE} = \frac{1}{g_m} \frac{\beta}{\beta+1} \tag{16-43}$$

Now we need to know, what is $\Delta VB/\Delta IE$?

$$IE = (\beta + 1)IB$$

which leads to:

$$\Delta IE = (\beta + 1)\Delta IB$$

By Ohm's law, $\Delta VB = Rs\Delta IB$, so:

$$\frac{\Delta VB}{\Delta IE} = \frac{Rs\Delta IB}{\Delta IE} = \frac{Rs\Delta IB}{(\beta + 1)\Delta IB} = \frac{Rs}{(\beta + 1)}$$

$$\frac{\Delta VB}{\Delta IE} = \frac{Rs}{(\beta + 1)} \tag{16-44}$$

$$\frac{\Delta VE}{\Delta IE} = \frac{\Delta VEB}{\Delta IE} + \frac{\Delta VB}{\Delta IE} = Rout \tag{16-40}$$

Combining information from Equations (16-43) and (16-44) into Equation (16-40), we arrive at:

$$Rout = \frac{1}{g_m}\frac{\beta}{\beta + 1} + \frac{Rs}{(\beta + 1)} \tag{16-45}$$

For $\beta \gg 1$:

$$\beta \approx (\beta + 1) \quad \text{and} \quad \frac{\beta}{\beta + 1} \approx 1$$

$$Rout = \frac{1}{g_m} + \frac{Rs}{\beta} \tag{16-46}$$

We can now model the emitter follower as shown in Figure 16-6 for AC signals. For $Rs = 0\ \Omega$, the gain is unity unless it is loaded to the outside world. This is so because the entire voltage across the emitter-base junction is just VEBQ, the quiescent bias turn-on voltage, and that turn-on voltage acts as a DC level shifter with no change in amplitude to AC signals from the base to the emitter. Note that equations for Rout in Equations (16-45) and (16-46) are applicable to NPN and PNP emitter follower circuits.

Using the Emitter Output Resistance to Determine Characteristics of a Common-Emitter Amplifier

Let's take a look at the common-emitter amplifier shown in Figure 16-7. It should be noted that in terms of an AC signal across the base-emitter junction, $\Delta VBE = \Delta Vbe$. To determine the AC collector current with a small signal voltage at the base, Vbe, we can use the transconductance formula for an NPN transistor:

$$\frac{\Delta IC}{\Delta VBE} = g_m$$

$$\Delta IC = \Delta VBE\ g_m \tag{16-47}$$

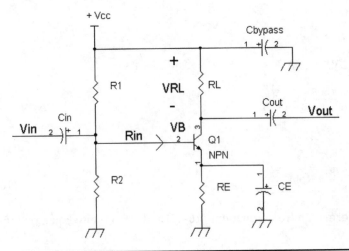

Figure 16-7 Common-emitter amplifier with its emitter AC grounded via capacitor CE.

The voltage drop across the collector resistor RL is then:

$$VRL = \Delta IC \; RL = \Delta VBE \; g_m \; RL = \Delta Vbe \; g_m \; RL$$

Note that VRL is taken from the (+) polarity at Vcc, an AC ground, and the (–) polarity on the collector. If we take the output voltage at the collector referenced to ground, the polarity inverts to:

$$Vout = -\Delta Vbe \; g_m \; RL \tag{16-48}$$

The AC signal $Vin = \Delta Vbe$, the small signal into the base-emitter junction of Q1. Therefore, the AC gain is:

$$Vout/Vin = -g_m \; RL \tag{16-49}$$

The AC input resistance of the grounded emitter amplifier is calculated when we can calculate the input voltage divided by the base current:

$$\beta \Delta IB = \Delta IC$$

$$\Delta IB = \Delta IC/\beta$$

$$Rin = \frac{\Delta VBE}{\Delta IB} = \frac{\Delta VBE}{\Delta IC/\beta} = \frac{\Delta VBE}{\Delta IC}\beta \tag{16-50}$$

The transconductance is defined as the change in collector current over the change in base emitter voltage:

$$\frac{\Delta IC}{\Delta VBE} = g_m \rightarrow \frac{\Delta VBE}{\Delta IC} = \frac{1}{g_m}$$

which leads to:

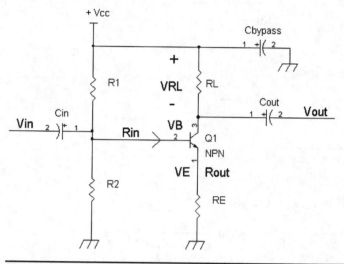

FIGURE 16-8   Common-emitter amplifier with a finite resistance to an AC ground for lower distortion.

$$Rin = \frac{\Delta VBE}{\Delta IB} = \frac{\Delta VBE}{\Delta IC}\beta = \frac{1}{g_m}\beta = \frac{\beta}{g_m} = Rin$$

$$(16\text{-}51)$$

In general, for signals whose frequencies span many octaves, a grounded emitter amplifier generates too much distortion for input signals exceeding 10 mV peak. A way to reduce distortion is to incorporate a finite resistance from the emitter to an AC ground, as shown in Figure 16-8. To calculate the gain, Vin = VB, and the input voltage source resistance Rs = 0 from Vin.

From Equation (16-45), the output resistance Rout from the emitter of Q1 is:

$$Rout = \frac{1}{g_m}\frac{\beta}{\beta+1} + \frac{Rs}{(\beta+1)}$$

$$(16\text{-}45)$$

However, Rs = 0, and therefore:

$$Rout = \frac{1}{g_m}\frac{\beta}{\beta+1}$$

The AC voltage at the emitter:

$$\Delta VE = \Delta Vin\frac{RE}{RE + Rout}$$

and the emitter current is then:

$$\frac{\Delta VE}{RE} = \Delta IE = \Delta VE\frac{1}{RE}$$

$$\Delta IE = \Delta Vin \ \frac{RE}{RE + Rout} \ \frac{1}{RE} = \Delta Vin \ \frac{1}{RE + Rout}$$

$$\Delta IC = \frac{\beta}{\beta + 1} \ \Delta IE$$

$$\Delta IC = \Delta Vin \ \frac{1}{RE + Rout} \ \frac{\beta}{\beta + 1} = \Delta Vin \ \frac{1}{RE + \dfrac{1}{g_m} \ \dfrac{\beta}{\beta + 1}} \ \frac{\beta}{\beta + 1}$$

$$\Delta IC = \Delta Vin \ \frac{1}{RE + \dfrac{1}{g_m} \ \dfrac{\beta}{\beta + 1}} \ \frac{\beta}{\beta + 1} \tag{16-52}$$

We can multiply by $g_m/g_m = 1$ to get:

$$\Delta IC = \Delta Vin \ \frac{1}{RE + \dfrac{1}{g_m} \ \dfrac{\beta}{\beta + 1}} \ \frac{\beta}{\beta + 1} = \Delta Vin \ \frac{g_m}{g_m RE + \dfrac{\beta}{\beta + 1}} \ \frac{\beta}{\beta + 1}$$

and the new transconductance is:

$$\frac{\Delta IC}{\Delta Vin} = \frac{g_m}{g_m RE + \dfrac{\beta}{\beta + 1}} \ \frac{\beta}{\beta + 1}$$

If $\beta \gg 1$, then:

$$\frac{\Delta IC}{\Delta Vin} = \frac{g_m}{g_m Re + 1}$$

This is a commonly described way of expressing the reduced transconductance due to the finite emitter resistance connected to an AC ground.

However, if we multiply both numerator and denominator by $1/g_m$, we get an expression that is easier to render quick calculations:

$$\frac{\Delta IC}{\Delta Vin} = \frac{1}{RE + \dfrac{1}{g_m}} = g_m'$$

For example, if ICQ = 1 mA, then $1/g_m = 26 \ \Omega$, and if RE = 470 Ω, which leads to a total of 496 Ω in the denominator:

$$\frac{\Delta IC}{\Delta Vin} = \frac{1}{496\Omega} = g_m'$$

The gain of a common-emitter amplifier with a finite resistance from the emitter to an AC ground is:

$$Vout/Vin = -g_m'RL = -\frac{1}{RE + \dfrac{1}{g_m}} \ RL = -\frac{RL}{RE + \dfrac{1}{g_m}} = Vout/Vin$$

We now turn to calculating the input resistance. Derived from Equation (16-52), we had:

$$\frac{\Delta IC}{\Delta Vin} = \frac{1}{RE + \frac{1}{g_m}\frac{\beta}{\beta+1}}\frac{\beta}{\beta+1}$$

The reciprocal of conductance $\Delta IC/\Delta Vin$ is resistance. This means that $\Delta Vin/\Delta IC$ is a resistance but not yet the input resistance that we want:

$$\frac{\Delta Vin}{\Delta IC} = \frac{RE + \frac{1}{g_m}\frac{\beta}{\beta+1}}{1}\frac{\beta+1}{\beta} = [RE + \frac{1}{g_m}\frac{\beta}{\beta+1}]\frac{\beta+1}{\beta}$$

$$\frac{\Delta Vin}{\Delta IC} = [RE + \frac{1}{g_m}\frac{\beta}{\beta+1}]\frac{\beta+1}{\beta}$$

$$(16\text{-}53)$$

To find the AC input resistance, we need to know the ratio of the input voltage to the input base current $\Delta IB = \Delta IC/\beta$. Thus:

$$\frac{\Delta Vin}{\Delta IB} = \frac{\Delta Vin}{\frac{\Delta IC}{\beta}} = \frac{\Delta Vin}{\Delta IC}\beta = \frac{\Delta Vin}{\Delta IB}$$

We can multiply both sides of Equation (16-53) by to get $\Delta Vin/\Delta IB$. Thus:

$$\frac{\Delta Vin}{\Delta IB} = [RE + \frac{1}{g_m}\frac{\beta}{\beta+1}]\frac{\beta+1}{\beta}\beta = [RE + \frac{1}{g_m}\frac{\beta}{\beta+1}][\beta+1]$$

$$\frac{\Delta Vin}{\Delta IB} = Rin = [RE + \frac{1}{g_m}\frac{\beta}{\beta+1}][\beta+1] = (\beta+1)RE + \frac{\beta}{g_m}$$

$$\frac{\Delta Vin}{\Delta IB} = Rin = (\beta+1)RE + \frac{\beta}{g_m}$$

$$(16\text{-}54)$$

In a practical example, if $\beta = 100$ with $RE = 1\ k\Omega$ and $ICQ = 1\ mA$, $g_m \sim 0.038$.

$$Rin = (100+1)(1\ k\Omega) + \frac{100}{0.038} = 101\ k\Omega + 2.6\ k\Omega = 103.6\ k\Omega$$

looking into the base of the transistor. As we can see, the resistor RE used for local negative feedback raises the input resistance by quite a bit, and often it results in the the majority of the input resistance with $(\beta + 1)$ versus the base-emitter resistance β/g_m.

Analyzing an Inverting-Gain Video Amplifier

There is one further point we need to add when analyzing amplifiers. In general, most active amplifying devices have current source outputs. When a load resistor is

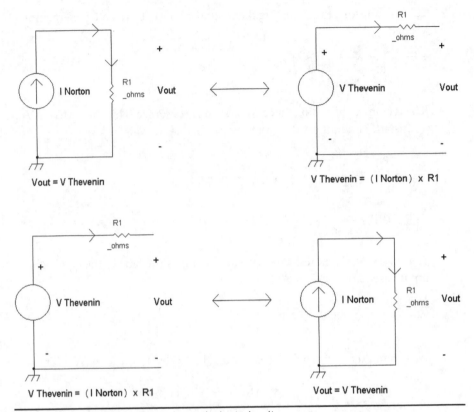

FIGURE 16-9   Norton and Thevenin equivalent circuits.

connected to the collector of a bipolar transistor or to the drain of a FET, generally, the output resistance at the collector or the drain is just the resistance of the load resistor. A current source driving current into a resistor forms a Norton circuit that is also equivalent to a Thevenin circuit. The output resistance in the Norton circuit is the same value as the Thevenin resistor (see Figure 16-9).

Let's now turn to the inverting-gain video amplifier from Chapter 13 (see Figure 16-10). First, we find the DC operating conditions so that we can determine the transconductances for Q2 and Q3. Generally, when the parallel combination of the base biasing resistors (R22 and R23) has a resistance that is less than 20× the 1-kΩ emitter-to-ground resistance of R25, we can assume that there is no DC base current that would pull down the voltage at the base of Q2.

In this case, for R22 = 10 kΩ and R23 = 2.7 kΩ:

$$R22||R23 = 10\ k\Omega||2.7\ k\Omega \approx 2.1\ k\Omega < 20 \times 1\ k\Omega = 20\ k\Omega$$

The voltage at the base of Q2 is then 9 volts [2.7 kΩ/(2.7 kΩ +10 kΩ)] = +1.9 volts DC. With a 0.6 volt turn-on voltage for Q2 and Q3, the emitter DC voltage at Q2 is 1.9 volts – 0.6 volt = 1.3 volts. With the high current gain β =100, $\beta \approx (\beta + 1)$, we can

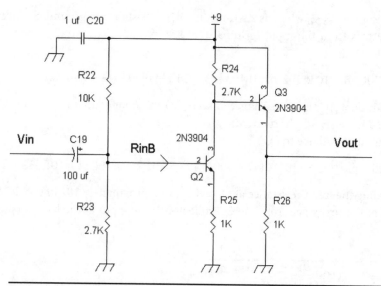

Figure 16-10   Video amplifier circuit Q2 and Q3 (originally Figure 13-21).

say that the emitter current and collector current of Q2 are the same. Likewise, the emitter and collector currents of Q3 are essentially the same.

With 1.3 volts at the emitter of Q2, the emitter current is then 1.3 volts/R25 = 1.3 volts/1 kΩ = 1.3 mA = IE2 = IC2, where IE2 and IC2 are the DC bias emitter and collector currents. Given 1.3 mA of collector current for Q2, the voltage drop across R24 (2.7 kΩ) is 1.3 mA(2.7 kΩ) = 3.5 volts. The Q2 collector voltage is then 9 volts – 3.5 volts = 5.5 volts DC.

Finally, we need to find the collector current of Q3. Because Q2's collector voltage is 5.5 volts, and the collector of Q2 is connected to the base of Q3, the base of Q3 also has 5.5 volts DC. Hence the Q3 emitter voltage is 5.5 volts – 0.6 volt (Q3 turn-on voltage) = 4.9 volts DC. The Q3 emitter current is then 4.9 volts/R26 = 4.9 volts/1 kΩ = 4.9 mA. Because the current gain is 100, which is >> 1, the collector current of Q3 is essentially the same as the emitter current. Thus IC3 = 4.9 mA.

Let's gather what we know so far:

IC2 = 1.3 mA $\rightarrow$ g$_m$Q2 = 1.3 mA/0.026 V = 0.050 mho   (or 50 mA per volt)

IC3 = 4.9 mA $\rightarrow$ g$_m$Q3 = 4.9 mA/0.026 V = 0.188 mho   (or 188 mA per volt)

β = 100

The input resistance RinB into the base of Q2 is:

(β +1)R25 + β/g$_m$Q2 = (101)1 kΩ + 100/0.050 mho = 101 kΩ + 2 kΩ = 103 kΩ

In reality, the input resistance as seen by Vin is a combination of the parallel resistances of R22 and R23 because both resistors go to an AC ground, and RinB is the input resistance to Q2, which is also an equivalent resistance to ground. So there are three resistors from the input terminal that are in equivalent parallel connection. The result is that:

$$R22 || R23 || RinB \approx 10 \text{ k}\Omega || 2.7 \text{ k}\Omega || 103 \text{ k}\Omega \approx 2.12 \text{ k}\Omega || 103 \text{ k}\Omega \approx 2.07 \text{ k}\Omega$$

is the equivalent AC signal resistance as seen by Vin. Input coupling capacitor C19 is considered to be an AC short circuit.

The input resistance to Q3 is:

$$(\beta +1)R26 + \beta/g_m Q3 = (101)1 \text{ k}\Omega + 100/0.188 = 101 \text{ k}\Omega + 532 \ \Omega = 101.532 \text{ k}\Omega$$

Disregarding the input resistance at Q3, which is approximately 102 k$\Omega \gg$ R24, the AC gain of the first stage from the Q2 collector to the input at the base is approximately:

$$-\frac{R24}{R25 + \dfrac{1}{g_m^2}} = -\frac{2.7 \text{ k}\Omega}{1 \text{ k}\Omega + \dfrac{1}{0.050}} = -\frac{2.7 \text{ k}\Omega}{1 \text{ k}\Omega + 20 \ \Omega} \sim -2.65$$

If the input resistance into Q3 is taken into account, the more accurate gain calculation is:

$$-\frac{R24 || 102 \text{ k}\Omega}{R25 + \dfrac{1}{g_m^2}} = -\frac{2.7 \text{ k}\Omega || 102 \text{ k}\Omega}{1 \text{ k}\Omega + \dfrac{1}{0.050}} = -\frac{2.63 \text{ k}\Omega}{1 \text{ k}\Omega + 20 \ \Omega} \sim -2.58$$

In both cases, they are very close. For a first approximation, we will just use approximately –2.65 for the voltage gain of Q2.

The output resistance RoutQ3 looking into the emitter of Q3 is approximately:

$$[R24/(\beta +1)] + (1/g_m Q3) \approx 2.7 \text{ k}\Omega/101 + 1/0.188 \text{ mho} \approx 26.7 \ \Omega + 5.3 \ \Omega \approx 32 \ \Omega = \text{RoutQ3}$$

The model for Q2 and Q3 is shown in the block diagram in Figure 16-11.

It should be noted that the hand calculations are usually sufficient to get a feel for the DC biasing currents and voltages along with the AC gain of the amplifier. For more accurate answers, one can calculate further with more exacting models, but at

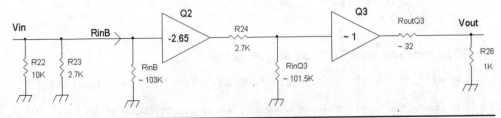

FIGURE 16-11   An approximate representation of the amplifier shown in Figure 16-10.

that point, it is probably easier to download a free circuit simulation software pack such as LTSPICE by Linear Technology at www.linear.com/designtools/software/.

If base currents are sufficiently significant as to affect the operating point of a previous resistive network such as the base current in Q3 pulling down the collector voltage of Q2, then the calculations become a lot more involved, and there are a few options one can take. One is to design the previous stage's resistance values low enough so that the base current of Q3 is insignificant, which is preferable, or the second option is to go through the longer calculation for the worse-case current gain and determine whether the amplifier will still work within expectations. The third is to build or simulate the circuit. When I designed the inverting video amplifier in Chapter 13, I deliberately chose smaller-value resistors for biasing (R22 and R23) and loading (R24) such that I did not have to worry about base currents affecting the bias point significantly.

Improvements or Variations of Previous Projects

Sometimes one can improve existing projects by replacing one part with a completely different part or by adding a few parts. We will start with an improvement of the superheterodyne radio that used a TA7642 chip (see Figure 16-12).

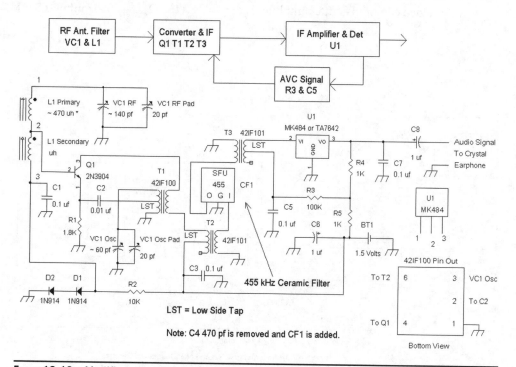

FIGURE 16-12 Modified circuit by replacing the 470-pf capacitor originally between the two IF transformers with ceramic filter CF1 (e.g., SFU455KU2-B0).

By adding almost any ceramic filter as shown, the selectivity of the AM radio goes up. A 455-kHz ceramic filter (e.g., Murata Part Number SFULA455KU2A-B0, which is available at www.mouser.com, or, alternatively, the SFULA455KU2B-B0) generally is equivalent to adding one or two more intermediate-frequency (IF) transformers. With this modification from the Bay Area, the radio was able to receive radio station KFI from Los Angeles and adjacent channel interference was were very well filtered out. Selectivity was noticeably improved, even though the SFULA455KU2A-B0 ceramic filter has a "normal" bandwidth of 10 kHz. Other 455-kHz ceramic filters are also available, and some have multiple sections for sharper selectivity. These can be tried as well.

Another modification that is relatively easy is to add gain in the IF system for an FM radio to increase its sensitivity. In the FM tuner using the TA2003 integrated circuit (IC), one improvement was to connect two 10.7-MHz ceramic filters in series for increased selectivity. We can add a common-emitter amplifier (R5, R6, Q1, R3, R4, and C10) in between the two filters, as shown in Figure 16-13, to make up for losses in the ceramic filters and thus provide a bit more sensitivity. Note that an extra supply bypass capacitor C11 is added to this amplifier supply line.

The gain of the amplifier assumes a 330-Ω input resistance into the second ceramic filter F2, so the effective collector load to Q1 is 330 Ω||330 Ω, or 165 Ω. The DC collector current is about 0.8 mA $\rightarrow$ g_m = 0.8 mA/0.026 V = 0.0307 mho. The gain of the added common emitter amplifier is then = $-g_m(165\ \Omega)$ = -0.0307 mho$(165\ \Omega)$ ≈ -5.

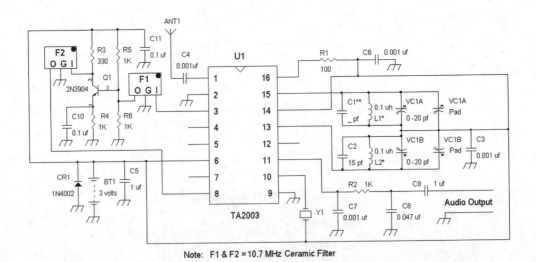

Note: F1 & F2 = 10.7 MHz Ceramic Filter

Y1 = 10.7 MHz Ceramic Resonator & Quadrature Detector

Figure 16-13   A one-transistor amplifier is placed between F1 and F2 for improved sensitivity.

Just for Fun ... A Wein Oscillator Using a 60-Watt Light Bulb

In Chapter 8, various types of low-voltage lamps were used as a nonlinear variable resistor to allow the Wein oscillator to achieve a steady-state-amplitude sine-wave signal without clipping. I later tried larger 120 volt bulbs starting from an 11-watt bulb to as much as 60 watts, and the success rate was not as good as using the lower-voltage bulbs. What happened was that the amplitude was not stable and tended to "breathe." It occurred to me that in some of the Wein oscillators, such as the HP 200, the lamp is connected between the cathode of the first tube and ground. The tube is DC biased for a particular cathode current, resulting in a slight DC current flowing through the lamp as well as (having) an AC signal across it. So, to improve the Wein circuit such that it worked with higher-voltage and higher-wattage lamps, I decided to add a prebias circuit (R6A to R6C) as shown in Figure 16-14 and the completed project in Figure 16-15.

The circuit works well enough to provide a reasonable sine-wave signal, and the prebiasing resistors R6A to R6C reduce or eliminate low-frequency amplitude modulation on the output signal. Note that a regulated 24 volt supply is recommended, such as using a raw 30 volt DC supply and connecting it to a 24 volt regulator chip such as the 7824 (TO-220 package). Biasing resistors R6A, R6B, and R6C are ¼-watt resistors and can be replaced with a single 1-watt resistor. Gain-control potentiometer VR1 may be 100 Ω when used with a standard 60-watt soft white light bulb. Some other 60-watt lamps may have higher resistances and VR1 → 200 Ω. However, because of the amount of power that is dissipated by VR1, it is recommended that at least a ½-watt or 1-watt variable resistor is used. Other types of bulbs may be used; see Table 16-1 for a summary. Note that R6(total R) = (R6A + R6B + R6C).

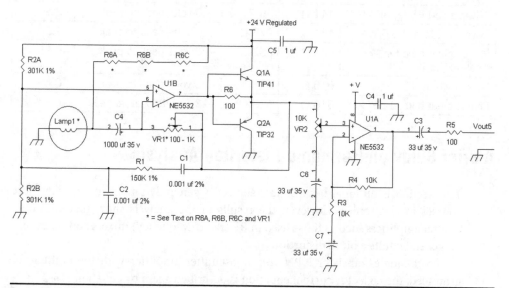

FIGURE 16-14   Wein oscillator with added DC current to provide better amplitude stability.

Figure 16-15 The 60-watt light bulb Wein oscillator with a Feit old style 60-W lamp installed in the socket with Philips 25-W tube and 11-W clear bulbs behind.

Table 16-1 Summary of Various Lamps Used for the Wein Oscillator

Bulb (W = Watts)	R6(Total R)	VR1	Vout at Q1A, Q2A Emitters	Distortion
Feit old style, 60 W	513 Ω	96 Ω	~20 volts peak to peak	~ 2%
Sylvania soft white, 60 W	441 Ω	56 Ω	~20 volts peak to peak	~3%
Sylvania standard, 25 W	454 Ω	275 Ω	~20 volts peak to peak	~1.2%
Philips clear tube for displays, 25 W	545 Ω	282 Ω	~20 volts peak to peak	~0.95%
GE soft white, 15 W	616 Ω	468 Ω	~20 volts peak to peak	~0.6%
Philips clear bulb, 11 W	616 Ω	610 Ω	~13.5 volts peak to peak	~0.6%

A Brief Study on Harmonic Distortion Analysis

This section is optional because this subject is generally taught in very advanced courses in electronics. However, the results of the analysis show that common-emitter amplifiers such as those used in RF (radio-frequency) mixers and fuzz boxes do generate quite a bit of distortion.

For a single-transistor NPN common-emitter amplifier with the emitter AC grounded, the collector current equation is described again by equation (16-27):

$IC_{npn} = ICQ\ e^{(Vbe/V_T)}$ (16-27)

The $e^{(Vbe/V_T)}$ function can be more characterized beyond a linear approximation as highlighted in boldface in the following:

$e^{(Vbe/V_T)} = 1 + a_1(Vbe/V_T) + \mathbf{a_2(Vbe/V_T)^2 + a_3(Vbe/V_T)^3} + \dots + \mathbf{a_n(Vbe/V_T)^n}$ (16-55)

where $a_1 = 1, \mathbf{a_2 = \frac{1}{2}, a_3 = \frac{1}{6}}, \dots, \mathbf{a_n = \frac{1}{n!}}$

NOTE $n! = (n)(n-1)(n-2)$ L (1). For example, $5! = (5)(4)(3)(2)(1) = 120$, and $3! = (3)(2)(1) = 6$.

If we are just interested in second- and third-order distortions from Equation (16-27), we plug in $a_1 = 1$, $\mathbf{a_2 = \frac{1}{2}}$, and $\mathbf{a_3 = \frac{1}{6}}$ into Equation (16-55) and combine it with Equation (16-27) to get:

$IC_{npn} = ICQ[1 + (Vbe/V_T) + \mathbf{(1/2)(Vbe/V_T)^2 + (1/6)(Vbe/V_T)^3}]$ (16-56)

For a sinusoidal input signal, we let the AC signal $= \mathbf{B\ cos(\omega t)} = Vbe$, and Equation (16-56) becomes:

$IC_{npn} = ICQ[1 + (B\ cos(\omega t)/V_T) + \mathbf{(1/2)(B\ cos(\omega t)/V_T)^2 + (1/6)(B\ cos(\omega t)/V_T)^3}]$ (16-57)

The harmonic distortion is the ratio of the amplitudes of the harmonic signal to the linear term signal $(B\ cos(\omega t)/V_T)$ that has the fundamental frequency, ω.

For the second harmonic term, let's take a look at:

$(1/2)(B\ cos(\omega t)/V_T)^2 = 0.5\ (B\ cos(\omega t))^2(1/V_T)^2$

and:

$(B\ cos(\omega t))^2 = (0.5)B^2[cos(2\omega t) + 1](1/V_T)^2$

Note that squaring the input signal results in a second harmonic and a DC term. In particular, we get a $(0.5)B^2\ cos(2\omega t)$ second-harmonic term. We can now say that:

$(1/2)[Bcos(\omega t)/V_T]^2 = 0.5(0.5)B^2[cos(2\omega t) + 1](1/V_T)^2$

The ratio of the second-harmonic signal to the fundamental-frequency signal is:

$0.5(0.5)B^2[cos(2\omega t)](1/V_T)^2/[B\ cos(\omega t)/V_T]$

but we are only interested in the amplitude $0.5(0.5)B^2(1/V_T)^2$ of the second-harmonic signal and the amplitude (B/V_T) of the linear term to determine the distortion. The second-harmonic distortion HD_2 is then:

$HD_2 = 0.5(0.5)B^2(1/V_T)^2/(B/V_T) = 0.25(B/V_T)$

$HD_2 = 0.25(B/V_T) = (\frac{1}{4})(B/V_T)$ (16-58)

Because $V_T \approx 0.026$ volt:

$$HD_2 \approx B/4(0.026\ V) \approx B/0.104\ V \approx B/100\ mV \tag{16-59}$$

What this says is that if the peak sinusoidal voltage B is even as small as 1 mV, then the second-order harmonic distortion is approximately $1/100 \approx 1$ percent. Equation (16-59) also says that HD_2 is directly proportional to the amplitude of the input signal, which equates to 1 percent per 1 mV peak signal. For example, if the input signal is 3 mV peak, we would expect about 3 percent second-harmonic distortion.

The third-harmonic analysis gets a bit more messy, so I will refer to Robert G. Meyer's EE140 and 240 notes from 1975 and 1976. Thus:

$$HD_3 = (1/24)(B/V_T)^2 \approx (1/24)(B/0.026\ V)^2 \tag{16-60}$$

Note that HD_3 is proportional to the square of the input voltage. A sine wave at B = 1 mV peak, $HD_3 \approx (1/24)(0.001/0.026)^2 = 0.00616$ percent. However, if B → 10 mV peak, $HD_3 \to 0.616$ percent, which is a dramatic increase.

In summary, for second- and third-order distortions we have:

$$HD_2 \approx \mathbf{B}/0.104\ V \tag{16-61}$$

$$HD_3 \approx (1/24)(B/0.026\ V)^2 = \mathbf{B^2}(1/24)(1/0.026\ V)^2 \tag{16-62}$$

where **B** is the peak input amplitude for a sinusoidal signal, and the base of the transistor is driven by a pure voltage source—a situation where there is no series impedance between the input voltage signal source and base.

From either Equation (16-61) or Equation (16-62), it would appear that if the input amplitude exceeds 1,000 mV peak, the amplitude of the second-harmonic distortion signal will be larger than the input signal by about 10×, and the third-harmonic signal will be larger than the input signal by approximately 61×. Is this possible? The answer is no because at some point with a large amplitude for Vbe, the polynomial approximation for $e^{(Vbe/V_T)}$ blows up and is no longer valid.

In fact, if a large-amplitude sinusoidal signal (e.g., 1,000 mV peak) is driving the base-emitter junction of a transistor, the collector output current will approach a narrow periodic pulse having the fundamental frequency. This means that the amplitudes of the fundamental frequency and harmonics will tend to be equal.

For example, in calculating second-order distortion with a polynomial, the inaccuracies begin when the peak amplitude of the input sine wave approaches one V_T or 26 mV peak for approximately 26 percent HD_2. Beyond 26 mV peak, a more accurate assessment of the second-order distortion can be achieved by using modified Bessel functions instead of the power–series/polynomial method.

Using Differential Gain as a Method to Determine Harmonic Distortion

Based on a general power-series expansion equation, Professor Robert G. Meyer derived formulas to calculate distortion products based on differential gains. The basis is that there is a relationship between variations in gain during the sinusoidal output swing. Using this method allows us not only to calculate distortions for bipolar transistors with emitters that are AC grounded, but it also works well in determining distortions in amplifiers with negative feedback.

It should be noted that any distortion analysis based on power-series expansion requires that the distortion be small and the input-output characteristic have a smooth curve without any abrupt jumps in value. For example, when the amplifier clips or exhibits crossover (aka deadband) distortion, this method is not applicable.

When we look at transconductances of the following:

$$\frac{\Delta IC}{\Delta VBE} = g_m = \frac{ICQ}{V_T}$$

for an AC-grounded emitter amplifier and:

$$\frac{\Delta IC}{\Delta Vin} = \frac{g_m}{g_m RE + 1} = g_m{}'$$

for a common-emitter amplifier with a finite resistance from emitter to ground, we see that the transconductances are a function of the DC quiescent collector current ICQ. As the signal swings above and below ICQ, the transconductance changes. In the case of:

$$\frac{\Delta IC}{\Delta VBE} = g_m = \frac{ICQ}{V_T}$$

the transconductane $\Delta IC/\Delta VBE$ is directly proportional to the collector current. With a form of negative feedback, the other case of:

$$\frac{\Delta IC}{\Delta Vin} = \frac{g_m}{g_m RE + 1} = g_m{}' = \frac{\dfrac{ICQ}{V_T}}{\dfrac{ICQ}{V_T} RE + 1}$$

the transconductance $\Delta IC/\Delta Vin$ is not as directly related to ICQ.

So here are paraphrased formulas based on Professor Meyer's EE140 notes:

$$HD_2 = (\tfrac{1}{8})|(A_{positive\ swing} - A_{negative\ swing})/(A_{no\ signal})|$$

$$HD_3 = (\tfrac{1}{24})|(A_{positive\ swing} + A_{negative\ swing} - 2A_{no\ signal})/(A_{no\ signal})|$$

where $A_{positive\ swing}$, $A_{negative\ swing}$, and $A_{no\ signal}$ may represent the gain in terms of current, voltage, or transconductance.

The really cool thing about using the differential-gain method is that it analyzes distortions in amplifiers where the current gain β is varying as a function of collector current. Also, this method is applicable to determining distortion in multiple-stage amplifiers with or without feedback, push-pull amplifiers including differential pairs, FET devices that exhibit the body effect, and so on.

Let's again take a common-emitter amplifier with its emitter AC grounded and test for second-order harmonics based on an AC signal of 1 mV peak. Thus:

$$IC_{npn} = ICQ \, e^{(Vbe/V_T)} \tag{16-27}$$

We need to calculate three collector currents, when Vbe = –1 mV, 0 mV, and +1 mV. The ICQ term will eventually cancel out and is independent of distortion in this case because there is no negative feedback. $V_T \approx 0.026$ V and $g_m = ICQ/V_T$. Thus:

$$HD_2 = (\tfrac{1}{8})|(g_{m(+)} - g_{m(-)})/(g_m)|$$

Since g_m is directly proportional to ICQ and HD_2 (including HD_3), it is the ratio of tranconductances, HD_2 (and HD_3) can be equivalently expressed as a function of ICQ. This leads to:

$$HD_2 = (\tfrac{1}{8})|(ICQ \, e^{(0.001/V_T)} - ICQ \, e^{(-0.001/V_T)})/(ICQ \, e^{(-0.000/V_T)})|$$

$$HD_2 = (\tfrac{1}{8})|(1.039211 - 0.962269)/(1)|$$

$$HD_2 \approx (\tfrac{1}{8})(0.077) = 0.96\%$$

From Equation (16-61), $HD_2 \approx B/0.104$ V, with B = 1 mV peak; thus $HD_2 \approx 0.001/0.104 \approx 0.96\%$. So there is a match between using the differential gain and power-series analysis.

Using differential-gain analysis again for calculating HD_3 with 1 mV peak signal, we can take the calculations we already have and change the sign from subtraction to addition, but do not forget to add a minus 2 times the zero signal gain, $-2ICQ \, e^{(-0.000/V_T)}$, to the numerator and change the $\frac{1}{8}$ factor to $\frac{1}{24}$ to get:

$$HD_3 = (\tfrac{1}{24})|(ICQ \, e^{(0.001/V_T)} + ICQ \, e^{(-0.001/V_T)} - 2ICQ \, e^{(-0.001/V_T)})/(ICQ \, e^{(-0.000/V_T)})|$$

$$HD_3 = (\tfrac{1}{24})|(1.039 + 0.962 - 2.00)| \approx 0.001479/24 \approx 0.00616\%$$

From the power-series equation (16-62), $HD_3 = 0.00616\%$ for a 1 mV peak sine-wave signal, which is in agreement with the differential-gain method.

For an amplifier with feedback, one way to use the differential-gain method is to determine the output swing, which is still sinusoidal in nature, and determine the changes in collector current. For example, a common-emitter amplifier with an unbypassed emitter resistor that has a collector current swing of ±10 percent ICQ will be set up in the following manner for calculating second-order distortion. Note that ICQ does not cancel out and does play a role in the amount of distortion (e.g., the higher the ICQ, the lower is the distortion). Thus:

$$HD_2 = |(A_{\text{positive swing}} - A_{\text{negative swing}})/(A_{\text{no signal}})| =$$

$$(\tfrac{1}{8})[\frac{\dfrac{1.1ICQ}{V_T}}{\dfrac{1.1ICQ}{V_T}RE + 1} - \frac{\dfrac{0.9ICQ}{V_T}}{\dfrac{0.9ICQ}{V_T}RE + 1}]/[\frac{\dfrac{ICQ}{V_T}}{\dfrac{ICQ}{V_T}RE + 1}]$$

From feedback equations via the Professor's Meyer's EE140 notes, the same common-emitter amplifier has the following characteristics of second-order distortion as a function of the input signal:

$$HD_2 = (\tfrac{1}{4})(B/V_T)/(1 + g_m RE)^2$$

But note that as the distortion is reduced by $(1 + g_m RE)^2$, the gain is reduced by **$(1 + g_m RE)$** via:

$$\frac{\Delta IC}{\Delta Vin} = \frac{g_m}{(g_m RE + 1)}$$

References

1. Robert G. Meyer, "EE140 Notes," University of California, Berkeley, CA, 1975.
2. Robert G. Meyer, "EE240 Notes," University of California, Berkeley, CA, 1976.
3. Allan R. Hambley, *Electrical Engineering: Principles and Applications, 2nd ed.* Upper Saddle River, NJ: Prentice-Hall, 2002.
4. Paul R. Gray and Robert G. Meyer, *Analysis and Design of Analog Integrated Circuits, 2nd ed.* New York: Wiley, 1984.

CHAPTER 17
Hacking, Inventing, and Designing

In the context of this book, the term *hacking* will be defined as modifications or investigations of hardware circuits and systems. For example, an electronic device is taken apart for investigation. The parts revealed within the device can save time in finding the right components for making other types of circuits. Throughout the book so far we have shown examples of hacks, such as taking apart a AA cell to make a flashlight or using a power rectifier (1N5400 series) as a varactor tuning diode. In this chapter, we will look at tearing apart a battery tester and modifying it for something else.

We will also touch on inventing and designing. The term *invention* includes a novelty of ideas. These ideas can be an improvement of existing ideas, or they can be wholly new. Also, ideas that lead to unexpected results are generally patentable. While showing some examples of inventions, we will also show the design aspects of such inventions.

Hacking

Let's start with a battery tester bought at Harbor Freight for about $3 (Figure 17-1). The mystery is how does the light-emitting diode (LED) that requires 1.7 volts to light up work when this device tests NiMh batteries that provide only 1.25 volts?

It looks like we will have to break open the battery tester to find out how this device works (Figures 17-2a and 17-2b). When the voltage is measured across C2 (see Figure 17-4a), whose positive terminal is connected to the cathode of diode D1, we get about 4 volts. The 33-µH inductor L1 provides an approximately 160 kHz pulse waveform to the anode of D1. We can use this battery tester as a DC-to-DC step-up converter. Although we can just obtain a switching regulator integrated circuit (IC), coil, and Schottky diode, by the time we purchase all these components, we would have spent more than $3 in time and transportation.

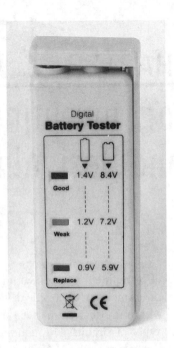

FIGURE 17-1   Three-LED battery tester for 1.5 volt and 9 volt batteries.

Because this device tests batteries, there are load resistors connected across a crude voltmeter that has three voltage ranges to denote "Replace," "Weak," and "Good" conditions of the tested battery. We will have to locate R1 and R4 marked "120" (12 Ω) connected to the 1.5 volt terminals of the battery tester. Once we remove them, we can tap out from C2 to provide approximately 3.5 volts to 4 volts direct current (DC) (see Figure 17-3).

The supply of approximately 3.5 volts DC is handy to power a white LED that normally turns on at about 3 volts. We should add a 47-Ω series resistor for current limiting to the white LED. Also with approximately 3.5 volts, we can power op amp and amplifier circuits. Low-voltage op amps such as the LM324 (quad), LM358 (dual), and the faster FET op amps TLC274 (quad) and TLC272 (dual) will work fine at about 3.5 volts. Other op amps, such as the single op amp rail-to-rail versions LMV118 or the Max4321, can be used.

But what if we need to boost the voltage further? We can use our friend the DC restoration circuit and double the original voltage Vout_1 = ~3.5 volts to about 7.5 volts DC for Vout_2 (see Figure 17-4b).

Of course, if we want more current, we can hack the LED flashlight in Figure 2-33 (Figure 17-5a). In this hack, we could use the printed circuit board or we can trace down which IC was used. The switching supply IC is a BL8530 with surface-mount connections (SOT-89). From the data sheet, we can build our own 1.5 volt to approximately 4.5 volt power supply (Figure 17-5b).

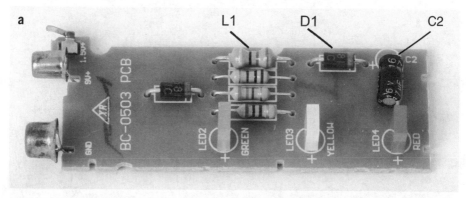

FIGURE 17-2 (*a*) Top side of the battery tester's circuit board showing a 33-µH inductor L1 and Schottky rectifier D1. (*b*) Bottom side of the device showing a multitude of surface-mount components including transistors, diodes, resistors, and an LM324 quad op amp U1.

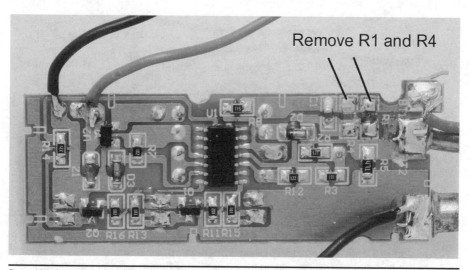

FIGURE 17-3 Surface-mount 12-Ω resistors R2 and R4 removed and two wires (*top left of board*) brought out to provide approximately 3.5 volts at approximately 30 mA.

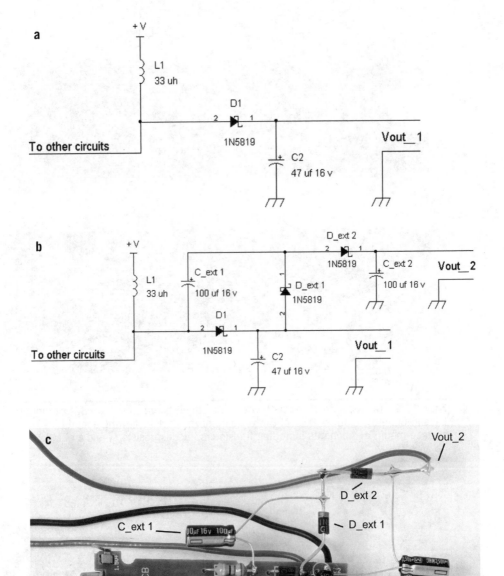

a

+ V

L1
33 uh

D1
1N5819

C2
47 uf 16 v

To other circuits

Vout__1

b

D_ext 2
1N5819

+ V

L1
33 uh

C_ext 1
100 uf 16 v

D_ext 1
1N5819

C_ext 2
100 uf 16 v

Vout__2

D1
1N5819

C2
47 uf 16 v

To other circuits

Vout__1

c

Vout_2

D_ext 2

C_ext 1

D_ext 1

C_ext 2

Figure 17-4 (*a*) Partial schematic of the battery tester's original circuit. (*b*) Partial schematic of the battery tester's switching converter that is converted for double voltage by adding external components D_ext 1, D_ext 2, C_ext1, and C_ext 2. (*c*) Modifications to the battery tester to provide about 7.5 volts DC from a 1.5 volt battery.

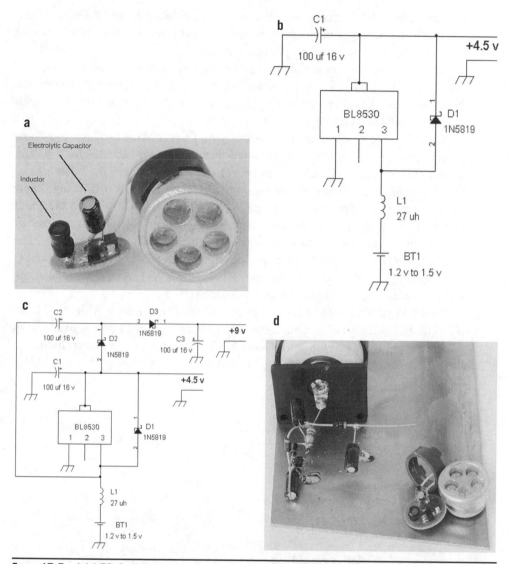

Figure 17-5 (*a*) LED flashlight taken apart to reveal a BL8530 IC. (*b*) Switching power-supply schematic. (*c*) Schematic of the voltage doubler circuit with the BL8530 chip. (*d*) Prototype circuit of the BL8530 1.5 volt to about 9 volt converter (voltage doubler) and the original LED flashlight taken apart.

Again, can we boost the voltage? If we use a DC restoration circuit that is part of a voltage doubler circuit, we should be able to boost the voltage (Figure 17-5*c* and *d*).

We now turn our attention to a couple of AM/FM radios. The first radio to examine is the RadioShack Catalog Number 1200586 radio, and we will take it apart (Figure 17-6*a*). There are only two Phillips screws on the back that are removed. One is seen on the back of the case above the "RadioShack" lettering, and the other is hidden inside the battery compartment. Remove the batteries, and the screw is located in the middle. Then we carefully pry open the case.

The first thing we should notice is just how few parts there are on the top side of the board. What's more intriguing is that there are two potentiometers, one for the audio level and the other for tuning. Also, notice on the top that there is a watch crystal (32.768 kHz) (see Figure 17-6b).

Let's now take a look at the other side (Figure 17-6c). While I will not show how to hack into this radio, it should be noted that the Silicon Labs 4820A10 IC can receive other bands, such as SW (shortwave) and FM (frequency modulation) stations below 88 MHz. What's interesting about this investigation is that we now know a part number for a dual low-voltage audio power amplifier IC (the TDA2822) and that this radio, built in 2014, uses no variable capacitors and no coils that are fixed or variable. In fact, this radio is based on software-defined radio (SDR) technology. The Silicon Lab IC has two input ADCs (analog to digital converters) to process I and Q radio signals. The intermediate-frequency (IF) filtering, AM and FM demodulation, and so on are done with digital signal processing (DSP).

Now let's take a look at a radio that we can modify, the Sony AM/FM Model ICF-S10MK2 pocket radio. This radio was produced in 2014 and is available at Walgreens and www.amazon.com (Figure 17-7a). There is only one screw hidden in the battery compartment to remove. Take out the lower AA battery from its holder

FIGURE 17-6 (a) Front side of the RadioShack radio.

Figure 17-6 (*b*) Front-side circuit board with two potentiometers and a cylindrical watch crystal. (*c*) Two surface-mount ICs, a 24-pin Silicon Labs 4820A10 AM/FM tuner chip and an 8-pin TDA2822 two-channel audio amplifier.

on the back, and remove the screw. Carefully pry apart the case, and note that the FM antenna's wire may get in the way and have to be temporarily desoldered from the circuit board. Also, the AM ferrite-bar antenna may be glued lightly to the case and can be separated so that the circuit board is free. Also, the two battery wires soldered to the circuit board will have to be temporarily desoldered.

Figure 17-7*b* shows the circuit board with two three-pin ceramic filters, one for AM (455 kHz) and the other for FM (10.7 MHz). The FM ceramic filter is located just above the Sony IC (integrated circuit). For this modification, we would like to improve the selectivity so that more stations (e.g., in between) can be tuned. We will replace the 10.7 MHz filter with a 110 kHz bandwidth filter, the Murata SFELF 10M7KAH0-B0, or the 180 kHz bandwidth filter, the SFELF10M7HA00. These filters are direct pin-to-pin replacements, and note the orientation of the filter in that the front side of the filter that has its printing is facing out toward the edge of the circuit board.

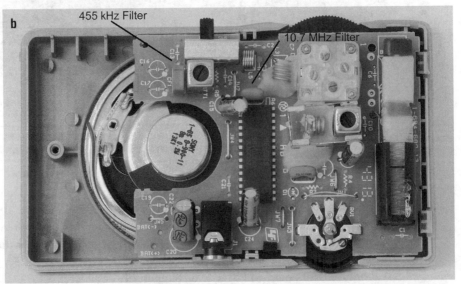

FIGURE 17-7   (*a*) Sony AM/FM Model ICF-S10MK2 radio. (*b*) FM ceramic filter that can be upgraded for higher selectivity.

Typically, most AM/FM radios have 10.7 MHz FM ceramic filters of wide bandwidth such as 230 kHz or 280 kHz. We can replace them with a 180 kHz or even a 110 kHz bandwidth filter. Ideally, if you can get a 150 kHz bandwidth filter, it will work optimally in that it can handle the full frequency deviation of the FM signal without substantial distortion while achieving good selectivity. In practice, though, satisfactory results were achieved with the 110 kHz bandwidth filter in terms of the demodulated signal with an acceptable level of audio distortion.

In terms of improving the AM band selectivity, there are not too many choices today. Most available 455 kHz filters that are a direct pin-to-pin replacement have the same 10 kHz bandwidth as the one that already comes with the radio. One can try multielement 455 kHz filters, but just remember that the input pin to the 455 kHz filter is closest to the band switch or to the unloaded capacitor C13.

Inventions and Patents

When one thinks about invention, Thomas Edison's light bulb comes into mind. Before the advent of incandescent electric lighting, there were gas street lamps and electric arc lights. So what constitutes an invention?

First, there are three types of invention:

- Utility
- Design
- Plant

Most people are more familiar with the *utility* patent, such as Thomas Edison's light bulb or George Westinghouse's air brake for trains. Almost all high-technology developments these days from circuits, to semiconductor devices (e.g., new types of transistors such as the FinFet), to computers, to software, and so on fall into the category of the utility patent. *Design* patent examples include distinctive industrial designs of products such as audio cables. For example, the company Monster Cable has many design patents on their various speaker cables. A *plant* patent example includes plants that are reproduced without genetic seeds via grafts or cuttings. In this chapter, we will be discussing utility patents.

There are various ways of coming up with an invention. One is solving a problem that no one else has been able to solve. For example, the safety pin solves a problem for clipping two pieces of clothing together. Another and just as common way to invent is to use something in one field to solve a problem in another field. For example, one can use ideas from audio circuits to solve problems in the bar-code reader field. Specifically, U.S. Patent 4,740,475 shows an invention for a digital bar-code reader for cards imprinted with bar codes. The cards are swiped through the bar-code reader. However, because some people may move the card through the reader at a slower than normal rate, the existing bar-code reader circuit may have insufficient signal. By using a trick found in a phono preamp RIAA equalization curve or in an analog tape recorder NAB equalization curve, the problem is solved. A bass boost equalization curve is added to the gain function of the amplifier that results in the capability to read bar codes at a speed 4× slower than the original circuit.

And yet another way to come up with an invention is to produce an unexpected result. A very well-known attorney once told me this hypothetical example: if someone finds out that aspirin can be used to make chickens grow faster, then that idea is patentable, the reason being that aspirin is commonly used as a pain reliever. Therefore, the concept of *new use* resulting in an unexpected consequence is patentable. The unexpected consequence of this hypothetical example is that the aspirin can be used to make chickens grow faster.

Another way of coming up with a patentable idea is to improve an existing invention. However, if the existing invention's patent is still enforceable, the original owner of that patent may have to be paid or you may have to be given permission to practice the improved patent.

We will examine three patent examples that will include the design philosophies behind them:

- A circuit using a ceramic filter as opposed to a ceramic resonator in a wide-frequency-deviation voltage-controlled oscillator
- An electronic voltage-controlled variable capacitor without using varactor diodes
- A Class AB output-stage amplifier circuit that uses no thermal feedback diodes or transistors to compensate for the quiescent current of the output transistors

Ceramic Filter Oscillator for Wider Frequency Deviation

Generally, oscillators are designed for a fixed frequency or for a variable range of frequencies. However, what if we want to make a variable-frequency oscillator that will always power up within a known range of frequencies? The problem with an inductor-capacitor oscillator is that the inductor and capacitor need to be chosen with tight tolerances. For example, if we want to build a 455 kHz variable-frequency oscillator, to ensure that the nominal frequency is within 5 kHz, both inductor and capacitor must be within about 1 percent tolerance. While 1 percent tolerance capacitors are available, 1 percent tolerance inductors are generally not common because they come in 5 percent varieties.

If we chose a ceramic resonator or a crystal, we will get very accurate frequencies. However, it is very difficult, if not impossible, to move a crystal oscillator's frequency beyond a 0.3 percent peak-to-peak range. Ceramic resonators allow for a wider range than crystals in terms of frequency deviation in a variable-frequency oscillator, but what if we want a wider range such as on the order of 0.5 percent peak to peak or more?

Ceramic filters are made to replace IF transformers that provide an IF bandwidth filter. They are not intended to be used as part of an oscillator circuit. In U.S. Patent 5,229,535, wider-frequency-deviation oscillator techniques are created via multiple crystals and ceramic filters (see Figures 17-8 and 17-9). Figure 17-8 shows multiple crystals tuned at slightly different frequencies to form a composite crystal band-pass filter that has a wider bandwidth than a single crystal. This approach achieved the desire result. However, it was desired to design the oscillator with a single element to replace custom-made crystals cut at slightly different frequencies. The idea of a ceramic filter became a possible solution because of its common availability (see Figure 17-9).

A basic premise for using a simple ceramic filter (element 141) is that its phase and amplitude responses are similar to those of a single LC (inductor-capacitor) band-pass filter. What we would like from the ceramic filter is a phase shift on the order of about +90 degrees below the center frequency and –90 degrees above the center frequency. If a multiple-section ceramic filter is used, there may be excessive phase shift, such as ±360 degrees, when the amplitude response is not attenuated, which will lead to oscillation at multiple frequencies. Figure 17-9 shows two inverting amplifiers, 142 and 143, which provide an amplifier with a 0 degree phase shift and voltage gain >1. There are two phase shifters in cascade, 144 and 145, that allow for about double the range of a single phase-shifting circuit. This then provides an increased ability to vary the oscillation frequency.

We now will take a look at practical implementations of ceramic filter oscillators. The first one is shown in Figure 17-10 as the top-level block diagram. An A-to-B dissolve or fader amplifier allows us to continuously mix from 100 percent negative phase signals to 100 percent positive phase signals via the control signal Vcontrol at the base of Q1. With a 50 percent mix, there are negative and positive phase signals that combine to have net effect of 0 degrees phase of shift at the output. Control voltage Vcontrol allows the output of the fader amplifier to provide a signal with adjustable negative and positive phase shifts, including a 0 degree phase shift. The output of the fader amplifier is connected to the input of the ceramic filter. At the output of the ceramic filter, a unity-gain amplifier sends a signal to the lag and lead networks. For the actual implementation, the lag network is a series inductor forming a low-pass filter with the input resistance of the A input. Positive phase via a lead network is provided by a small-value series capacitor to provide a high-pass filter with the B input. An output of the oscillator may be taken from the output of the unity-gain amplifier Q4.

Now let's look at the circuit implementation shown in Figure 17-11. The control voltage to the base of Q1 is provided at the slider of potentiometer VR1. By varying the base voltage of Q1, a mixture of lagging and leading signals via L2 and C4 is

US005229735A

United States Patent [19]

Quan

[11]   Patent Number: **5,229,735**

[45]   Date of Patent: **Jul. 20, 1993**

[54]   **WIDE FREQUENCY DEVIATION VOLTAGE CONTROLLED CRYSTAL OSCILLATOR HAVING PLURAL PARALLEL CRYSTALS**

[75]   Inventor: **Ronald Quan**, Cupertino, Calif.

[73]   Assignee: **Macrovision Corporation**, Mountain View, Calif.

[21]   Appl. No.:   **860,643**

[22]   Filed: **Mar. 30, 1992**

[51]   Int. Cl.^{5} H03B 5/36; H03H 9/00; H04N 9/455

[52]   U.S. Cl. .... 331/116 R; 358/19; 331/162; 331/177 V; 333/189

[58]   Field of Search .... 331/162, 116 R, 116 FE, 331/177 V; 358/19, 20; 333/187, 188, 189, 190, 191, 192

[56]   **References Cited**

U.S. PATENT DOCUMENTS

3,170,120	2/1965	Jensen et al.	333/188 X
3,358,244	12/1967	Ho et al.	331/162 X
4,063,194	12/1977	Helle	331/116 R X
4,994,764	2/1991	Peters	331/158 X

FOREIGN PATENT DOCUMENTS

1045480	12/1958	Fed. Rep. of Germany	331/162

OTHER PUBLICATIONS

Science Manual—Video Tape Recorder, IVC–900, International Video Corp., Feb. 1972, pp. 4–30.
Handbook of Filter Synthesis, A. I. Zverev, John Wiley & Sons, Inc., 1967, pp. 425, 426, 446, 452, 486–487.
"Survey of Crystal Oscillators", Roger Harrison, *Ham Radio*, Mar. 1976, pp. 10–22.

Primary Examiner—Siegfried H. Grimm
Attorney, Agent, or Firm—Skjerven, Morrill, MacPherson, Franklin & Friel

[57]   **ABSTRACT**

A wide frequency deviation voltage controlled crystal oscillator includes a multiple section bandpass filter containing a plurality of crystal or ceramic elements. One embodiment includes a plurality of crystals which are connected into parallel conduction paths. The crystals have resonant frequencies which are separated by small, selected intervals so that the oscillator may be adjusted to a wider range of frequencies than prior art oscillators. Alternatively, the same result may be achieved by using crystals with the same resonant frequency and connecting each crystal to a capacitor having a selected value. In another embodiment a multiple section crystal or ceramic bandpass filter is substituted for the parallel conduction paths. The principles of the invention are applicable to a number of different devices, including the ringing circuits used in color televisions to perpetuate the color bursts which appear in the blanking intervals of the incoming TV signals.

57 Claims, 19 Drawing Sheets

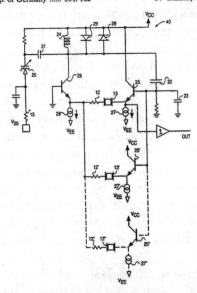

FIGURE 17-8   Front page of the patent with multiple crystals to emulate a wider-band filter suitable for a wider-frequency-deviation crystal oscillator.

U.S. Patent

July 20, 1993

Sheet 19 of 19

5,229,735

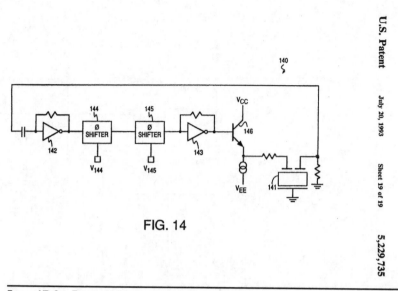

FIG. 14

FIGURE 17-9 Example of the use of a ceramic filter for a wide-frequency-deviation oscillator.

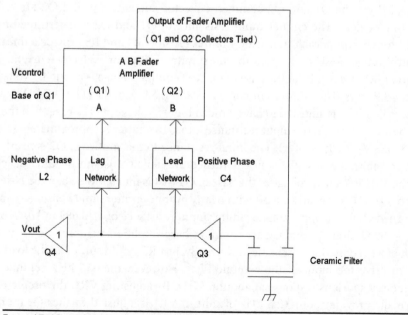

FIGURE 17-10 A 455 kHz ceramic filter oscillator.

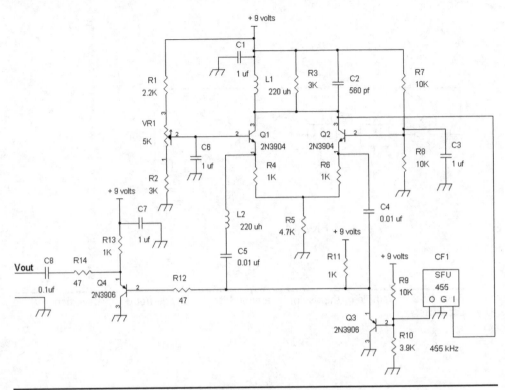

FIGURE 17-11 A 455 kHz ceramic filter voltage-controlled oscillator.

provided via the Q1 and Q2 summed collector currents. Q1 and Q2 are current steering devices. The emitter input resistance of Q1 and Q2 is determined by the collector current of each transistor. Emitter resistors R4 and R6 provide a linearized control range. Resistor R5 sets the maximum emitter or collector current. With approximately 4.5 volts at the base of Q2, the emitter voltage is about 4.5 volts – 0.6 volt = 3.9 volts. The emitter tail current is about 3.9 volts/(1 kΩ + 4.7 kΩ) ≈ 0.684 mA. When VR1 is adjusted to have a much higher voltage at Q1's base than the voltage at the base of Q2, the input resistance into Q1's emitter is approximately $1/[0.684$ ma/0.026 V$] = 1/g_m ≈ 38$ Ω. The input resistance to the emitter of Q2 is nearly infinite resistance because Q2 is nearly turned off while Q1 is almost fully turned on. When VR1 is set to a voltage at the base of Q1 that is much lower than the base voltage of Q2, Q1 is at nearly cut off with nearly infinite emitter input resistance, and Q2 is turned on for an approximate emitter input resistance of $1/[0.684$ mA/0.026 V$] = 1/g_m ≈ 38$ Ω. Both Q1 and Q2 are common-base amplifiers that provide the positive gain of >1 required for oscillation. Normally, just R3 would suffice as the load resistor to drive the input of the ceramic filter. However, the 455 kHz ceramic filter includes a spurious response at about 6 MHz. By adjusting VR1, the oscillator will normally provide about ±2.5 kHz of shift. If VR1 is adjusted further for more frequency shift at 455 kHz, a 6 MHz oscillation will occur. A wide-bandwidth, low-Q band-pass filter L1 and C2 provides attenuation at 6 MHz while passing 455 kHz signals with very little effect.

If the input resistance to the ceramic filter is ignored because the bandwidth of the parallel LCR filter L1, R3, and C2 is much wider than the ceramic filter's bandwidth, the Q is approximately $2\pi(455\text{ kHz})(R3)(C2) = 2\pi(455\text{ kHz})(3\text{ k}\Omega)(560\text{ pF}) \approx 4.8$. The bandwidth of the LCR filter is $455\text{ kHz}/Q = 455\text{ kHz}/4.8 \approx 94\text{ kHz} >> 10\text{ kHz} = $ bandwidth of the ceramic filter CF1 (e.g., Murata SFULA455KU2A-B0).

Base bypass capacitors C3 and C6 ensure that the bases of Q1 and Q2 are AC grounded at 455 kHz to ensure that Q1 and Q2 are (AC) grounded amplifiers for the lag and lead networks L1 and C4. DC blocking capacitor C5 is used so that the DC voltage at the emitter of Q3 does not couple directly DC-wise into the emitter of Q1. Although the output of the oscillator may be taken from the emitter of Q3, an extra emitter follower stage Q4 provides isolation from the outside world so as to minimize any effect of external loading to the oscillator.

For a 10.7 MHz wide-frequency-deviation oscillator, see Figure 17-12. For this oscillator, about ±30 kHz of frequency deviation is provided. The 10.7 MHz ceramic filters of bandwidths from 100 kHz to 230 kHz worked satisfactorily (e.g., Murata SFELF 10M7KAH0-B0 or SFELF 10M7GA00-B0). A 350 kHz wider-bandwidth

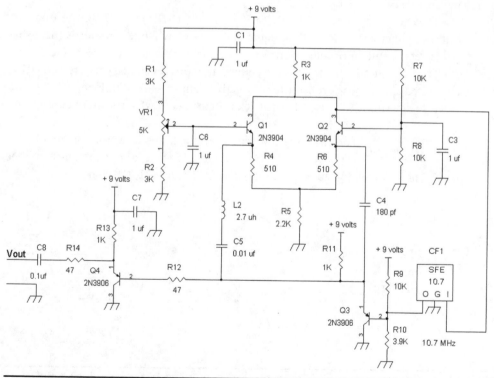

FIGURE 17-12   Ceramic filter 10.7 MHz voltage-controlled oscillator.

ceramic filter proved unreliable in providing a single stable frequency. Because there are low amounts of spurious response from the 10.7 MHz ceramic filters, an LC band-pass filter is not required. Thus only a resistive load R3 is required. For a narrower range of frequency deviation, a 20 kHz ceramic filter (Murata Part Number SFKLF 10M7NL00-B0) can be used.

Note that the values for L2, C4, R3, R4, R5, and R6 are different in Figure 17-12 from those in Figure 17-11. A higher quiescent current was required for Q1 and Q2 when operating at 10.7 MHz versus 455 kHz.

A Non-Varactor Electronic Variable Capacitor

Varactor tuning diodes are limited in their capacitance, and sometimes an alternative electronically controlled capacitor is desired. To provide some background as to how a nonvaractor electronic variable capacitor works, we need to briefly examine some simple feedback circuits (Figure 17-13).

The circuit in Figure 17-13a shows a voltage source V connected to a resistor R1. If we want to determine what the equivalent resistance seen by the voltage source is, we can measure the current I. For example, the voltage across R1 is V, and the resulting current is $V/R1 = I$.

The equivalent input resistance is $Rin = V/I = V/(V/R1) = R1$. By inspection, one can see right away that Rin will be equal to R1, but we will find out shortly that using V/I is very useful for finding equivalent resistances, capacitances, and inductances.

The next question is, what is the equivalent input resistance when resistor R1 is connected as a feedback resistor for a negative-gain amplifier? In Figure 17-13b, we will also determine Rin via V/I. The trick is figuring out what the input current I is.

> **NOTE** For all of the inverting-gain amplifiers shown in Figure 17-13b through 17-13d, the input resistance is infinite into the amplifier itself. Therefore, all the current I will be flowing through R1, C1, and L1.

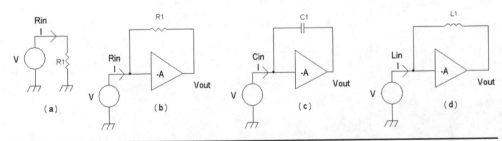

FIGURE 17-13 Circuits, (a) through (d), with each an input voltage V and a resulting input current I.

For the circuit in Figure 17-13b, the input of the amplifier has a voltage V at the input, and thus the output voltage Vout = –AV. The voltage across R1 then is V – (–AV) = V + AV = V(1 + A). The current I is then the voltage across R1 divided by its resistance R1. We have I = V(1 +A)/R1, and we can substitute this into Rin = V/I = V/[V(1 + A)/R1] = 1/ [(1 + A)/R1] = R1/(1 + A)= Rin.

What this says is that if the gain A = 0, Rin = R1 because Vout will be 0 volt or ground potential. But if A = 1, Rin = R1/2. For example, if R1 = 10 kΩ and –A = –1 so that A = 1, Rin = 10 kΩ/(1 + 1) = 10 kΩ/(2) = 5 kΩ. Having a feedback resistor with a finite-gain amplifier results in having a lowered equivalent resistance. Put another way, if V = 1 volt and –A = –1, then Vout is –1 volt, and there are (1 volt – –1 volt) = 2 volts across R1, or 2× more current flowing through R1 than if R1 were grounded and had 1 volt across it. The output of the amplifier causes more current to be drained at the input and thus simulates a lower input resistance. Because the resistor current is increased by a factor of (1 + A), the equivalent resistance as seen by the input source V is R1/(1 + A). In Figure 17-13b, Rin = R1/(1 + A). This means that if we were able to vary the gain of the inverting amplifier, we could change the input resistance without having to change R1's resistance value.

Now let's examine what happens when there are feedback capacitances and inductances when the voltage source V provides a sinusoidal signal in Figure 17-13c and d. For a capacitor, the impedance is $Z_{C1} = 1/j\omega C1$. The voltage across C1 is then (V – Vout) = V – (–AV) = (1 + A)V = voltage across C1. The AC current due to the voltage across C1 is $I = (1 + A)V/ Z_{C1}$.

The input impedance Zin = V/I, which will eventually tell us what the input capacitance Cin is. Thus:

$$Zin = V/I = V/(1 + A)V/Z_{C1} = 1/(1 + A)/ Z_{C1} = Z_{C1}/(1 + A) = (1/j\omega C1)/(1 + A) =$$

$$\frac{1/(j\omega C1)}{1 + A} = \frac{1}{j\omega C1(1 + A)} = \frac{1}{j\omega[C1(1 + A)]} = Zin$$

If the inverting gain is 0, A = 0, so Zin = $1/(j\omega C1)$, which tells us that Cin = C1.

But for any nonzero gain A, we see a multiplied capacitance because the input impedance is:

$$Zin = \frac{1}{j\omega[C1(1 + A)]}$$

This means that the input impedance is purely capacitive and that capacitance is:

$$Cin = C1(1 + A)$$

which means that there is an equivalent capacitance referenced to ground that is a multiplied version of C1. For example, if C1 = 100 pF and –A = –99 so that A = 99, then:

$$Cin = (1 + 99)100 \text{ pF} = 10,000 \text{ pF}.$$

Also, by changing the gain of the inverting amplifier, we can provide the equivalent of a variable capacitor. This makes sense because there is more AC current flowing through C1 when there is an out-of-phase signal at Vout. The *unexpected* increase in current is equivalent to having an increased capacitance.

In musical effects devices such as a wah-wah circuit, a foot pedal variable resistor changes the gain of an inverting amplifier. A feedback capacitor is used such that a variable capacitor that depends on the foot pedal position changes the frequency and phase response of the wah-wah circuit. Generally, an inductor is connected to the inverting amplifier's input and ground to form a parallel LC circuit, where the C comes from a configuration similar to Cin shown in Figure 17-13c.

Before we get to the voltage-controlled variable capacitor, let's take a quick look at Figure 17-13d. The input impedance is Zin = V/I, where the impedance of an inductor is $Z_{L1} = j\omega L1$. Again, we have Vout = −AV. Thus the voltage across L1 is (V − Vout) = [V − (−AV)] = (1 + A)V. Thus:

$$I = (V - Vout)/ Z_{L1} = (1 + A)V/ Z_{L1}$$

$$V/I = Zin = V/(1 + A)V/Z_{L1} = 1/(1 + A)/Z_{L1} = Z_{L1}/(1 + A) = j\omega L1/(1 + A)$$

$$= j\omega[L1/(1 + A)]$$

$$Zin = j\omega[L1/(1 + A)]$$

Zin is purely inductive and has a new inductance value of L1 divided by (A + 1), which is similar to resistances. In other words, the input inductance Lin = L1/(1 + A). For example, if −A = −19, then A = 19, and if L1 = 100 μH, then Lin = 100 μH/(1 + 49) = 2 μH.

Again, note that if the inverting amplifier changes gain, then we can provide a variable inductor that is referenced to ground. Now the question is, what type of amplifier do we use if we want to make an electronically controlled variable resistor, capacitor, or inductor?

First, we have to determine how large the input signals are. For audio signals, the typical amplitudes are greater than 100 mV peak. For some radio-frequency (RF) signals, the levels can be less than 1 mV peak. In Chapter 16, we examined the distortion characteristics of a common-emitter amplifier. For an AC-grounded common-emitter amplifier, the gain or transconductance is directly proportional to the collector current, which allows a very large range of transconductances, but a 1 mV peak sinusoidal voltage into the base-emitter junction already produces 1 percent second-order distortion. What if we decide to apply local feedback via a finite resistance RE to the common-emitter amplifier to lower the distortion? What happens to the gain or transconductance g_m' with changes in collector current?

$$g_m' = g_m/(1 + g_m RE) = 1/[(1/g_m) + RE)]$$

In Chapter 16, we saw that second-order distortion goes down by a factor of $1/(1 + g_m RE)^2$ for any given input signal. For example, if the input signal is a 10 mV peak sine wave, without feedback or RE = 0 equivalent via large emitter bypass capacitor to ground, the distortion would be 10 percent second-order harmonic distortion. If g_m = 0.0384 mho for 1 mA of DC collector current and RE = 200 Ω, then the second-order harmonic distortion would be

$$10\%/[1 + (0.0384)(200)]^2 = 10\%/75 = 0.13\% = HD_2.$$

Now let's see if we can control the gain by increasing the collector current twofold from 1 mA to 2 mA. Note that g_m = 0.0384 mho for 1 mA DC and $1/g_m$ = 26 Ω. At 2 mA, g_m = 0.0768 mho, and $1/g_m$ = 13 Ω. Note that we are using RE = 200 Ω. Thus:

$$g_m' = 1/[(1/g_m) + RE)]$$

At 1 mA, g_m' = $1/[(1/g_m) + RE)]$ = $1/[(26\ \Omega) + 200\ \Omega]$ = 0.0044 mho. At 2 mA, g_m' = $1/[(13\ \Omega) + 200\ \Omega]$ = 0.0047 mho.

Increasing the collector current 100 percent (from 1 mA to 2 mA) only increased the transconductance, $1/[(1/g_m) + RE)]$, or equivalent gain from 0.0044 mho to 0.0047 mho, which is about 6 percent. Now we are in a "catch 22" situation because while we can lower the distortion in the common-emitter amplifier, we also linearize gain function such that there is hardly any gain variation when the collector current changes greatly. Of course, this is what we should expect when we lower distortion of an amplifier since the differential gain variation is proportional to second-order distortion; lowering this distortion also lowers the differential gain.

I suppose that one can try using a grounded source or cathode FET or vacuum tube to achieve greater gain variation with acceptable distortion characteristics. But FETs need to be selected because their transconductances vary by quite a bit even within the same part number, and tubes are not as convenient to use because they generally require higher voltages.

One solution to this problem is to use half of a voltage multiplier gain cell as a voltage-controlled amplifier (Figure 17-14). The input capacitance equation from the patent is shown in Figure 17-15.

To provide a voltage-controlled variable capacitance looking into C_m, transistor 16 is a common-emitter amplifier with an emitter local feedback resistor to ensure low distortion characteristics. The collector signal current is connected to the emitters of transistors 12 and 14. A voltage-control voltage V_C steers the signal current from lower transistor 16 to output transistor 14 or transistor 12, or some mixture of signal current is sent to both transistors. The signal current is proportional to the equation shown in Figure 17-15. When V_C = 0, half the signal current goes to transistor 14, and the other half goes to transistor 12. If V_C is tipped more positive in voltage, more signal from transistor 16's collector current will be diverted to transistor 12 than to transistor 14. If V_C is tipped more negative in voltage, more signal current from transistor 16 will be diverted to transistor 14 than to transistor 12. Thus transistor 14 provides a gain control of the signal current, and along with load resistor

United States Patent [19]

Quan

[11]	**Patent Number:**
[45]	**Date of Patent:**

4,516,041

May 7, 1985

[54] **VOLTAGE CONTROLLED VARIABLE CAPACITOR**

[75] Inventor: **Ronald Quan**, Cupertino, Calif.

[73] Assignee: **Sony Corporation**, Tokyo, Japan

[21] Appl. No.: **443,250**

[22] Filed: **Nov. 22, 1982**

[51] Int. Cl.^{3} **H03K 5/00**; G06G 7/18
[52] U.S. Cl. **307/494**; 307/490;
307/498; 307/590; 307/320; 307/262; 307/510;
307/514; 307/597; 330/282; 328/127; 334/15;
357/14
[58] Field of Search 307/490, 494, 498, 320,
307/590, 592, 597, 603, 606, 262, 510, 514;
357/14; 328/127, 155; 334/15; 330/282, 107;
361/277

[56] **References Cited**

U.S. PATENT DOCUMENTS

3,944,945	3/1976	Corte et al.	331/78
4,191,899	3/1980	Tomczak	307/303
4,223,271	9/1980	Furukawa	328/127
4,396,890	8/1983	Kato et al.	330/86

FOREIGN PATENT DOCUMENTS

H04B1/16 7/1982 PCT Int'l Appl. 455/233

Primary Examiner—Stanley D. Miller
Assistant Examiner—Richard Roseen
Attorney, Agent, or Firm—Philip M. Shaw, Jr.

[57] **ABSTRACT**

A voltage controlled capacitor is created by connecting a capacitance between the output of an amplifier whose gain is voltage controllable and an inverting input of the amplifier.

6 Claims, 1 Drawing Figure

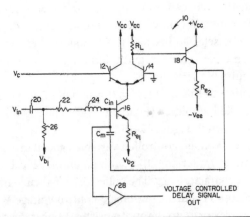

Figure 17-14 Front page of U.S. Patent 4,516,041 showing a voltage-controlled variable capacitor.

$$C_{in} = C_m \left[\left[\frac{g_{m1}R_L}{g_{m1}R_{e1} + 1} \right] \frac{1}{1 + e^{V_c/KT/q}} + 1 \right]$$

where

R_L = collector load resistance
R_{e1} = first emitter resistance
V_c = control signal voltage

Figure 17-15 The variable capacitance described by the equation shown earlier, with an enlarged view of the schematic in Figure 17-16.

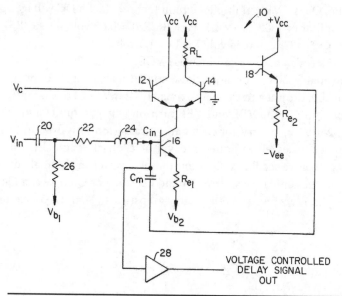

Figure 17-16 A larger schematic of the voltage-controlled capacitor.

RL, they form a voltage-controlled inverting-gain amplifier. With capacitor C_m, the capacitance Cin is in the form of:

Cin = (1 + A)C_m

where the gain A is a function of control voltage VC. Transistor 18 is an emitter-follower amplifier to provide a low-impedance output stage that drives Cm, which is essential so that the variable capacitance Cin is minimized in resistive losses.

Class AB Output Stage with Quiescent Bias Using a Servo Control

When designing a push-pull output stage, one of the most important circuits is the bias output circuit. Under a no-signal condition, the output transistors are set to a quiescent collector or drain current for bipolar transistors and MOSFETs. Should the bias current be set too low, crossover distortion appears. If the bias current is set too high, steps must to taken so that the output transistors do not self-destruct due to thermal runaway when the DC quiescent current increases until the power dissipation of the transistor becomes too high and exceeds its heat and power ratings.

A common way of providing thermal feedback to an output current bias circuit is to mount biasing diodes or transistors to a heat sink where the output transistors are installed (see Figure 17-17). As shown in Figure 17-17 this audio power amplifier includes output transistors Q4 and Q5 that are typically mounted on a large heat sink. The bias transistor Q3 sets the output quiescent currents of Q4 and Q5 via resistors R5 and R6. Q3 is a VBE multiplier circuit that provides a DC voltage across its collector and emitter of $VCE_{Q3} \approx VBE_{Q3}(1 + R6/R5) \approx 0.65$ volt$(1 + R6/R5)$. For example, with R6 = R5, $VCE_{Q3} \approx 0.65$ volt$(1 + 1) \approx 1.3$ volts.

By reducing R5's resistance, we can increase VCE_{Q3}. When the bias for Q4 and Q5 is set to approximately 50 mA or more, these transistors heat up, and the turn-on voltages for Q4 and Q5 drop at a rate of approximately –2 mV/°C. For example, if the rise in temperature is 30°C, the VBE and VEB turn-on voltages for Q4 and Q5 will change by 30 × (–2 mV) = –60 mV for each output transistor. This drop in turn-on voltage of the output devices left without some type of feedback eventually will cause an ever-increasing positive-feedback situation that will result in thermal runaway. By mounting Q3 to the same heat sink as Q4 and Q5, such as Q3 between Q4 and Q5, thermal feedback is achieved. As the output transistors Q4 and Q5 heat up, so

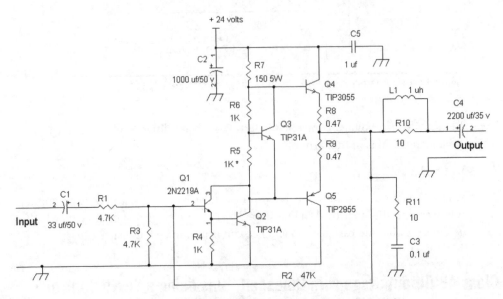

Notes: Q2 should be attached to a small heatsink.

Q3, Q4, and Q5 should be attached to a common heatsink.

Figure 17-17 A class AB audio amplifier.

does Q3. This then causes the turn-on voltage of Q3, VBE_{Q3}, to also decrease and prevent thermal runaway via $VCE_{Q3} \approx VBE_{Q3}(1 + R6/R5)$. It should be noted that emitter resistors R8 and R9 also help in reducing thermal runaway, but their values must be kept low so that the load receives maximum power.

> **NOTE**   Because the tabs or cases of Q3, Q4, and Q5 are the collector terminals, care must be taken so that all three transistors are electrically insulated from the metal heat sink.

One question is, can we set the bias of output transistors another way? What was learned in Chapter 13 was that in a video amplifier, we can use a servo or feedback system to ensure that the blanking level of a video signal is set to a reference voltage such as 0 volt regardless of small amounts of DC offsets that may vary in the video signal (see Figure 17-18).

A feedback clamp circuit forces the "back porch" voltage of a video signal to a desired voltage set by the 50 kΩ potentiometer (pin 2) connected to R11 in Figure 17-20. A back porch sampling pulse (BPSP) signal activates a sample-and-hold circuit (S/H ckt.), U3C and hold capacitor C7 to store the back porch level of the video signal. The stored voltage is connected to pin 5 of U2B and it is compared with a reference voltage at R11 via pin 2 of the 50 kΩ potentiometer and U2B. An integrator amplifier U2B via pin 7 sends an error voltage to R6 such that the output at U1B has a predetermined back porch level for video signal connected to output resistor R7.

The question is, can we adapt something from video clamp circuits for a new use in amplifiers to set up the bias of the output transistors? Video signals include peri-

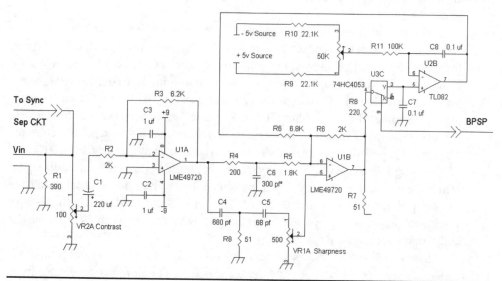

FIGURE 17-18   A feedback clamp circuit for a video amplifier.

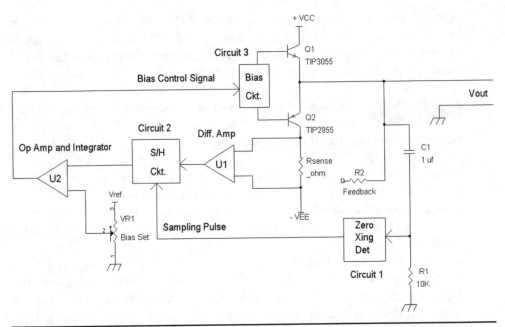

FIGURE 17-19 A block diagram of a servo system similar to a video clamp circuit that sets the output device's quiescent currents.

odic levels such as blanking levels after the trailing edge of horizontal sync pulses. However, audio signals are generally not periodic in nature. For example, music starts and stops with different notes depending on the program material. So can we adapt some type of video clamp circuit to a class AB amplifier? The answer is yes. See Figure 17-19.

A differential amplifier U1 with a collector-current sense resistor Rsense forms a voltage indicative of Q2's collector current. This voltage is amplified by U1 and fed to the input of the sample-and-hold circuit. The collector current of Q2 changes as the audio signal at Vout is supplied to the load, such as a loudspeaker. However, when Vout is nearly zero in AC voltage, the collector currents of Q1 and Q2 are approximately the same. High-pass filter C1 and R1 passes audio signals from about 16 Hz and up. Because AC signals have to cross 0 volt, we would like to sense the collector current of Q2 only during times when the AC signal at Vout is nearly 0 volt AC.

By using a zero crossing detector, circuit 1 in Figure 17-19, when the audio signal is near 0 volt, the zero crossing detector circuit outputs a high-logic signal to activate the sample-and-hold circuit (circuit 2). The sample-and-hold circuit then stores voltage values only during periods when the AC signal crosses 0 volt DC or when the audio signal is indicative of silence.

These stored voltage values are then voltages indicative of the output transistor's collector current when the Vout's AC voltage is zero. The stored voltage value from the sample-and-hold circuit is then compared with a reference voltage set by VR1. An error amplifier that includes an integrator circuit sends a signal to a bias circuit to vary the DC bias for the collector currents of Q1 and Q2.

It should be noted that when there is no audio signal, the zero crossing detector always outputs a logic high, and the sample-and-hold circuit is continuously sampling Vout. With this circuit, emitter series resistors such as R8 and R9 shown in Figure 17-17 are not needed, but they can be included.

Now look at Figure 17-20, which shows the servo-controlled biasing circuit. The output transistors 12 and 14 are tied together at their emitters. A high-pass filter 36 is connected to a zero crossing detector that includes comparators 30 and 32. These two comparators form a window comparator circuit that outputs a logic-high voltage over a narrow range of voltages such as between –20 mV and +20 mV, and it outputs 0 volt for voltages greater than +20 mV and less than –20 mV. The output terminals of the comparators are connected to control sampling switch 34.

A sense resistor 38 develops a voltage proportional to the collector current of transistor 14. This voltage is amplified by differential amplifier 40. The output of the differential amplifier is connected to the input of sampling switch 34. Switch 34's output is fed to a hold capacitor and is buffered by voltage-follower amplifier 48, which is connected to inverting amplifier 50. The output of amplifier 50 is fed to an integrator amplifier 52. A reference voltage to set the collector bias current is provided by potentiometer 54 into the (+) input of integrator 52. The output of the integrator amplifier is connected to zener diode 56 to ensure that at startup or power on the biasing circuit does not latch and cause the output transistors to be burned out. Zener diode 56 allows the bias current to start at zero before being increased to the set collector current. The collector biasing circuit includes a variable-current-source circuit via common-emitter amplifier transistor 24, and the voltage developed across resistor 22 then biases the bases of transistors 12 and 14.

When this circuit was built, the quiescent bias current was set and did not change even when the output devices were raised in temperature via a heat gun. However, the circuit did fail, but that was because the heat gun had melted the solder connections, and the transistor terminals eventually became open circuits.

We have now shown three examples of inventions:

- The use of a ceramic filter in an oscillator instead of a resonator
- Application of a differential pair voltage-controlled amplifier for a voltage-controlled capacitor that handles large signals without distortion
- Taking a page from video clamp circuits and varying the circuit a little with a zero crossing detector resulting in the design of a new output stage (the quiescent current of the output devices is set without the use of thermal feedback elements such as diodes or a transistor attached to the heat sink where the output devices are installed)

United States Patent [19]

Quan

[11]	**Patent Number:** **4,458,213**
[45]	**Date of Patent:** **Jul. 3, 1984**

[54] **CONSTANT QUIESCENT CURRENT, CLASS AB AMPLIFIER OUTPUT STAGE**

[75] Inventor: **Ronald Quan**, Cupertino, Calif.

[73] Assignee: **Sony Corporation**, Tokyo, Japan

[21] Appl. No.: **449,133**

[22] Filed: **Dec. 13, 1982**

[51] Int. Cl.³ H03F 3/30
[52] U.S. Cl. 330/267; 330/265; 330/296
[58] Field of Search 330/265, 267, 268, 270, 330/273, 274, 290, 296

[56] **References Cited**

U.S. PATENT DOCUMENTS

4,306,109 12/1981 Kondov .... 330/274 X

Primary Examiner—James B. Mullins
Assistant Examiner—Steven J. Mottola
Attorney, Agent, or Firm—Philip M. Shaw, Jr.

[57] **ABSTRACT**

A circuit maintains the quiescent collector current of an output transistor amplifier constant by sensing the voltage developed across a resistor connected in series with the output transistor's collector and, using a differential amplifier, produces a first signal representative thereof which is sampled and held by a circuit controlled by a zero voltage crossing detector sensitive to the amplifier output voltage. The sampled and held signal is inverted and integrated with respect to an adjustable reference voltage to produce a second signal. The second signal controls the current passed by a transistor connected in a voltage divider network supplying the base bias voltage to the output transistor.

3 Claims, 1 Drawing Figure

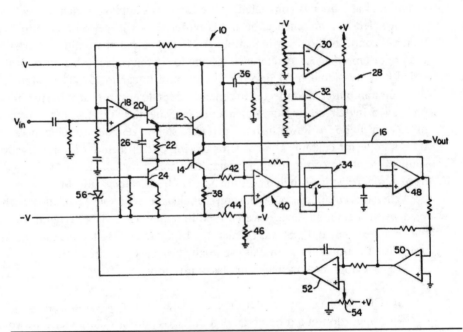

FIGURE 17-20   A Class AB output-stage amplifier with a zero crossing detector to set the output collector bias current.

Other Examples of Designing

We have discussed how relaxation oscillators work. So let's try a variation on that theme.

Modulating the Power-Supply Pin of an Oscillator to Provide FM

In some of the circuits mentioned using a hysteresis oscillator such as a 74HC14 inverter gate with a resistor and capacitor, it was recommended that the power-supply voltage should be regulated (e.g., +5 volts). However, we can vary the power-supply voltage to the 74HC14 chip to convert it to a VCO (see Figure 17-21).

The relaxation oscillator including a 74HC14 uses R6 and C6 to determine the center oscillating frequency. To vary the oscillation frequency, one can replace C6 with a varactor diode or a variable capacitor that is electronically controlled. In this circuit, we modulate the power pin 14 of U3A to change the oscillation frequency. Op amp U1A's output at pin 1 nominally has about 4 volts DC and can vary above and below 4 volts once Vin, an AC signal, is present. Resistor R8 isolates the output of op amp U1A from bypass capacitor C5 to ensure that the op amp does not oscillate due to loading directly to a capacitive load. Output pin 2 of U3A then has both AM and FM. Zener diode ZD1 and transistor Q1 work as positive and negative clipping circuits so that pin 14 of the 74HC14 chip does not suffer from over- or under-

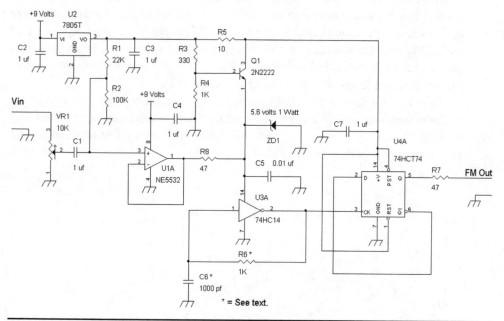

Figure 17-21   VCO made from modulating the power-supply pin of a 74HC14 inverter gate.

voltage. A divide-by-two in frequency flip-flop circuit U4A provides a 50 percent duty-cycle FM signal that also has a stable amplitude signal that is free of AM effects.

An LC RF Matching Network

The concept of an RF matching network using inductors and capacitors is to mimic a transformer for the frequencies of interest. For example, if we want to match a 100-Ω source resistance with a 10 kΩ system, ideally, we would want the equivalent of a transformer with a turns ratio 1/N. Thus:

$$1{:}N = \sqrt{\frac{R\_low}{R\_high}} = \sqrt{\frac{100}{10K}} = 1{:}10$$

That is, if the turns ratio of a transformer from its primary winding to its secondary winding is 1:10 and the secondary winding is connected to a 10 kΩ system, the primary winding of the transformer will have an equivalent resistance of 100 Ω. Thus, if the signal generator's source resistance is 100 Ω, we will have achieved an optimal power match because the input resistance looking into the primary winding is also 100 Ω.

For an RF matching network to achieve matching from 100 Ω to 10 kΩ, we need to step up the voltage somehow. One solution is to use an LC network, as shown in Figure 17-22. In Figure 17-22, at resonance, we want to choose the values of L1 and C1 such that the voltage source Vlow with source resistance R_low matches the equivalent resistance Rs when C1 is loading down with R_high. If R_high were an open circuit, we would have a series LC circuit, and at resonance, the output voltage of C1 would depend on the driving resistance R_low. For example, if R_low → 0 Ω and R_high → infinite resistance, the voltage across C1 will be very large. This is so because huge currents will flow through L1 and C1 at the resonant frequency where the combination series impedance of L1 and C1 is 0 Ω.

If we load C1 down with a finite resistance R_high, the voltage across C1 at the resonant frequency will be lower because we no longer have as huge currents flowing through L1 and C1. Having resistor R_high in parallel with C1 wrecks the perfect cancellation of impedances with L1.

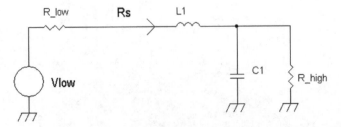

Figure 17-22   A low-pass LC network with Q > 1 to provide stepped-up voltage to R_high.

The question is, how do we find the correct matching network such that when R_high is across C1, Rs looks like R_low, and the voltage at C1 is stepped up by a ratio of $1:N = \sqrt{R\_low/R\_high}$? We could start by cranking out impedance formulas for the capacitors, resistors, and inductor with complex impedances, or we could take a look at the circuit backwards and equate the series and parallel LC filter's Q (see Figures 17-23 and 17-24).

We will now use the magic of Q, which will be:

$$Q_{\text{series res}} = (\omega_r)L1/Rs$$

$$Q_{\text{parallel res}} = (\omega_r)Rp(C1)$$

where $\omega_r = 1/\sqrt{L1C1}$ = resonant frequency in radians per second. By squaring ω_r, we have:

$$(\omega r)^2 = 1/L1C1$$

and by equating $Qs_{\text{eries res}} = (\omega_r)L1/Rs = Q_{\text{parallel res}} = (\omega_r)Rp(C1)$:

$$(\omega_r)L1/Rs = (\omega_r)Rp(C1)$$

Divide by ω_r both sides to get:

$$L1/Rs = Rp(C1)$$

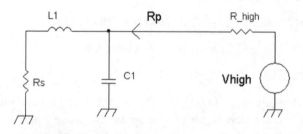

FIGURE 17-23   Looking backwards, we want to find an equivalent Rs such that Rp = R_high.

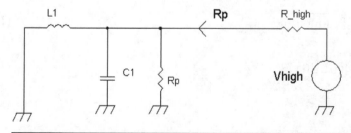

FIGURE 17-24   A resistor in series with L1 is equivalent to a lossless L1 but with an equivalent parallel resistor Rp across C1 and L1.

which leads to:

L1/C1 = RpRs

L1 = RpRsC1, but from $(\omega_r)^2 = 1/L1C1$, C1 = $(1/L1[(\omega_r)^2])$, which leads to a substitution for C1:

L1 = RpRs$(1/L1[(\omega_r)^2])$ = RpRs/L1$[(\omega_r)^2]$

L1 = RpRs/L1$[(\omega_r)^2]$

Multiply by L1 on both sides to get:

$(L1)^2$ = RpRs/$(\omega_r)^2$

Now take the square root of both sides to get:

$$L1 = [\sqrt{RpRs}]/(\omega_r) \tag{17-1}$$

$$C1 = (1/L1[(\omega_r)^2]) \tag{17-2}$$

Alternatively by using the formula, $\omega_r = 1/\sqrt{L1C1}$, we also get:

$$C1 = 1/(\omega_r)\sqrt{RpRs} \tag{17-3}$$

For example, suppose that we want to match 100 Ω = R_low = Rs to load into 10 kΩ = R_high = Rp, and suppose that the frequency we are interested is approximately 724 kHz, or ω_r = 2π(724 kHz). Then:

L1 = $[\sqrt{(100)(10K)}]/[2\pi(724\ k\Omega)]$ = 1,000/2π(724 kΩ) = 220 μH= L1

With the resonant frequency ω_r and L1 known, we can solve for:

C1 = 1/L1$[(\omega_r)^2]$ = 1/(220 μH)$[2\pi(724\ kHz)]^2$ = 220 pF= C1

It should be known that Equations (17-1), (17-2), and (17-3) are accurate when (Rp/Rs) ≥ 10. However, a correction factor can be applied to these equations for more accurate calculations for the matching network in Figure 17-22.

The correction factor is $\sqrt{1 - \frac{Rs}{Rp}}$ and this results in:

$$L1 = [\sqrt{1 - \frac{Rs}{Rp}}][\sqrt{RpRs}]/(\omega_r) \tag{17-4}$$

$$C1 = [\sqrt{1 - \frac{Rs}{Rp}}](1/L1[(\omega_r)^2]) \tag{17-5}$$

alternatively:

$$C1 = [\sqrt{1 - \frac{Rs}{Rp}}][1/(\omega_r)\sqrt{RpRs}] \tag{17-6}$$

NOTE The low-impedance side is at inductor L1, and the high-impedance side is at capacitor C1. The matching network looks like Figure 17-22 with L1 and C1.

References

1. United States Patent 5,229,735, "Wide Frequency Deviation Voltage Controlled Crystal Oscillator Having Plural Parallel Crystals," Ronald Quan, filed March 30, 1992.
2. United States Patent 4,516,041, "Voltage Controlled Variable Capacitor," Ronald Quan, filed November 22, 1982.
3. United States Patent 4,458,213, "Constant Quiescent Current, Class AB Amplifier Output Stage," Ronald Quan, filed December 13, 1982.
4. Chris Bowick, John Blyler, and Cheryl Ajluni, *RF CIRCUIT Design, 2nd ed.* Amsterdam, The Netherlands, Newnes, 2008.
5. David Pressman, *Patent It Yourself, 13th ed.* Berkeley, CA: Nolo, 2008.

CHAPTER 18

Troubleshooting and Final Thoughts

We will close out this book with some tips in troubleshooting. When we think of troubleshooting a circuit, this can mean getting the circuit to work in terms of tracing an error in wiring, but troubleshooting also can relate to improving a circuit that was not designed optimally in the first place. Sometimes a circuit may not work at all in one type of construction, but it works perfectly when built on a different board. So troubleshooting involves both detective work and the builder's flexibility to construct the circuit again in a different environment.

Assembling the Circuit

One of the ways to ensure that all the connections are made during construction of a circuit is to print a copy of the schematic. As each wire or component is soldered or connected, use a highlighter (marker) pen (e.g., yellow, pink, or green) to mark off each of the completed connections. Sometimes, with more complex projects, one may forget to hook up a power pin of an integrated circuit (IC) or ground a lead of a capacitor or resistor. The marking off of the schematic helps the builder keep track of what's been wired. It is not uncommon to "finish" wiring a board just to find out in the schematic that one or two lines have not been highlighted and thus not connected.

When the Circuit Is Completed and Ready for a Test ... Before Powering It On

If the circuit seems ready to power up, then first make sure that all the electrolytic capacitors are wired in the correct polarity. Incorrect connection of electrolytic capacitors can result in excessive current flowing into the electrolytic capacitor. A power supply with a current-limit adjustment feature comes in handy. For example, if the circuit is low power, you can set the current limit to about 50 mA. Otherwise, you can connect a 1 Ω to 10 Ω, ¼-watt series resistor between the power supply and the circuit. If there are any short circuits or excessive power drains, you can determine current by measuring the direct-current (DC) voltage across the series resistor and dividing the voltage by the resistance.

If you are testing an audio power amplifier, one old trick is to monitor the output terminals with a voltmeter *without* connecting the amplifier to a load. Connect a variac transformer to the power outlet, and hook up the variac autotransformer to the amplifier. The variac is first adjusted all the way down to 0 volt. Then slowly turn up the voltage to the full alternating-current (AC) line voltage. If the output terminals generate a huge DC offset voltage, quickly shut off the variac.

Another method as told to me by John Curl is to connect a standard 100-watt incandescent light bulb in series with the power supply to the power amplifier. If there are any shorted output transistors, the lamp will light up, and its resistance will rise due to heating of the filament and limit the DC current into the amplifier.

DC Bias Conditions

For determining the DC bias conditions of a circuit that does not include oscillators, measure various bias points with a voltmeter.

> **NOTE** Measuring for a DC voltage where there is an AC voltage riding on top of the DC voltage usually results in erroneous measurements. It is better to use an oscilloscope instead for this case.

For NPN transistors, if the circuit is an amplifier, voltage across the base-emitter junction VBE has to be approximately 0.6 volt to approximately 0.7 volt. The range of collector-to-emitter voltage VCE is for VCE > 0.2 volts, and VCE is typically >1 volt. And for PNP transistors, the emitter-base voltage is in the 0.6 volt to 0.7 volt range, with the emitter-collector voltage VEC > 0.2 volt, typically >1 volt.

If for some reason the VCE for an NPN or VEC for a PNP transistor is less than 0.2 volt, one can try reducing the resistance value of the collector load resistor or increasing the associated emitter resistor. If that does not help, take out the transistor, and with an ohmmeter, measure any short circuit across the collector and emitter.

Transistors that are blown commonly have the following characteristics:

- The collector-emitter (or the collector-base) junction has become shorted to less than 10 Ω when measured with an ohmmeter both ways. By *both ways*, we mean to measure across the collector emitter with the positive probe on the collector and negative probe on the emitter and then measure again with the probes reversed. Some transistors that are fine (e.g., not shorted) have a diode across the collector and emitter and will measure in the approximately 15 Ω range depending on the ohmmeter in one direction but close to infinite resistance when the probes are reversed.

- The base-emitter junction has opened up, and the VBE voltage in the circuit exceeds 0.7 volt. With an ohmmeter and the transistor taken out, measure with the X1 resistance setting. A good transistor will measure typically less than 100 Ω in one direction and near infinite resistance when the probes are reversed. A blown base-emitter junction of the transistor will measure infinite resistance in both directions. In some Class AB amplifiers, where a biasing transistor is used as a VBE multiplier circuit to set the quiescent current of the output devices, the VBE multiplier circuit will fail if the base-emitter junction is blown or open circuited. This usually results in too much voltage across the bases of the output devices, resulting in excessive quiescent current that leads to failure of the output devices in the Class AB amplifier.
- It is rare, but it is possible that the base-emitter junction can be shorted, and if an ohmmeter shows low resistance in both directions, the transistor is bad.

For op amps with negative feedback used as an amplifier, the (+) input terminal's DC voltage should force the (−) input terminal to the same voltage. For example, if the (+) input of an op amp is +3.5 volts DC, the (−) input terminal should also read +3.5 volts DC with a voltmeter. Note that the voltmeter should have a very high input resistance such as 10 MΩ, which is more common with FET analog voltmeters or digital volt-ohmmeters (VOMs) that cost more than $20 (e.g., as of the year 2014).

Choosing the Correct OP Amps

In many cases, a single-supply voltage powers a circuit that includes op amps. Both dual- and single-supply op amps are capable of working correctly with single or dual (+) and (−) supplies. The distinction for a single-supply op amp generally means that it can work off +5 volts on the supply, whereby the input stage can sense down to 0 volt, and the output terminal can still be operational at close to 0 volt. Some dual-supply op amps may work with +5 volts, but the input stage needs to be biased to about 2 volts to 2.5 volts, and the output stage will work within a narrow range around 2.5 volts.

For unipolar signals such as a photodiode preamp with the cathode of the photodiode connected to the (−) input of the op amp and the (+) input grounded, the output will have to swing above 0 volt. A single-supply op amp is recommend over a dual-supply op amp in this case.

One can perform a Google search for rail-to-rail op amps, which generally will work at +5 volts or with (±) power supplies. If the supply voltages are above 6 volts, usually one can get away with using dual-supply op amps by biasing the (+) input terminal to half the supply voltage.

Another consideration is that many of the newer op amps will have a maximum supply voltage of 10 volts to 16 volts across the power terminals. For example, high-speed op amps (as of 2014) generally have a +10 volt limit (or ±5 volts), such as the AD8038 and AD8005, and the TLC272 series op amps have approximately a +16 volt maximum (or ±8 volts) supply voltage. Many of the older op amps, such as the NE5532, RC4558, LM4562, and AD797, can handle 30 volts or ±15 volts safely.

What if you need more current output capability from an op amp? Most op amps have output short-circuit protection with current limiting to protect the output transistors. This current limit is typically about 10 mA to 15 mA, such as those found in the LM1458, LM358, and TL082. For higher output capability in the 25 mA or more range, one can use the NE5532, LM4562, LM833, or NJM4560 op amps.

Logic Gates with Increased Speed and Output Current

In some instances, a logic gate such as an inverter, flip-flop, etc. may be required to drive more than the input of another logic circuit. For example, a logic circuit may drive a light-emitting diode (LED) or a loudspeaker or even be used as a "final" power amplifier (PA) for a low-power radiofrequency (RF) transmitter. The 74AC or 74ACT series logic chips not only possess high speed (>50 MHz) but also provide output currents in excess of 30 mA compared with the lower-speed and lower-output current of the 74HC or 74HCT series logic chips.

Decoupling Capacitors

Typically, decoupling capacitors of 0.1 μF to 1 μF are placed at the power pins and ground in ICs. Ceramic or film capacitors are generally preferred, although sometimes electrolytic types can be used. Without them, ICs such as amplifiers can oscillate or digital ICs can output spurious pulses.

For other circuits, such as discrete transistor amplifiers or oscillators, usually there are at least two power-supply decoupling capacitors, which can include two or more ceramic or film capacitors or, alternatively, electrolytic capacitors. Preferably a larger-value electrolytic capacitor (e.g., at least 4.7 μF) is paralleled with a 0.01 μF to 1 μF ceramic or film capacitor. Electrolytic capacitors have equivalent internal series inductors, and at frequencies greater than 100 kHz, the impedance of the capacitors starts to flatten out and then rise. Ideally, a capacitor should have an impedance characteristic that continues to drop as the operating frequency rises.

The smaller-value ceramic or film capacitors have smaller equivalent inductors compared with electrolytic capacitors. When a ceramic or film capacitor is paralleled to an electrolytic capacitor, the parallel combination will have a dropping impedance characteristic as the operating frequency increases.

What to Do If an Amplifier Oscillates

In some cases, just the type of breadboard can cause some amplifiers to oscillate. For example, the superstrip protoboard has been found to cause certain LM386 audio amplifiers to oscillate. The capacitances between each row or each track were enough to cause an unsatisfactory operation of the LM386. A solution was to mount the LM386 IC on a vector board or to a copper-clad breadboard.

For a high-speed op amp or a transistor amplifier circuit, usually a 47 Ω to 100 Ω resistor connected in series to the (+) input of the op amp or to the base of the transistor should be enough to stop a parasitic oscillation. Of course, circuit layout is important, and input and base leads should be kept short.

When the output terminal of an op amp or amplifier is connected to the outside world, which includes capacitances from cables, to avoid oscillation, it is best to add a series 47 Ω to 100 Ω resistor between the output of the op amp or amplifier and the output connector. When working with low-voltage circuits of less than 40 volts DC of supply voltage, often the parasitic oscillation is of a high frequency, and if you have an oscilloscope set to 100 MHz analog bandwidth and probing the output, the location of the oscillation can be determined by touching parts of the circuit.

> **NOTE** To avoid injury, do not touch higher-voltage circuits such as those that operate at greater than 40 volts.

Low-Frequency Oscillation

In some oscillator circuits, such as transformer-coupled types, often an AC coupling capacitor is included, such as the oscillator circuit in Figure 10-41. A condition known as *squegging* occurs when the AC coupling capacitor is generally too large in value and produces a low-frequency on-off modulation of the oscillator's desired signal. For example, if the oscillator's frequency is set to 455 kHz, as shown in Figure 18-1, capacitor C3, if too large in value (e.g., >1 μF), can cause the 455 kHz signal to turn on and off at a rate much lower than 455 kHz (see Figure 18-1).

One way to eliminate the squegging is to reduce the value of coupling capacitor C3 such that the oscillation signal is stable in amplitude. Typically, to avoid squegging, C3's value is between 0.0047 μF and 0.012 μF. Larger values of C3 generally lead to the squegging problem.

Another type of parasitic low-frequency oscillation is *motor-boating*, which can be caused by an amplifier circuit with a series of RC power-supply filter circuits (see

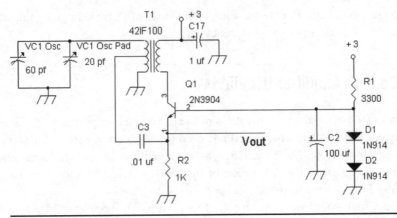

FIGURE 18-1 AM radio local oscillator circuit with AC coupling capacitor C3.

Figure 18-2). In Figure 18-2, the original value of C1 was 0.05 µF, which avoided motor-boating at Vout. However, to improve the low-frequency response, C1 was increased to 0.22 µF. This caused a positive-feedback effect due to the voltage change at C7's positive terminal, effectively a shift in the power-supply voltage into vacuum tubes V1A and V1B.

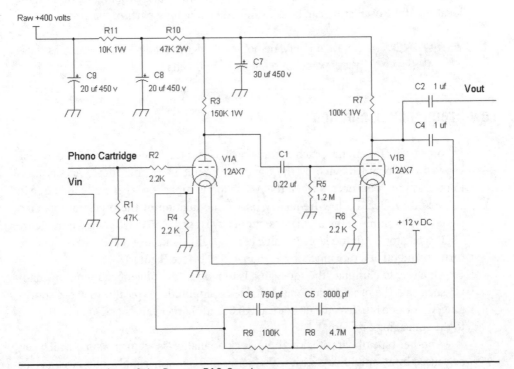

FIGURE 18-2 A variant of the Dynaco PAS 3x phono preamp.

What happens is that the amplifier on startup pulls current and causes a delayed action of the power-supply voltage ramping up via R11 with C8 and R10 with C7. This causes the amplifier to reduce in current drain, and the supply voltage rises at C7 but then starts to increase in current and cause the voltage at C7 to decrease. This decreasing and increasing of voltage at C7 then provides a form of relaxation oscillation. The oscillation mimics the sound of a motor boat. If the voltage to the amplifier is made "stiffer," the low-frequency oscillation is eliminated.

A fix to motor-boating can be done via a voltage regulator. Note that many older tube preamplifiers can suffer from motor-boating if modifications are done to increase low-frequency response. But often a simple voltage regulator will solve the problem (see Figure 18-3).

NOTE   The circuits shown in Figure 18-2 and 18-3 use high voltages that can at least cause severe injuries, if not worse. Please do not attempt to build such circuits if you are not careful and familiar with high-voltage circuits.

The 600 volt regulator transistor Q1 should typically be a power MOSFET with a TO-220 case capable of handling several amps of drain current, such as a Vishay Siliconix IRFIB6N60APBF. The 150 volt zener diodes ZD1 and ZD2 connected in

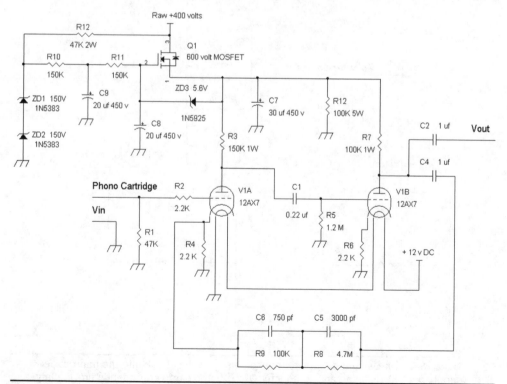

FIGURE 18-3   Voltage regulator circuit including ZD1, ZD2, and Q1 to eliminate motor-boating.

series form a 300 volt reference voltage that also reduces much of the ripple voltage from the raw +400 volt power supply. Further filtering of the 300 volt reference voltage is provided by R10/C9 and R11/C8. MOSFET Q1 is configured as a source-follower circuit that provides a low-impedance voltage source of about 300 volts. Zener diode ZD3 is connected to the gate and source of Q1, which provides current limiting and reverse-breakdown protection. Also note that R11 may be increased to 330 kΩ or even to 680 kΩ, if desired, for more power-supply ripple reduction.

Decoupling Sources from Injecting Noise into the Power Rails

One way to isolate noise from the power pins of circuits such as oscillators or logic gates is to use inductors and capacitors, as shown in Figure 18-4a. In some cases, larger-value inductors may be too expensive or may take up significant board space. If a slight voltage loss is acceptable, such as operating at 4.3 volts instead of 5.0 volts, an active inductor circuit that includes a transistor, resistor, and capacitor may be used (see Figure 18-4b).

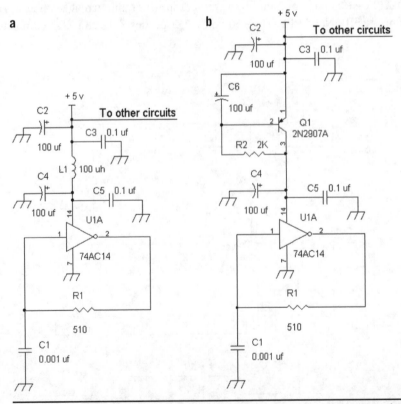

Figure 18-4 (a) Using inductors to reduce noise coupling back into the main power-supply rail. (b) Using an active inductor Q1, R2, and C6 to equivalently provide larger inductance values.

In Figure 18-4*b*, transistor Q1 acts as a current source for AC signals. At DC or 0 Hz, capacitor C6 can be treated as an open circuit. With a high-current-gain transistor (e.g., $\beta > 50$), there are negligible base currents leading to close to a 0 volt drop across R2. This results in equivalently tying the collector and base together to form a diode, which leads to a 0.7 volt drop (e.g., 4.3 volts at the collector of Q1) and an output resistance of approximately 1/gm. For example, if the current drain is about 2 mA, gm = 2 mA/0.026 V = 0.0768 mho, and 1/gm = 1/0.0764 mho $\approx$ 13 Ω.

However, at higher frequencies, where C6 acts as an AC short circuit or a battery, the output resistance at the collector of Q1 is 2 kΩ via R2. This means that the noise from pin 14 of U1 has to couple through R2, a 2 kΩ resistor, before injecting noise into the 5 volt rail. Thus R2 provides isolation between the noise source from pin 14 of U1 and the 5 volt rail. One can increase R2 to a higher resistance value such as 10 kΩ, but there may be appreciable voltage drop across R2, and the voltage at the collector of Q1 will be less than 4.3 volts. If the current gain of the transistor is higher (e.g., $\beta > 200$), usually R2 can be increased in value to avoid incurring a larger voltage drop across it.

When an Oscillator Does Not Oscillate

For oscillators that should oscillate but do not, here are some suggestions:

- If the oscillator uses a transformer, try reversing *either* the outer primary or the secondary leads to ensure positive feedback.
- If the oscillator circuit has the correct phase for positive feedback and still does not oscillate, increase the collector current to provide more gain.
- Change the values of the positive-feedback network to provide more feedback.
- If the frequency of oscillation is beyond 20 MHz, try a higher-frequency transistor. For example, although general-purpose transistors such as the 2N3904 or 2N4124 can be used for a 100 MHz oscillator, often the collector current needs to be increased. A higher-frequency transistor such as the MPSH10 or 2N5179 should work better at lower collector currents.

See Figures 18-5*a* and 18-5*b*.

In Figure 18-5*a*, if the circuit fails to oscillate, one can try reversing the primary windings, pins 5 and 8, or reversing pins 1 and 4 of the secondary windings. Note that the low-side tap (T1, pin 6 to pin 8), which has the least measured resistance compared with the other primary windings, is usually coupled to the emitter of Q1. The output signal of the oscillator may be taken at pin 6 of T1 or at the emitter of Q1.

If the Colpitts oscillator in Figure 18-5*b* fails to start up, one can decrease the value of R3 from 1.5 kΩ to a resistance value as low as 330 Ω to provide more gain or transconductance in Q1. Alternatively, with R3 = 1.5 kΩ, the amount of positive

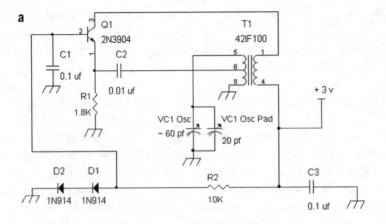

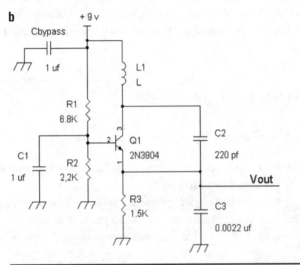

FIGURE 18-5   (*a*) A transformer-coupled oscillator. (*b*) An example Colpitts oscillator circuit.

feedback can be increased by providing more signal from the collector to the emitter. The approximate amount of voltage dividing is $C2/(C2 + C3) \approx 1/11$ when $C2 = 220$ pF and $C3 = 2,200$ pF (0.0022 μF). One can try increasing $C2$ and decreasing $C3$. For example, try making $C2 = 390$ pF and $C3 = 390$ pF. If the oscillation frequency is greater than 20 MHz, change the transistor to a higher-frequency one, such as an MPSH10 or 2N5179.

It should be known that there are losses due to the inductors as well that can prevent an oscillator from starting up. Pertaining to different frequencies of oscillation, there are practical values for inductors. For example, in the range of 500 kHz to 3 MHz, one can use inductor values from about 1,000 μH to about 100 μH. For 3 MHz to 30 MHz, one can try inductor values from 100 μH to1 μH. And for frequencies from 30 MHz to 150 MHz, using inductor values from about 1 µH to 0.05 µH is a good start. In all these ranges, the larger inductance values should be used for the lowest frequency. For example, at 3 MHz, start at about 100 uH, and adjust to lower values as desired.

Reducing External Noise

Suppose that there is an audio mixer circuit that has been designed with high-value resistors in the 500 kΩ range and there is hum pickup from the AC line via fluorescent lamps or other external factors. One way to reduce hum pickup is to place the mixer circuit in a metal enclosure and connect the enclosure to a ground connection in the audio mixer circuit. But suppose that an enclosure for shielding out the external noise sources is not available? What's an alternative?

We can lower the values of the resistors by hundredfold or more and reduce the noise by modifying the mixer circuit from 500 kΩ to 5 kΩ resistors (see Figure 18-6). Because many of the audio signal sources have low-impedance outputs, having an audio mixer circuit with lower input resistances, as shown in Figure 18-6, where the potentiometer's resistance value is changed from 500 kΩ to 5 kΩ, is a solution to reducing external noise without loading down (too much) the audio sources.

When the Circuit Kind of Works

Sometimes when a circuit is built, the performance may not be up to par because one or two of the transistors have been inadvertently wired backwards, such as having emitter and collector leads reversed. When the collector and emitter leads are

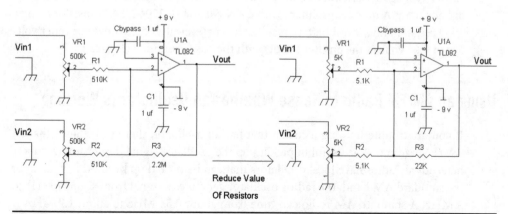

FIGURE 18-6   Simple summing amplifier scaled to lower resistances to reject external noise sources.

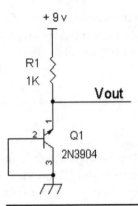

+9v

R1
1K

Vout

Q1
2N3904

FIGURE **18-7**   An example of using an NPN silicon transistor for a zener diode.

reversed, there is some transistor amplifying action, but the "reverse" current gain β is usually 1 or less versus when the transistor is wired correctly, when the forward current gain β is generally greater than 20. A consequence of having the collector and emitter leads reversed is that an amplifier may still work somewhat, but the DC base currents will be very large and will generate errors in the base DC biasing resistive network. Also, a reversed connection of the emitter and collector leads to a lower breakdown voltage of the transistor.

For example, did you know that if you want to make a cheap zener diode out of a transistor, just reverse bias the base and emitter terminals. Leave the collector terminal unconnected, or connect the collector to the base. For example, with an NPN transistor, ground the base lead, and connect a 1 kΩ to 10 kΩ resistor to the emitter. The other side of the resistor goes to a 12 volt or 9 volt power supply. A reference voltage between 5 volts and 8 volts will be available at the emitter reference to ground.

The actual breakdown voltage of the base-emitter junction will vary, and this "zener" voltage should not be construed as something reliable in terms of zener voltage accuracy. A quick experiment with a KN3904 or 2N3904 NPN transistor using a 1 kΩ resistor and a 9 volt battery resulted in a base-emitter breakdown or zener voltage of 7.64 volts at the emitter (Vout) with the base grounded (Figure 18-7).

Using an AM/FM Radio to Sense Whether an Oscillator Is Working

If you are troubleshooting a circuit that has an oscillator, there is a chance that an AM/FM radio will be useful in picking up the oscillator signal. For example, superheterodyne radios have local oscillators that can be picked up by another radio. For the standard AM band, the radio's oscillator generates a signal from about 1 MHz to 2 MHz. A standard AM radio can tune from about 0.54 MHz to about 1.7 MHz. If you are repairing an AM radio, you can quickly find out if the converter or oscillator circuit is working by setting the dial to about 600 kHz and using another radio to listen for a drop in hiss at about 1,055 kHz. Some radios, such as the one mentioned in Chapter 17, the Sony ICF-S10MK2, has an LED signal-strength indicator, which also can be used to sense oscillator signals from another superheterodyne radio. An FM radio has a local oscillator with a range of about 99 MHz to 119 MHz, and an FM radio tunes from 88 MHz to 108 MHz, which would be able to pick the local oscillator signal from the other radio.

It should be known that with a lower-frequency oscillator, generally harmonics of the fundamental frequency are also generated. For example, a 100 kHz oscillator will generate signals that occur at frequencies of N × 100 kHz, where N is a natural number greater than or equal to 1, such as 1, 2, 3, …, 10, 11. Therefore, an AM radio will be able to pick up the 100 kHz oscillator at its harmonic signals at 600 kHz, 700 kHz, 800 kHz, 900 kHz, 1 MHz, 1.1 MHz, 1.2 MHz, and so on.

Using Low-Dropout Voltage Regulators

In some instances when a 12 volt DC power supply is built with a voltage regulator, the raw DC voltage drops just enough to cause the 7812 voltage regulator to pass some of the ripple voltage at its output. When the raw supply's voltage falls to less than 13.7 volts (e.g., the rated regulated voltage plus 1.7 volts), the 7812 circuit fails to regulate the input voltage. A low-dropout voltage regulator such as the LM2940 requires only 0.5 volt above the regulated voltage or ≥ 12.5 volts at the input to provide a regulated 12 volt output. For the low-dropout regulator chips, bypass capacitors are generally a little larger in value. It is recommend that at least a 33 µF capacitor be connected across the output terminal and ground, and at least a 1 µF capacitor connected across the input terminal and ground.

Low-dropout regulator chips are available in different output voltages. For example, see the following links from Texas Instruments:

- www.ti.com/ldo
- www.ti.com.cn/cn/lit/ds/symlink/lm2940-n.pdf
  (for positive voltage regulators)
- www.ti.com.cn/cn/lit/ds/symlink/lm2990.pdf
  (for negative voltage regulators)

Back to Test Equipment

If I were to choose four pieces of test gear, they would be

- A good DVM that measures AC signals with a 2 volt or less full scale and measures beyond 10 MΩ in resistance with a transistor tester and frequency counter up to at least 1 MHz

- An oscilloscope that costs less than $1,000, such as the Rigol 1000Z Series (e.g, DS1104Z four-channel 100 MHz scope at $830); or for about $400 you can try the Rigol 1102E dual-channel 100 MHz scope
- A function generator up to at least 1 MHz
- An RF signal generator up to 100 MHz

One can make an audio signal tracer by using an amplified computer speaker. Also, by adding a germanium or Schottky diode in series with the hot or tip lead, demodulation of AM signals can be provided.

Summary on Troubleshooting

The art of troubleshooting generally takes a lot of experience, and as one gets to build or repair or modify existing circuits, the troubleshooting will become easier. The reader should always be aware that almost any design can be improved on in terms of reliability, performance, and so on.

One way to learn troubleshooting is to obtain some test equipment, build some circuits, and for example, probe with an oscilloscope if possible. Troubleshooting is like being a detective using a process of elimination. Usually, we start with whether the components have power and ground voltages, and later we look for the actual signals.

For example, there was the case of using a CD4046 phase-lock loop circuit. There are two phase detectors that one can select from—an exclusive OR logic circuit (phase comparator I) or a positive-negative pulse circuit that charges a capacitor (phase comparator II). According to the specification sheet, a typical AC signal at the input terminal 14 is 200 mV for both phase comparators. It was found that at 200 mV, phase comparator I indeed worked fine. However, phase comparator II required twice the signal level (400 mV) at pin 14.

Final Thoughts on the Book

When I mapped out this book in mid-2013, I had many ideas to choose from. Some of these were concepts that are no longer relevant to today's generation, such as learning the resistor color code and working with coils. For example, it's easier just to measure the resistor with an ohmmeter and build projects without coils or transformers. I learned quite a bit about what is more important today while mentoring some students at Stanford (thank you, Professor Bob Dutton!) in 2013 and 2014. So I decided to include some of the basics but also present projects that were not as nostalgic. In Chapter 11, for example, one of the FM radio projects uses no variable capacitors but instead a 2014 state-of-the-art Silicon Labs SDR (software defined radio) chip.

It's been a pleasure to start this book with electricity projects involving LEDs, batteries, and lamps and then to progress to an intermediate level of electronics, including amplifiers and feedback systems. Finding more advanced projects up to Chapter 13 took quite a bit of selection as to what type of AM and FM radio projects to present along with a good system of circuits illustrated as the last big project, the video processor and scrambler.

In terms of math, I decided to just concentrate on lines and polynomials. Speaking of polynomials, I saw this "smart" answer on *expanding* the following:

$$(a + b)^2 = (a \qquad + \qquad b)^2$$

Of course, if I were the teacher, I would have given the student half credit because the answer "should have" been *expanded* this way:

$$(a + b)^2 = \left(a \qquad\qquad + \qquad\qquad b \right)^2$$

I would have told the "smart" student that if you really want to expand $(a + b)^2$, at least expand it in two dimensions (horizontally and vertically)! I would have also said, "Why think only in one dimension?" Just kidding!

> **NOTE**   Of course, the real answer is $(a + b)^2 = a^2 + 2ab + b^2$.

Now I will present one last figure for this book. It's a picture of the simple LED flashlight with two NiMh AA cells (low self-discharge types) that has been on since I wrote Chapter 4 (see the "original" Figure 4-11) during the first week of November 2013. Well, it is still on after over six months, and it probably will continue to give useful lighting past the last half of 2014 (see Figure 18-8).

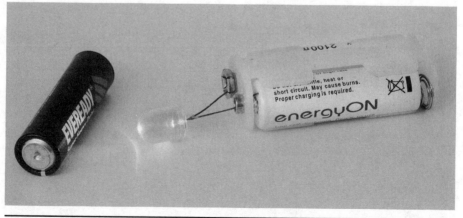

FIGURE 18-8   Simple LED flashlight lit for over six months continuously.

Lastly, the material in this book has been more diverse than my first book, *Build Your Own Transistor Radios*, and I hope that at least some parts of *Electronics from the Ground Up* has resonated with the reader.

Thank you.

Ron Quan
20 May 2014

Postscript: As of September 15, 2014, the LED flashlight in Figure 18-8 is still giving off usable light. It has been on for over 10 months.

APPENDIX A
Parts Suppliers

Transistors, FETs, Diodes, LEDs, Photodiodes, ICs, Resistors, Capacitors, Crystals, Batteries, Inductors, Transformers, Tools, and Soldering Equipment

- Mouser Electronics, www.mouser.com

- Digi-Key Corporation, www.digikey.com

- Anchor-Electronics, www.anchor-electronics.com

- HSC Electronic Supply, www.halted.com/commerce/index.jsp?czuid =1338011465538 (may have to purchase items at their store(s); call to confirm)

- MCM Electronics, www.mcmelectronics.com

- Jameco Electronics, www.jameco.com

- Amazon, www.amazon.com (almost every type of electronic parts and kits as well)

- Joe Knows Electronics, www.joeknowselectronics.com/?page_id=1271 (other types of electronic parts as well)

- RadioShack, www.radioshack.com

- Frys Electronics, www.frys.com

- Linear Integrated Systems, www.linearsystems.com (for low-noise transistors and FETs)

Vacuum Tubes

- Antique Electronic Supply, www.tubesandmore.com

Ceramic Resonators and Filters (455 kHz and 10.7 MHz)

- Mouser Electronics, www.mouser.com

- Digi-Key Corporation, www.digikey.com

- HSC Electronic Supply, www.halted.com/commerce/index.jsp?czuid =1338011465538 (may have to purchase items at their store(s); call to confirm)

- eBay, www.ebay.com (search terms are *ceramic resonators* and *filters*)

Oscillator Coils, IF Transformers, and Audio Transformers

- Mouser Electronics, www.mouser.com (Xicon transformers and coils)

- Abra Electronics Inc., www.abra-electronics.com (Xicon IF and audio transformers; for IF transformers, search under 42IF)

- Electronix Express, www.elexp.com (search under audio and IF transformers)

Antenna Coils

- Scott's Electronics (hard-to-find electronics parts), www.angelfire.com/electronic2/index1/loopstick.html (for ferrite antenna coils)

- MCM Electronics, www.mcmelectronics.com (search for *loop antenna*, the AM radio type, for example, see http://electronics.mcmelectronics.com/?N =&Ntt=loop+antenna)

- Antique Electronic Supply, www.tubesandmore.com (the search term is *antenna coil*; as of 2014, the only antenna coil available has an inductance of 240 μH, which will match a 330-pF to 365-pF variable capacitor)

- Amazon, www.amazon.com

- eBay, www.ebay.com

Variable Capacitors

- Scott's Electronics, www.angelfire.com/electronic2/index1/index.html (hard-to-find electronics parts) www.angelfire.com/electronic2/index1/Variable141pf.htm (for a 140-pF/60-pF dual variable capacitor) www.angelfire.com/electronic2/index1/266pf.html (for a twin 266-pF/266-pF variable capacitor) www.angelfire.com/electronic2/index1/335PF-Polyvaricon.html (for a twin 335-pF/335-pF variable capacitor)

- eBay, www.ebay.com [search under the terms *polyvaricon*, *variable capacitor*, *tuning capacitor*, or *crystal radio parts*; note that some of the variable capacitors on eBay are timer capacitors, which do not have sufficient capacitance (e.g., <100 pF) for the radios in this book]

- Antique Electronic Supply, www.tubesandmore.com (365-pF variable capacitor and others at www.tubesandmore.com/products/capacitors?filters =Type%3DVariable)

Crystal Earphones

- Scott's Electronics, www.angelfire.com/electronic2/index1/index.html (hard to find electronics parts)

- eBay, www.ebay.com (search under the terms *crystal earphone*, *crystal radio*, and *crystal radio parts*)

Science Kits, Cool Things, and Everything Else

- Evil Mad Scientist Laboratories, www.evilmadscientist.com [components such as resistors, LEDs, displays, and also many do-it-yourself (DIY) kits, including a discrete transistor version of the 555 timer by Eric Schlaepfer]

- Elenco, www.elenco.com/product/educational (all sorts of educational kits)

- Curcio Audio Engineering, www.curcioaudio.com (tube amplifier kits)

- Amazon, www.amazon.com

- Jameco Electronics, www.jameco.com

- Frys Electronics, www.frys.com

APPENDIX B

Online Learning Resources

- Dave L. Jones, www.eevblog.com (lots of cool videos on evaluation of test equipment and much, much more)

- Jeri Ellsworth, www.youtube.com/user/JeriEllsworthJabber (has many videos on science and electronics)

- The Kahn Academy on YouTube, www.youtube.com/user/khanacademy (has many great videos on math and science)

- For the advanced hobbyist, try the MIT OCW site for televised classes, http://ocw.mit.edu/courses/find-by-department/

- Element14 electronics and technology videos on YouTube, www.youtube .com/user/element14

APPENDIX C
Free Software

- LTSPICE is a circuit simulation program available at www.linear.com /designtools/software (you can download this simulation program to run on your Windows or Mac operating system). This program is quite useful for analyzing circuits in terms of DC biasing and AC analysis.

- You can convert your computer via its soundcard audio inputs into a spectrum analyzer via the Spectran software program at http://digilander .libero.it/i2phd/spectran.html.

- A program to run on your computer pertaining to software-defined radios (SDRs) is available at http://winrad.org/. You can search the Web for various SDR front-end circuits to build, which will be connected to your left and right audio inputs on your computer. Winrad will process the signals to receive AM, FM, single-sideband, and CW (Morse code) signals.

Index

References to figures are in italics.

555 timer circuits, 237–239

AA cell batteries, 7
    converting into a penlight, 16–21
    See also batteries
AAA cell batteries, 8
    See also batteries
AC, converting to DC for power supplies,
    107–113
AC circuits, analysis, 416–423
AC resistance, 121
acid core solder, 64
air dielectric variable capacitors, 266–267
aliasing, 338
    See also resolution
alkaline batteries, 10
    See also batteries
AM. *See* amplitude modulation
AM radios, compared to FM radios,
    301–303
AM signals, 251–256
    mixing and beating signals, 278–279
    single sideband signals (SSBs), 265–266
    suppressed carrier AM signals, 258–263
    vestigial sideband (VSB) AM signals,
    256–258

American Wire Gauge (AWG), 5
AM/FM radios
    hacking, 465–469
    using to sense whether an oscillator is
    working, 504–505
amplifiers, 38–39, 48–51
    automatic level control (ALC) amplifiers,
    174–176
    building the LM386 audio amplifier on a
    vector board, 71–74
    common-emitter amplifiers, 39, 288,
    443–447
    common-mode, 191–192
    emitter-follower amplifiers, 40, 438–443
    integrated circuits as, 44–45
    inverting-gain video amplifiers, 447–451
    linear amplifiers, 380
    low-voltage vacuum tube amplifiers,
    212–213
    microphone preamplifier circuits, 195–200
    nonlinear amplifiers, 381
    polynomials used in determining
    distortion effects of negative feedback
    in amplifiers, 395–401
    voltage-controlled common emitter
    amplifiers, 177–178

517